Energiepolitik und Klimaschutz
Energy Policy and Climate Protection

Reihe herausgegeben von
L. Mez, Berlin, Deutschland
A. Brunnengräber, Berlin, Deutschland

Diese Buchreihe beschäftigt sich mit den globalen Verteilungskämpfen um knappe Energieressourcen, mit dem Klimawandel und seinen Auswirkungen sowie mit den globalen, nationalen, regionalen und lokalen Herausforderungen der umkämpften Energiewende. Die Beiträge der Reihe zielen auf eine nachhaltige Energie- und Klimapolitik sowie die wirtschaftlichen Interessen, Machtverhältnisse und Pfadabhängigkeiten, die sich dabei als hohe Hindernisse erweisen. Weitere Themen sind die internationale und europäische Liberalisierung der Energiemärkte, die Klimapolitik der Vereinten Nationen (UN), Anpassungsmaßnahmen an den Klimawandel in den Entwicklungs-, Schwellen- und Industrieländern, Strategien zur Dekarbonisierung sowie der Ausstieg aus der Kernenergie und der Umgang mit den nuklearen Hinterlassenschaften.
Die Reihe bietet ein Forum für empirisch angeleitete, quantitative und international vergleichende Arbeiten, für Untersuchungen von grenzüberschreitenden Transformations-, Mehrebenen- und Governance-Prozessen oder von nationalen „best practice"-Beispielen. Ebenso ist sie offen für theoriegeleitete, qualitative Untersuchungen, die sich mit den grundlegenden Fragen des gesellschaftlichen Wandels in der Energiepolitik, bei der Energiewende und beim Klimaschutz beschäftigen.

This book series focuses on global distribution struggles over scarce energy resources, climate change and its impacts, and the global, national, regional and local challenges associated with contested energy transitions. The contributions to the series explore the opportunities to create sustainable energy and climate policies against the backdrop of the obstacles created by strong economic interests, power relations and path dependencies. The series addresses such matters as the international and European liberalization of energy sectors; sustainability and international climate change policy; climate change adaptation measures in the developing, emerging and industrialized countries; strategies toward decarbonization; the problems of nuclear energy and the nuclear legacy.
The series includes theory-led, empirically guided, quantitative and qualitative international comparative work, investigations of cross-border transformations, governance and multi-level processes, and national "best practice"-examples. The goal of the series is to better understand societal-ecological transformations for low carbon energy systems, energy transitions and climate protection.

Reihe herausgegeben von
PD Dr. Lutz Mez
Freie Universität Berlin

PD Dr. Achim Brunnengräber
Freie Universität Berlin

Weitere Bände in der Reihe http://www.springer.com/series/12516

Achim Brunnengräber
Maria Rosaria Di Nucci
Ana María Isidoro Losada · Lutz Mez
Miranda A. Schreurs
Editors

Challenges of Nuclear Waste Governance

An International Comparison
Volume II

Springer VS

Editors
Achim Brunnengräber
Freie Universität Berlin
Berlin, Germany

Maria Rosaria Di Nucci
Freie Universität Berlin
Berlin, Germany

Ana María Isidoro Losada
Freie Universität Berlin
Berlin, Germany

Lutz Mez
Freie Universität Berlin
Berlin, Germany

Miranda A. Schreurs
Bavarian School of Public Policy
Technical University of Munich
Munich, Germany

This volume is a contribution of the Environmental Policy Research Centre (Forschungszentrum für Umweltpolitik – FFU), Freie Universität Berlin to the project "Multi Level Governance-Perspective on Management of Nuclear Waste Disposal. A Comparative Analysis: Actors, Instruments and Institutions" that was part of the inter-disciplinary ENTRIA research platform funded by the German Federal Ministry of Education and Research (BMBF/02S9082B; 2013-2017).

Energiepolitik und Klimaschutz. Energy Policy and Climate Protection
ISBN 978-3-658-21440-1 ISBN 978-3-658-21441-8 (eBook)
https://doi.org/10.1007/978-3-658-21441-8

Library of Congress Control Number: 2018937637

Springer VS

Printed on acid-free paper

This Springer VS imprint is published by the registered company Springer Fachmedien Wiesbaden GmbH part of Springer Nature
The registered company address is: Abraham-Lincoln-Str. 46, 65189 Wiesbaden, Germany

Contents

Preface VII
Lutz Mez

Abbreviations IX

I. Introduction

The Technical, Political and Socio-Economic Challenges of Governing Nuclear Waste
A Comparative Perspective 3
Maria Rosaria Di Nucci, Ana María Isidoro Losada, Miranda Schreurs, Achim Brunnengräber and Lutz Mez

II. "The Big 4"- China, Russia, South Korea and Japan

A Long Way Off
Nuclear Waste Governance in China 25
Eva Sternfeld

A Profitable Business Strategy?
Spent Nuclear Fuel and Radioactive Waste Management in Russia 51
Felix Jaitner

Transitioning Away from Opaque Governance
Nuclear Waste Policy in South Korea 73
Sung-Jin Leem with Miranda Schreurs

False Premise, False Promise
Governance and Management of Nuclear Waste in Japan 95
Lila Okamura

III. A Missed Opportunity: The Case of the Eastern European Countries

Postponed Policy
Nuclear Waste Governance in Hungary 123
Zsuzsanna Koritár

Progress on Nuclear Waste Management
Nuclear Waste Governance in Lithuania 139
Povilas Poskas

Delays in Finding a Solution
The Governance of Nuclear Waste Disposal in Slovakia 161
Peter Mihók

An Arranged Marriage
Nuclear Waste Governance and Nuclear Energy in Slovenia and Croatia 183
Leo Šešerko

Too much to Handle
Radioactive Waste Management in a Post Nuclear Accident State: Ukraine 211
Olexi Pasyuk

IV. Nuclear Waste Governance in Canada and the Global South

A Twinned Approach
The Challenges of Nuclear Waste Governance in Canada 247
Cindy Vestergaard

A Democratic Deficit
The Challenges of Nuclear Waste Governance in Argentina and Brazil 273
Moïra Jimeno

Disposal and Contamination
Nuclear Waste Governance in South Africa 295
David Fig

Contributors 329

Preface

This book is the second edited volume of three volumes comparing nuclear waste governance in various countries developed during the research project, A Multi Level Governance-Perspective on the Management of Nuclear Waste Disposal: A Comparative Analysis of Actors, Instruments and Institutions. The project was carried out at the Environmental Policy Research Centre (FFU), Freie Universität Berlin, and is part of the interdisciplinary research initiative, ENTRIA: Disposal Options for Radioactive Residues. Interdisciplinary Analyses and Development of Evaluation Principles. The ENTRIA project brings together twelve German universities and one Swiss partner to examine nuclear waste management in relation to technical options, decision making processes, risk and safety as well as social challenges.

This book continues the examination of the modes of governance that countries are developing to address the storage and disposal of high-level radioactive waste and nuclear spent fuel. As a follow-up of Volume I, the analysis starts with "the big four": China, Russia, South Korea and Japan. Subsequently, case studies on five East-European Euratom member states are presented: Hungary, Lithuania, Slovakia, Slovenia and Croatia as well as Ukraine. Finally, country reports on Canada, Argentina and Brazil, and South Africa complete the overview.

Following the pattern of Volume I, the progress these countries have made and the obstacles they face are discussed by looking at their regulations, technology choices, safety criteria, monitoring systems, compensation schemes, institutional structures, and approaches to public involvement. The chapters included in this book identify the primary stakeholders in the debate and their interests, the responsibilities and authority of different actors in relevant decision-making processes, and the value systems that are influencing their different national policy choices. The views and expectations of different communities regarding participatory decision-making and compensation and the steps that have been or are being taken to promote dialogue and constructive problem-solving are also considered.

The German Federal Ministry for Education and Research provided the core funding for the ENTRIA project. The editors are grateful for this support. Special thanks go to Dörte Themann, whose engagement has been key to the success of this book project. The texts were skilfully proofread by Jessica Wallach. We would like to thank Britta Göhrisch-Radmacher at Springer VS for her support. Any mistakes are the responsibility of the authors and editors.

Berlin, Dezember 2017 Lutz Mez for the editorial team.

Abbreviations

AEB	Atomic Energy Board (South Africa)
AEC	Atomic Energy Corporation (South Africa)
AECB	Atomic Energy Control Board (Canada)
AECL	Atomic Energy of Canada Limited
AEPC	Atomic Energy Promotion Commission (South Korea)
AESJ	Atomic Energy Society of Japan
AGE	Ezeiza Waste Management Area (Argentina)
AMD	Acid Mine Drainage
ANC	African National Congress
Angarsk	IUEC Joint stock company Angarsk International Uranium Enrichment Centre (Russia)
ANRE	Agency for Natural Resources and Energy (Japan)
ANSTO	Australian Nuclear Safety Organization
APM	Adaptive Phased Management (Canada)
ARAO	Slovenian National Agency for Radioactive Waste Management
ARN	Nuclear Regulatory Authority (Argentina)
ARSO	Slovenian Agency for Environment
ASECQ	Spent Fuel Dry Storage System (Argentina)
BINE	Beijing Institute of Nuclear Engineering
BNFL	British Nuclear Fuels Limited
BN-600, BN-800	Bystrych Nejtronach (Fast Breeder)
BRIUG	Beijing Research Institute of Uranium Geology
CAE	Ezeiza Nuclear Center (Argentina)
CAEA	China Atomic Energy Authority
CANDU	Canada Deuterium Uranium
CAREM	Central Argentina de Elementos Modulares
CDFR	China Demonstration Fast Reactor
CEAA	Canadian Environmental Assessment Act
CEFR	China Experimental Fast Reactor
CEPA	Canadian Environmental Protection Act
CEZ	Chernobyl Exclusion Zone
ChNPP	Chernobyl NPP
CIAE	China Institute of Atomic Industry
CIRP	China Institute for Radiation Protection

CLTSF	Centralized Long-Term Storage Facility for Spent Sources (Ukraine)
CNEA	National Commission of Atomic Energy (Argentina)
CNEN	National Nuclear Energy Commission (Brazil)
CNEN	National Commission for Nuclear Energy (Mexico)
CNEPP	Comprehensive Nuclear Energy Promotion Plan (South Korea)
CNFF	Central Nuclear Financial Fund (Hungary)
CNL	Canadian Nuclear Laboratories Ltd
CNNC	China National Nuclear Corporation
CNPE	China Nuclear Power Engineering Company
CNS	Council for Nuclear Safety (South Africa)
CNSC	Canadian Nuclear Safety Commission
CRL	Chalk River Laboratories (Canada)
CSFSF	Centralized Spent Fuel Storage Facility (Ukraine)
CTDN	Nuclear Technology Development Center (Brazil)
DAD	Decide-Announce-Defend
DGD	Deep Geological Disposal
DGR	Deep Geological Repository
DME	Department of Minerals and Energy (South Africa)
DPJ	Democratic Party of Japan
EBRD	European Bank for Reconstruction and Development
EC	European Commission
EFPY	Effective Full Power Years
EIA	Environmental Impact Assessment
EIS	Environmental Impact Statement
EMR	Department of Energy, Mines, and Resources (Canada)
ENRN	National Nuclear Regulatory Body (Argentina)
EPA	Environmental Protection Agency
ERI	Energy Research Institute (China)
EU	European Union
Euratom	European Atomic Energy Community
EW	Exempt Waste
FA	Fisheries Act (Canada)
FCM	Fuel Content Materials
FEPC	Federation of Electric Power Companies of Japan

FOIA	Freedom of Information Act (Slovakia)
FUNAM	Environment Defense Foundation (Argentina)
FYP	Five Year Plan (China)
GEN	Energija d.o.o.Slovenian electrical power company (Slovenia)
GS	Geological Storage
HAEA	Hungarian Atomic Energy Authority
HEP	Hrvatska Elektroprivreda (Slovenia)
HLW	High-Level Waste
IAEA	International Atomic Energy Agency
IBRAE	Nuclear Safety Institute of the Russian Academy of Science
ICSRM	Industrial Complex for Solid Radioactive Waste Management (Ukraine)
ILW	Intermediate-Level Waste
INES	International Nuclear Event Scale
IPEN	Nuclear Energy Research Institute (Brazil)
IRRS	Integrated Regulatory Review Service
IRSN	Institut de radioprotection et de sûreté nucléaré (Slovenia)
ISF	Intermediate Spent Fuel Storage
ISFSF	Interim Spent Fuel Storage Facility
ISTC	The International Science and Technology Center in Moscow (Russia)
ITT	Isotope Information Association (Hungary)
I&Aps	Interested and Affected Parties
JAEA	Japan Atomic Energy Agency
JAEC	Japan Atomic Energy Commission
JAERO	Japan Atomic Energy Relations Organization
JAPC	Japan Atomic Energy Company
JAVYS	Nuclear and Decommissioning Company ("Jadrová a vyraďovacia spoločnos") (Slovakia)
JESS	Company "Jadrová energetická spoločnosť Slovenska" (Slovakia)
JNFL	Japan Nuclear Fuel Limited
KAERI	Korean Atomic Energy Research Institute
KHNP	Korea Hydro and Nuclear Power Company
KINS	Korea Institute of Nuclear Safety
KNF	KEPCO Nuclear Fuel (South Korea)
KORAD	Korea Radioactive Waste Agency

LEI	Lithuanian Energy Institute
LILW	Low-and Intermediate-Level Radioactive Waste
LL-LILW	Long-lived Low- and Interim-Level Radioactive Waste
LLRW	Low-Level Radioactive Waste
LLRWMO	Low-Level Radioactive Waste Management Office (Canada)
LLW	Low-Level Waste
LPD	Liberal Democratic Party
LRW	Liquid RAW
LRWTP	Liquid Radioactive Waste Treatment Plant (Ukraine)
MACH	Chubut Anti-Nuclear Movement (Argentina)
MCC	Mining and Chemical Combine, federal state unitary enterprise (Russia)
MEP	Ministry of Environmental Protection (China)
METI	Ministry of Economy, Trade and Industry (Japan)
MII	Ministry of Industry and Information (China)
Minatom	The Russian Ministry for Atomic Energy
MoF	Ministry of Finance (China)
MOP	Ministry of the Environment and Spatial Planning (Slovenia)
MOTIE	Ministry of Trade, Industry and Energy (South Korea)
MOX	Mixed Oxide Fuel
MPS	Ministry of Public Security
MSIP	Ministry of Science, ICT and Future Planning (South Korea)
MZI	Ministry of Infrastructure (Slovenia)
NA-SA	Nucleoeléctrica Argentina
NCSTIND	National Commission for Science, Technology and Industry for National Defense (China)
NDRC	National Development and Reform Commission (China)
NEA	Nuclear Energy Agency
NEA	National Energy Administration (China)
NECSA	South African Nuclear Energy Corporation
NERSA	National Energy Regulator of South Africa
NETI	Nuclear Environment Technology Institute (South Korea)
NFWA	Nuclear Fuel Waste Act (Canada)
NIMBY	Not in my Backyard
NINT	Northwest Institute for Nuclear Industry and Technology (China)
NJF/NNF	National Nuclear Fund (Slovakia)

NNR	National Nuclear Regulator (South Africa)
NNSA	National Nuclear Safety Administration (China)
NORM	Natural Radioactive Materials
NPP	Nuclear Power Plant
NRA	Nuclear Regulatory Authority (Japan)
NRA	Nuclear Regulatory Authority (Slovakia)
NRCan	Natural Resources Canada
NRWDI	National Radioactive Waste Disposal Institute (South Africa)
NRWR	National Radioactive Waste Repository (Hungary)
NSA	Nuclear Safety Account (Ukraine)
NSC	New Safe Confinement (Ukraine)
NSCA	Nuclear Safety and Control Act (Canada)
NSP	National Spatial Plan (Slovenia)
NSR	Near Surface Repository
NSSC	Nuclear Safety and Security Commission (South Korea)
NUMO	Nuclear Waste Management Organization of Japan
NWMO	Nuclear Waste Management Organization (Canada)
NYMTIT	Western-Mecsek Social Information Association (Hungary)
OAB	National Atomic Energy Committee (Hungary)
OECD	Organisation for Economic Co-operation and Development
OPG	Ontario Power Generation (Canada)
OS	Object Shelter (Ukraine)
PAIA	Promotion of Access to Information Act (South Africa)
PBMR	Pebble Bed Modular Reactor (South Africa)
PECOS	Public Engagement Commission on Spent Nuclear Fuel Management (South Korea)
PEGRR	Strategic Plan of the National Program for Radioactive Waste Management (Argentina)
PHAI	Port Hope Area Initiative (Canada)
PHWR	Pressurized Heavy Water Reactor (Argentina)
PIMCU	Uranium mining company Priargunsky Industrial Mining and Chemical Union (Russia)
PLEX	Plant Lifetime Extension
RAN	Russian Academy of Science
RATA	Radioactive Waste Management Agency (Lithuania)
RAW	Radioactive Waste

RFI	Request For Information
RI	Radioisotope
RMB	Renmin Bi (Chinese Yuan)
Rosatom	Russian State Atomic Energy Corporation
RosRAO	Enterprise for Radioactive Waste Management, Russia
RPC	Radiation Protection Centre (Lithuania)
RW	Radioactive Wastes
RWDS	Radioactive Waste Disposal Site
RWTDF	Radioactive Waste Treatment and Disposal Facility (Hungary)
SAR	Safety Analysis Report (Lithuania)
SASAC	State Owned Assets Supervision and Administration Commission (China)
SCC	Joint stock company "Siberian Chemical Combine" (Russia)
SE	"Slovenské elektrárne" company (Slovakia)
SEA	Strategic Environmental Assessment
SEZA	State Exclusion Zone Agency (Ukraine)
SF	Spent Fuel
SFA	Spent Fuel Assemblies
ŠFLJEZ/SNIDF	State Fund for Decommissioning of Nuclear Power Installations and Management of Spent Nuclear Fuel and Radioactive Waste (Slovakia)
SIP	Swedish International Project for Nuclear Safety
SIP	Shelter Implementation Plan (Ukraine)
SKB	Svensk Kärnbränslehantering AB (Swedish Spent Fuel Management Agency)
SNSA	Slovenian Nuclear Safety Administration
SRC RIAR	Joint stock company "State Research Centre of the Russian Federation - Research Institute of Atomic Reactors"
SRS	Spent Radiation Sources
SSE	Sustainable Sources of Energy
SSE CEMRW	Specialised State Enterprise Central utility to manage radioactive waste (Ukraine)
TEIT	Social Control and Information Association (Hungary)
TEPCO	Tokyo Electric Power Company
TETT	Social Association for Control and Information (Hungary)
TFMs	Tailings Management Facilities
TRU	Transuranic Waste

UAC	Unión de Asambleas Ciudadanas (Argentina)
Ucor	Uranium Enrichment Corporation (South Africa)
UECC	Joint stock company Ural Electrochemical Combine (Russia)
ÚJD	Slovak Nuclear Regulatory Authority (Úrad jadrového dozoru) (Slovakia)
URL	Underground Research Laboratory
URSJV	Administration of the Republic of Slovenia for Nuclear Safety
URSVS	Administration of Republic of Slovenia for Radiation Protection
USC Radon	Ukrainian State Corporation Radon
US NRC	US Nuclear Regulatory Commission
USS RWM	Unified State System for Radioactive Waste Management (Russia)
VATESI	State Nuclear Power Safety Inspectorate (Lithuania)
VLLW	Very Low Level Waste
VSLW	Very Short Lived Waste
VVER	see WWER
WNA	World Nuclear Association
WWER	Water- Water Energetic Reactor
ZMOS	Association of towns and communities of Slovakia

I. Introduction

The Technical, Political and Socio-Economic Challenges of Governing Nuclear Waste

A Comparative Perspective

Maria Rosaria Di Nucci, Ana María Isidoro Losada, Miranda Schreurs, Achim Brunnengräber and Lutz Mez[1]

This is the second volume examining high level radioactive waste (HLW) disposal processes around the world. Volume I examined European and North American cases (Belgium, the Czech Republic, Finland, France, Germany, Italy, the Netherlands, Spain, Sweden, Switzerland, the United Kingdom, and the United States). Volume II looks at: Argentina, Brazil, Canada, China, Croatia, Japan, Hungary, Lithuania, Russia, Slovakia, Slovenia, South Korea, South Africa, and the Ukraine.

The country studies reveal just how multifaceted the "wicked problem" of HLW is. It will take hundreds of thousands of years before natural decay processes will reduce radioactivity levels to a point where they are no longer dangerous to humans or other species of fauna and flora. This makes HLW management a wicked problem, that is a highly complex issue which is challenging to solve because of its many interrelated technical, material, social and political dimensions. A further complicating factor is the time dimension involved with radioactive waste management as no matter which option is followed future generations will be impacted.

Where there is some agreement among actors, whether they are supportive or opposed to nuclear energy, is that current approaches to HLW management cannot continue indefinitely. Longer term solutions must be found. Currently 50 countries have spent nuclear fuel (SNF) stored in pools at reactor sites or in caskets in central interim sites, awaiting reprocessing or disposal.

---

1 This article is part of the work of the Environmental Policy Research Centre (FFU) / FU Berlin within the ENTRIA project funded by the German Federal Ministry of Education and Research (BMBF/ 15S9082B): "Multi Level Governance-Perspective on management of nuclear waste disposal. A Comparative Analysis." ENTRIA (www.entria.de) is the acronym for "Disposal options for radioactive residues: Interdisciplinary analyses and development of evaluation principles".

A. Brunnengräber et al. (eds.), *Challenges of Nuclear Waste Governance*,
Energiepolitik und Klimaschutz. Energy Policy and Climate Protection,
https://doi.org/10.1007/978-3-658-21441-8_1

Volume I highlighted the challenge of developing heuristic tools to analyse the complexity of nuclear waste management (Brunnengräber et al. 2015). In both volumes, we use nuclear waste governance (NWG) as a lens to analyse how actors with at times deeply conflicting interests influence nuclear waste policies. We are especially interested in how conflict is managed and whether and how public participation is encouraged. While there is a large body of literature on technical issues concerning nuclear waste repositories complemented by socio-technical works on risk, there are only a few publications related to NWG, including research on acceptance, acceptability and participation in the siting process (Flueler 2006; Strandberg and Andrén 2010; Cotton 2017).

NWG is influenced by local, national, and international factors and institutions. The political and legal systems of a country as well as their nuclear history shape siting processes. Various actors with different interests and expectations seek to influence outcomes. In all countries, the handling of nuclear waste has simultaneously technical, social, and political implications. The rationale for taking a comparative perspective is obvious.

There is still no country in the world with an operating repository for the final disposal of HLW and the paths followed are varied. Some countries favour direct disposal of spent fuel and radioactive waste, others pursue disposal after reprocessing. Not all countries have the appropriate geological conditions for a deep geological disposal. For all options the costs represent a big challenge. For the countries with small nuclear programmes, the financial resources needed for the construction and operation of a disposal facility to host relatively limited volumes of spent fuel are excessive.

The nuclear waste produced in the years of military nuclear programme activities (nuclear weapons, navy reactors, military research, reprocessing plutonium plants, etc.) is also an extremely important issue, which in all countries, independent of their political setting, is subject to secrecy or non-transparency. These wastes are referred to as "legacy wastes" and represent a liability not covered by conventional nuclear waste legislation and funding arrangements. In some countries, military activities triggered ambitions for a closed nuclear fuel cycle strategy and reprocessing.

The main focus in the research on final disposal of nuclear waste still lies on the technical and scientific problems associated with final disposal. There is a tendency in part of the scientific community to claim that the technical feasibility of final disposal is not a major problem, only the social conflicts and the lack of acceptance are factors decisively preventing progress. Studies stress the remaining technical uncertainties and gaps in knowledge, for example on corrosion, geological formations and chemical effects. In the last five to ten years, however, a decisive paradigm shift has taken place, and greater attention has been devoted to socio-scientific issues in the field of nuclear waste disposal as well as to questions of participation and transparency (Solomon et al. 2010; Bergmans et al. 2015).

A broad social dialogue and greater involvement of the public as well as inclusive participation forms in the siting procedure are increasingly regarded as being the key for a conflict minimising approach to the problem of disposal (NEA 2015, NEA 2010; Brans et al. 2015; Di Nucci et al. 2017). Solutions must be scientifically robust and must achieve the highest technical standards, but they must also be publicly acceptable (Blowers 2016). Years of harsh controversies in countries around the globe have shown that finding a long-term deep geological disposal (DGD) for nuclear waste requires iterative learning and the re-adjustment of strategies. In almost all democratic countries, governments have started learning from their past efforts and mistakes in following Decide-Announce-Defend (DAD) approaches. More and more, governments are realising how counterproductive top-down approaches can be. Indeed, many changes in public participation procedures have occurred since the 1990s. Although there may still be resistance to more participatory and open planning structures, new and more democratic modes of governance are now widely recognized as necessary for moving forward with siting. In most OECD countries, legal and institutional frameworks for governing radioactive waste are in place. Licensing requirements and procedures for site selection and safety criteria have been established, and the responsibilities of stakeholders defined (Brunnengräber et al. 2015). The country cases we analysed reveal how siting decisions and processes are not only affected by geological, geographical and hydrological conditions.They are also shaped by the nature of the political and legal systems, formal and informal rules and procedures, culture, political constraints, technical skills, the stock of knowledge, public acceptance and, not least, a country's nuclear history. The way in which competing information and knowledge is processed and put to use by different actors in different political and cultural contexts also plays an important role.

In most countries, despite decades of efforts, science, society and politics are struggling to deal with this wicked problem. Nuclear waste governance is confounded by regulatory challenges, the conflicting values and preferences of stakeholders and still unresolved socio-technical and political issues. Due to political and economic costs, the high risks involved, the "known unknowns" and the "unknown unknowns" associated with nuclear waste management, policy makers have limited interest in governing the DGD-problem. This is not an issue widely discussed in the media or that is well understood by the public in most countries.

It is understandable why the search for site repositories is such a risky undertaking, irrespective of the political system in which they occur. In most of the countries analysed in this volume, the site selection procedures either have not started yet, are on-going or are controversial and mired in deadlock. The countries examined represent a heterogeneous sample in terms of their energy policies, nuclear programme size, share of nuclear power in domestic power generation, political decisions regarding whether or not to phase out nuclear energy, and state of advancement of decisions and procedures related to the siting process for nuclear waste repositories. Great variance exists with regard to their relevant national institutional, legal and industrial frameworks. There are also marked geographical, demographic, socio-economic, cultural, and political differences.

This is thus similar in some ways to what Adam Przeworski and Henri Teune (1970) defined as a most different system analysis. Despite their seemingly different political and economic systems and cultural predispositions, the countries studied all share a common challenge: what to do with HLW? Yet, it is understanding how differences in political, economic, and social structures and cultures influence NWG approaches that interest us in this research. We thus are interested in the more nuanced differences to be found in different country contexts and not simply in why NWG management is so difficult.

The process that leads to a selection of clay, salt or granite as host-rock for DGD is by no means only technical. Numerous socio-technical factors come into play in selecting a site. Regulatory structures, legal frameworks and at least in democratic states, social acceptability play a pivotal role in siting processes. As in many other cases concerning socio-technical issues, the core dimensions of the problem are closely interrelated. These are: the material nature of the problem and its possible consequences; the high technological requirements regarding controllability and safety risks (technical problems); and the social conflicts, conflict situations and political practices that are reflected in the behaviour of actors, institutions, organizations, the state or international organizations. In the case of disposal of radioactive waste, there are additional specific problems. Apart from the fact that the period to be taken into consideration reaches a million years, the risk dimension includes further problems: technical, geological and social measures must be taken, however their long-term robustness cannot be tested for such a lengthy time scale. Moreover, the fears of local populations, which cannot be minimized, considered unrealistic or unjustified, must be faced and dealt with.

The comparative approach helps us to identify common challenges facing governments in dealing with this wicked problem. It also shows that there are no one-size fits all solutions to NWG and that decision making approaches are strongly influenced not only by science and technology, but by softer social, political, and economic factors. The comparison facilitates in identifying which systems are doing better in terms of developing open, transparent, and robust decision making approaches and where lesson learning can occur. It also points out areas which could cause challenges or problems in the future.

Given the importance of finding solutions to HLW management that maximize safety for not only this but future generations, we are interested in how different political systems are doing at promoting debate and transparency in site selection. What steps have been taken to address HLW and what lessons have been learned about NWG over time? How are scientific, political, and technical disagreements dealt with? Who is involved in decision making processes and how are conflicts being addressed?

Lessons learnt

There are of course various scientific, technical and geological differences across the countries we consider. There are differences, for example, in terms of geological conditions (the prevalence of crystalline rock, clay or salt formations) and geological stability (is the country prone to earthquates or not). These factors certainly influence the designs for barriers and waste encapsulation. These are not the focus of this study. Rather we are interested in those factors which have influence on decision making procedures chosen and how deliberation and participation are encouraged or discouraged. Indeed the siting procedures and participation and deliberation processes present rather unique features in each national case.

Comparative and multi-level social and political analyses help to shed light on the efforts being made and the difficulties associated with trying to find socially, politically, economically, and technologically acceptable strategies for HLW storage and/or disposal. The progress the countries analysed have made and the obstacles they face are discussed in detail in the case studies. This is done looking at their regulations, technology choices, safety criteria, monitoring systems, compensation schemes, institutional structures, and approaches to public involvement.

There are some common patterns in the governance processes and approaches to NWG in the countries examined in this volume, but also many important differences. The spectrum of the countries under scrutiny encompasses not only cases with substantially different government types, economic development levels, geographical size and governance styles but also with very different perceptions and understanding of the role that the social control of technology and participation in procedural matters should play.

Size matters

Size matters when it comes to nuclear waste management. The smaller and more densely populated a country, the higher the likelihood that there will be opposition to DGD sites for the obvious reason that they are more likely to be near to population centers. The transport of HLW to the nuclear waste management sites is also more likely to travel near or through population centers. Canada, China, and Russia have vast territories, large sections of which are only sparsely populated. This does not eliminate the potential for Not In My Backyard (NIMBY) style protest, but it does reduce it. In comparison, relatively small countries like Lithuania, Croatia, or Slovakia may have particularly challenging times finding DGD sites domestically.

Government – civil society relations

The experience of the Nordic countries analysed in our first Volume (Brunnengräber et al. 2015) reveals that a key element in successful siting procedures is public trust in governmental institutions and a willingness to delegate the negotiation of agreements to them, as this is perceived to be in the community's interest. The countries examined in this volume vary significantly in terms of the relationship between the state and civil society. Whereas Canada bares some similarities to the Nordic countries, China and Russia are both nuclear weapons states with on-going conventional nuclear energy programs. While Russia has a long history with conventional nuclear energy and the Chernobyl nuclear accident occurred in the former Soviet Union (USSR), China had a late start with its development of NPPs. Thus, Russia obtains a far greater share of its electricity from nuclear energy than does China. Despite the Fukushima nuclear accident, both countries are continuing to expand their conventional nuclear industries. Both also have limited experience with public participation in nuclear waste management and there are various constraints on non-governmental organisations. The limited experience with civil society participation in NWG is similar in several of the central and eastern European countries examined in this volume. What this means for assuring a system of checks and balances in NWG needs to be considered.

Other countries analysed here, like Argentina, Brazil, and Canada have smaller nuclear energy programs but retain support for conventional nuclear energy. Surprisingly, the Ukraine which is the site of the world's worst nuclear disaster, has also not abandoned its nuclear energy program.

Even in countries with longer democratic traditions, the levels of trust citizens have in their governments related to nuclear questions as well as to the extent that public participation is encouraged and valued in nuclear waste management is not always high. This can be seen in the cases of Japan and South Korea. Both countries obtained large shares of their electricity from nuclear energy prior to the Fukushima nuclear accident. Most of Japan's NPPs remain off-grid still six years after the Fukushima accident. As of the summer of 2017, in South Korea there is a government which has called for an end to Korea's nuclear expansion plans.

The Fukushima nuclear disaster has forced major changes in Japan's approach to NWG. There has been a strengthening of checks and balances, a restructuring of institutional responsibilities, and moves to make decision making more participatory and transparent. The huge amounts of waste resulting from the nuclear accident, the volcanic nature of the island, and the lack of trust in the industry and government make the NWG challenge in Japan particularly large. South Korea has used compensatory approaches to win support for disposal sites for low- and medium level wastes, but has had trouble winning local support for a DGD site. Its nuclear policies are currently in flux.

In the central and eastern European countries where democratic traditions are relatively young or non-existent, there is still limited transparency and procedural opening despite requirements to this effect emanating from the Euratom Directive.

Responsibilities for nuclear waste management

There are some common patterns concerning the subdivision of responsibilities between waste producers and waste management organisations. A common feature of many national governance structures is the functional separation between "operators" and "regulators" in charge of overseeing safety requirements and standards. The level of independence of regulators can, however, be an issue as was learned in the case of Japan. After the Fukushima nuclear crisis regulatory institutions were restructured and responsibilities altered. There are, moreover, frequently prominent differences with respect to the ownerhip structures of the implementing organisations, which are sometimes state agencies and at other times in private hands. What these differences mean for the effectiveness and eventual safety of NWG need to be considered.

DGD – deep geologic disposal as a favoured path

Deep geologic disposal represents the preferred path as the most adequate way of disposing of HLW, independent of whether the host rock is crystalline, clay or salt. DGD is considered by the great majority of scientific and technical experts as the best available option that does not place enduring burdens on future generations. This paradigm is not without some critique, however. In some countries, there is a debate concerning whether there should be permanent closure of the final repository or options for depositing radioactive wastes in such a way that they are readily retrievable from the repository in the future should technologies for reducing radioactivity be developed. In yet other smaller countries with small volumes of waste there may be concerns about whether they are in a position to build DGD facilities at all and whether it makes sense for them to do so.

Long-term interim storage facilities

All countries, including those in this study, store their HLW in interim facilities which are used for initial cooling of the radioactive materials and because of the lack of alternative disposal options. In most cases, there have been substantial delays in selecting suitable and publically acceptable sites. To date none of the countries examined in this volume have constructed a DGD repository although they differ in the extent to which steps have been taken in this direction. Several states have opted for interim storage options that will be used for the next decades or even centuries, putting decisions about DGD sites on ice. In these cases, radioactive waste is stored above ground in specially constructed facilities which will allow retrieval for the length of the interim storage which might be decades or even centuries. Such interim storage decisions, however, transfer risk and decisions about what to do with the waste in the long term to future generations. The major criticisms of this approach concern the security risk in the case of terrorist attacks (from the outside) and the safety risk in the case of leakages of radioactivity and other yet unknown problems (from the inside).

Procedural and distributional justice, voluntarism, and compensation

Neither state control nor steering mechanisms of the kind that in the past shaped the ideas of political-administrative systems in many countries nor a (feasible) technical solution alone can represent good recipies for coping with the siting process. In most countries there is still considerable public mistrust regarding nuclear waste management. Governments have made too many false promises or shifted too often the waste question into the future. Therefore, siting is a broad societal project that needs coming to terms with the past and a broad societal dialogue. Voluntary search processes have been the preferred path in countries that are in an advanced stage of planning a repository or are already constructing one (Di Nucci et al. 2017). An essential advantage of voluntary siting approaches vis-à-vis D-A-D (Decide-Announce-Defend) strategies is the consent of the affected population. Often this process is associated with compensatory measures. These however are sometimes criticized as an act of bribery especially since possible compensation for the burdens of a repository are not communicated and negotiated in a transparent way.

In many countries, latent and open conflicts that have grown over decades have contributed to the "clumsy solutions" which have so far been taken (Verweij and Thompson 2011). Strong conflict lines have influenced actual decision making processes and these reach far into the future. Waste siting decisions have both inter- and intra-generational implications. Postponing decisions leave future generations to deal with problems they have not generated. There has been surprisingly little discussion of the inter- and intra-generational justice problems associated with nuclear waste management in the countries examined in this volume although there are hints of it in some. These tend to come in the form of compensation payments to host communities or plans for how compensation should be financed in the case of an accident.

One potential justice argument, for example, could be that the beneficiaries of nuclear plants in terms of jobs and business revenue should also bear the highest responsibility for the construction of a repository. However, in the case of a DGD repository, the site must be selected with regard to the most secure barrier functions, depending on the host-rock. This makes it rather unlikely that a nuclear power plant site will have the necessary conditions to make it a favourable location for a repository. Another perspective would be that in the past, communities hosting nuclear facilities have borne increased environmental and health risks for the benefit of society as a whole and now other areas should have to take over this responsibility and host a repository. In both cases, compensatory payments might be justified for reasons of fairness.

The multi-generational character to permanent disposal of nuclear waste motivates flexibility while the "wickedness" of the problem urges a comprehensive, continually adjusted and reviewed approach. These debates do not appear to play a strong role to date in the countries examined in this volume.

The nuclear-industrial complex

At the global level, the nuclear industrial complex is declining along with the decline in operational nuclear power plants and in their production of electricity since 2006 (Schneider and Froggatt 2016). The industry "is suffering from the cumulative impacts of the world economic crisis, the Fukushima disaster, ferocious competitors and [their] own planning and management difficulties" (Schneider and Froggatt 2012: 5). In several of the countries in this study, however, NPPs continue to be used and constructed and plans have been set for further expansion. This is the case, for example, in both China and Russia and was the case in South Korea until the election of Moon Jae-In. Whether Moon will be able to realize his aims for a decline in nuclear energy dependency in South Korea remains to be seen. In Japan's case the future of nuclear power in the country remains contested. In Japan, only a small number of NPPs are in operation although the government would like to restart more reactors. Whether it will succeed to return nuclear to a sizeable share of the country's electricity mix or will continue to be hampered by anti-nuclear activitists, the courts, and local government opposition remains to be seen. Argentina, Brazil, Canada and South Africa also aim to maintain or expand their nuclear energy sytems. Many of the eastern European countries examined here still have operating NPPs and have the added challenge of high dependency levels on nuclear for their electricity supply despite aging facilities. This makes the countries in this study particularly important for deeper analysis of the NWG plans, strategies, and tools.

Sophisticated: new forms of (robust) governance

The development of geological repositories for radioactive waste will take place over many decades. Societal developments over such a long time are not predictable and should be open to progress in science and technology, to evolving societal demands and to fixing potential implementation errors. In most of the countries examined in Volume I, nuclear waste management debates are no longer confined to scientific and techno-political actors, but also include many other relevant stakeholders, including civil society and social movements. The major actors involved include waste producers, waste management organisations, regulatory bodies, civil society, and policy makers at the national, regional and local levels. This is less obviously the case in many of the countries examined in Volume II. While this can be found to some extent in Canada and more recently Japan and South Korea, citizen involvement in NWG is more limited in many of the other countries examined.

One concern about nuclear waste management in countries with limited open debate is whether the conditions exist for reflexive governance, that is the ability to modify or reverse a decision and take a different course of action as long as a repository has not been sealed. Political decisions and the reversibility of decisions are the two sides of the same coin.

In many of the countries in this volume, government-led public discussions are strongly criticized as mere formalities. Although there may be resistance to more participatory and open planning structures, new and more democratic approaches to NWG will likely become necessary, even if the process is often slow and cumbersome.

Risks and uncertainties

Decisions concerning waste management require allocating risks and benefits to different regions, different generations, and social groups. As the US National Research Council noted, many of these decisions were (but still are) "linked to the national debate over the role of nuclear energy and the future of nuclear weapons" (NRC 1984: 1). Russia is a nuclear weapons state that is also pursuing reprocessing. There are concerns that a nuclear weapon state could use the spent fuel from other countries for military purposes (proliferation). This has raised concerns in Europe regarding the ties between the nuclear power programs in central and eastern European countries, like Hungary and Russia. In other countries in this study, the deployment of nuclear energy is linked to the narrative of a clean energy source and promoted as a good way to combat greenhouse gas emissions. This is the case with the Intergovernmental Panel on Climate Change as well as NGOs in Canada and the USA (IPCC 2014, chapter 7). Yet in others, the long term reliance on nuclear energy is beginning to be questioned not only for safety reasons, but also because of the problem of HLW management. In some cases, hopes are placed on finding a safer solution or even with the claim that for future generations waste could be a resource.

Problems are "socially and ideologicaly produced" and their (possible) solution depends on how the problem is framed. The definition of problems also depends on the potential solution being considered or taken. Categories such as "right" or "wrong" do not apply to this kind of issue and there is no ideal solution.

On the contents and contributions in this Volume

Following the same scheme as in Volume 1, the 14 country studies included in this book describe national nuclear waste inventories and the legal and institutional framework within which NWG occurs. The primary stakeholders in the debate and their interests, responsibilities and authority in relevant decision-making processes are discussed. Consideration is given to the value systems that are influencing the different national policy choices. The views and expectations of different stakeholders regarding participatory decision making and compensation and the steps that have been or are being taken to promote dialogue and problem-solving are also considered.

Various organisational logics could be used in looking for patterns across cases. Here the editors decided to group the case studies as follows:

a) Countries with large nuclear programmes and hence large volumes of nuclear waste (China, Russia, South Korea and Japan);

b) Central and eastern European countries which democratized in the post-Soviet era (Hungary, Lithuania, Slovak Republic, Slovenia, Croatia, and the Ukraine);

c) Canada and economically powerful countries in the Global South (Argentina, Brazil, and South Africa with some discussion of Mexico).

"The Big 4"- China, Russia, South Korea and Japan

In Chapter 2 *Eva Sternfeld* analyses nuclear waste governance in China where the nuclear power sector is growing faster than in any other country in the world. Most spent fuel is stored at reactor sites; the rest is transported to offsite interim storage facilities. The author purports that China has ambitions to realise a closed nuclear fuel cycle strategy with reprocessing. China first started looking for a DGD for HLW starting in the mid 1980s with a focus on the Beishan area in Gansu in northwest China as well as in Xinjiang Uygur Autonomous Region. Final site selection is still pending. Planned construction of an underground research laboratory (URL) has been delayed.

The existing legislation regarding the treatment of nuclear waste is still uneven and unsystematic, as China lacks a comprehensive National Atomic Law that regulates the obligation for NPP operators to treat and dispose of nuclear waste. The transportation of spent fuel via public highways, a distance of over 4,000 km from the NPPs in the eastern provinces to the designated interim storage facilities and later possible final repository site, appears to represent both the bottleneck and the Achille's heel of the Chinese nuclear waste strategy, as Eva Sternfelder shows in her chapter. Although nuclear safety issues have been discussed to some extent by the Chinese public in the aftermath of the Fukushima accident and in recent years, protests against NPPs and related industries have been reported, the issue of nuclear waste has received scant attention in the public debate. This is expected to change as soon as nuclear waste transports to the treatment facilities start increasing.

According to *Felix Jaitner,* the Russian nuclear industry has undergone a deep reorganisation in the last years. The Russian Federation has established the regulatory foundations in order to define radioactive waste and provide solutions for its disposal. The selected approach includes both near-surface disposal and disposal of radioactive waste and SNF in DGD. Apart from the development of the necessary infrastructure, increased reprocessing of spent nuclear fuel and the development of a closed fuel cycle are the paths followed. Jaitner reflects on the guiding principle in restructuring the nuclear industrial complex and concludes that these were not driven by the desire for new governance modes, increased democratic control or the implementation of ecological standards, but by the logic to adjust Russia´s nuclear industry according to the requirements of the market economy. In fact it is no longer possible to sell a NPP turnkey if a solution to the waste problem is not part of the deal. Thus, the overall rationale of this restructuring-process could be to secure the role of nuclear energy on the domestic market and to enhance Rosatom´s international expansion. Radioactive waste management is regarded as a decisive factor influencing the future success of the nuclear complex. The new approach to radioactive waste management appears to be driven by the valorisation of radioactive waste and spent fuel (with reprocessing and fuel banks). This approach is based on Russia's goals of having a closed nuclear cycle where there is reprocessing of spent fuel, deep well storage and deep geological and near-surface disposal. The foundation of an international fuel bank of low-enriched uranium to assure fuel supply to countries without their own fuel cycle facilities or repatriation of spent fuel for Soviet-built NPPs (the case in Belarus, Hungary, and Vietnam) is one of the pillars of the Russian strategy.

Sung-Jin Leem and Miranda Schreurs discuss the current status and limits of Korean nuclear waste management and governance. They argue that because the temporary spent fuel storage facilities at the sites of NPPs will reach full capacity in the near future, the disposal of spent fuel has become a matter of great concern in South Korea. Considering the ecological and social impacts of a HWL-repository, a societal consensus on nuclear waste disposal is needed. However, the participation of citizens and stakeholders in NWG decision-making processes and their access to information related to nuclear power were restricted. Following the failed attempts to find a nuclear waste repository site, legal and institutional instruments to engage the public were introduced. Anti-nuclear NGOs refused to participate in the conservative-government led public discussions and criticized them as a mere formality. With the election of Moon Jae-In, open dialogue on nuclear-related issues is being encouraged. According to Sung-Jin Leem and Miranda Schreurs, President Moon's more open, inclusive and transparent approach to decision making represents an important sign of the major changes underway in South Korea's energy policies.

Lila Okamura describes the situation before and after the Fukushima catastrophe. Prior to the Fukushima accident, the country was the world's third largest producer of nuclear energy. This huge industry and research sector was built on the premise that Japan would recycle its spent fuel. The author asserts that it is upon this unstable foundation that Japan attempted to establish its nuclear disposal concept. Discussion on DGD in Japan was restrained by the assumption of spent fuel recycling. This "vicious circle" hindered the attempt of a number of governments to deal with the disposal problem effectively. The lack of transparency, the absence of alternatives, and the confusion surrounding nuclear waste management in the country has meant that the debate has never really reached the public arena and new forms of governance have not really been discussed. Nuclear policy decisions have been taken in a top down fashion; local communities and civil society have very limited power. A major aggravation is that the waste from the Fukushima site and other types of waste were outside the scope of existing legislation. Okamura assumes that uncertainties will remain until the government decides which direction to take, and this in turn renders the search for a site even more complex. Okamura asks whether Japan will have the will, the courage, the means, and the time to take on the formidable challenge of nuclear waste disposal.

A Missed Opportunity: The Case of the Eastern European Countries

Directive 2011/70/Euratom required all member states to submit a report on the implementation of their national programmes for the safe management of spent fuel and radioactive waste by August 2015. The EU Members States have to comply with the regulations of Euratom. These requirements could have represented a chance to adjust the national structures of the eastern European member states to the requirements of the Euratom Directive, but the contributions in this sections seem to indicate that this was a missed opportunity.

In Hungary nuclear waste originates mostly from the four VVER-440 reactors at the Paks site. Although establishing a national HLW disposal site is a national task, the issue of SNF is tied to agreements with and shipment of SNF to the Russian Federation. *Zsuzsanna Koritár* explains that at the time of plant construction, in the 1970s, an agreement was made between Hungary and the Soviet Union that spent fuel should be transported to the USSR for reprocessing. The last shipment took place in 1998, but since then there has been no definitive decision regarding Hungary's NWG strategy. The reference scenario foresees a domestic DGD, however the option of shipping spent fuel to the Russian Federation for reprocessing still remains open. SNF is currently stored in an interim storage facility next to the Paks NPP. A complex screening procedure for a potential waste disposal site (hosting either HLW after decommissioning or spent fuel) started in 1999. As a result, six geological formations were "recommended for further research". The most favoured is the Boda Siltstone Formation (clay) in the Western Mecsek Mountain, in the southern part of Hungary. Public participation has been limited. There has been some engagement of the municipal associations and local governments, whose primary current task is to provide information to the public. Koritár argues that the national programme for radioactive waste management is inspired by a wait-and-see approach. The Hungarian national programme was prepared by the government and notified to the European Commission in August 2015. Its final version was adapted by the government in August 2016. According to the plans, the DGD should start operation in 2064.

Lithuania possesses only one NPP – the Ignalina plant. Both of its units are permanently shut down and under decommissioning. Unlike many other countries with NPPs, Lithuania never operated a research reactor. During the Ignalina NPP operation from 1983 until 2010, roughly 21,500 spent nuclear fuel assemblies were accumulated. *Povilas Poskas* reports that part of this is accomodated for 50 years in a facility for dry interim storage in casks whilst the rest will be transferred to a new facility still under construction. Since 2003, two near surface repositories have been under construction near the Ignalina NPP. The Lithuanian SNF disposal programme is still in the initial site investigation and preliminary facility design stage. Poskas underlines that in the site selection process the Lithuanian authorities considered public acceptance and good relations with neighbouring countries almost as important as geological criteria. Possibilities for disposal of SNF in a geological repository were analysed between 2001 and 2004 with support of Swedish experts. A repository in crystalline rocks has been proposed and a related generic safety assessment for this repository concept has been performed. There are plans to continue investigation of clay and rock formations with support of safety assessments. Poskas emphasizes that existing knowledge has largely been gained from international cooperation.

Slovakia is the focus of *Peter Mihók*'s chapter. After its establishment in 1993, Slovakia had to cope with new challenges in the governance of nuclear waste disposal, mainly with regard to final disposal of SNF. The Slovak nuclear waste disposal programme shares a part of its history with the Czech Republic. As the links between the two countries remained strong even after their separation, the author assumes that the Slovak authorities closely followed the development in the Czech Programme. Mihók points out that the major challenges were because the SNF generated in the territory of the former Czechoslovakia was to be disposed of on the territory of the Czech Republic due to the more suitable geological conditions there. Moreover there was, and still is a certain dependence on the Russian Federation because of cooperation agreements between the former Czechoslovakia and the former Soviet Union. The establishment of a national agency for nuclear waste envisaged in the Slovak legislation has not been realised and relevant roles were 'entrusted' to the state-owned company JAVYS only in 2011. All these factors can be considered as a ground for the delayed restart of the national SNF geological repository development programme, which was suspended in 2004. Despite the govern-ment's request to restart this programme as soon as possible, as affirmed in the Nuclear Back End Strategy in 2008, only some desk research activities have been carried out since 2013. The plan for SNF final disposal was reported in the joint National Policy and National Programme document (prepared in order to comply with the Council

Directive 2011/70/Euratom), approved in 2015. The author points out that further delays also affected the adoption of a law on compensation to municipalities affected by nuclear waste facilities. Mihok also criticizes the lack of financing for HLW management, the failure to enforce relevant laws and the failure to establish a legal entity to take responsibility for the final disposal of SNF.

Leo Šešerko examines forms of governance related to nuclear waste in Croatia and Slovenia. At the time the plans for the first Yugoslavian NPP, Krško were drafted in the 1970s, there was no mention of nuclear waste. After the disintegration of Yugoslavia in 1990, a contract was drawn up between Slovenia and Croatia dividing ownership of the NPP on a 50:50 basis. The agreement extended to the nuclear waste which was generated. This agreement is the origin of both cooperation and entanglements about nuclear waste. Until now, Slovenia and Croatia have decided to postpone the construction of a final repository for spent nuclear fuel. In describing the strategy for spent fuel management adopted in 1996, Šešerko notes that cooperation has proven difficult as both countries strive to take on as little of the costs associated with the spent fuel as possible. The author argues that the two states have had opposing interests from the beginning and that the present situation can be risky. He asks whether there might end up being two separate final repositories.

Twenty years after the start of operations, the payments being made into two separate decommissioning funds have not been coordinated. There are not sufficient funds to fully cover the expected costs of radioactive waste management and decommissioning. There have also been conflicting opinions among the owners regarding compensation payments to the surrounding population. The relation between the state institution responsible for handling all waste, preparing strategies, developing technical capacity as well as operating facilities for storage or disposal and the NPP company as a primarily profit-driven company is not synergetic.

Olexi Pasyuk posits that the consequences of the Chernobyl accident continue three decades later to define the Ukraine's radioactive waste management situation. Today, radioactive waste at the Chernobyl NPP site and surrounding exclusion zone constitutes over 98 percent of the country's total solid radioactive waste. SNF is excluded from this figure as it has a special legal status and is not considered to be radioactive waste. Following Ukraine's independence from the former Soviet Union, the country's nuclear waste management system has changed repeatedly. There is still no clear delineation of responsibilities or distribution of roles among involved institutions. Experts recognize the need for clarity and have made proposals to centralise the management system. EU and International Atomic Energy Agency (IAEA) funding has enabled research on which waste management system would be most suitable for the Ukraine. The research has examined DGD, regulatory system improvements and physical infrastructure. Adaptation of the Ukrainian standards and practices to the European standards as well as the development of legal, institutional and scientific structures will be accelerated in view of the EU-Ukraine Association Agreement. Because of the military conflict with Russia, Ukraine lost control of its research reactor in Sebastopol and nuclear waste repository in Donetsk. The scale and the cost of the Chernobyl nuclear waste disaster overshadows the problems of the waste accumulated at Ukraine's additional 15 operating commercial reactors. It also complicates the governance of radioactive waste. Pasyuk argues that Ukraine's waste management system suffers from the government's focus on solving day-to-day tasks rather than focusing on long-term objectives. This absence of strategy is particularly problematic for radioactive waste management.

Nuclear Waste Governance in Canada and the Global South

Cindy Vestergaard introduces and analyses the Canadian case. Over the past decade, Canada has been implementing an "Adaptive Phased Management" (APM) approach to handling its radioactive waste, involving interim storage at reactor sites and plans for eventual final disposal in a centralised, DGD. There are still many unanswered questions in this strategy, ranging from location to cost to long-term acceptance and even viability. Vestergaard posits that APM has evolved over almost five decades of debate, study, and public discourse. It has been shaped by the communities involved and scientific and technical studies based on past and ongoing inquiries within Canada and abroad. The author describes this process as a flexible step-by-step, phased approach supported by public engagement and scientific research at each interval along the way. The timeline for implementation is extended over a long period. She warns that the process will span generations, taking potentially upwards of two centuries to fully implement. It will require identifying a willing host community through to the construction, monitoring and eventual closure of the repository.

In Argentina and Brazil, the governance of nuclear waste remains an unresolved issue as *Moïra Jimeno* describes. Some regulations were issued after the restoration of democracy, like Law 25018 which was enacted in Argentina in 1998, and Law 10308 which was enacted in Brazil in 2001, but there are still no long-term disposal strategies for nuclear waste in either country. SNF is stored in interim facilities located at the NPPs. Argentina is experiencing a nuclear renaissance and there are NPPs under construction and plans for more to be built in the next years. The difficulty of finding a solution to HLW management resides in the limited involvement of relevant actors in the process of selecting a nuclear disposal strategy. The national nuclear energy commissions in both countries, together with their national governments, have exercised centralized control over nuclear issues, including nuclear waste, leaving almost no room for social participation and new forms of governance.

Jimeno claims that national legislation for the classification of radioactive wastes as well as nuclear disposal and storage and licensing are well developed. However, problems are related to implementation and there have been delays in obtaining licences and authorizations. Currently, even decisions on disposal sites for very low, low, and intermediate-level waste repositories are experiencing delays. This does not bode well for finding a solution for final disposal of HLW.

David Fig uncovers the history of nuclear waste in South Africa, focusing on the legal and institutional structures and controversies. As a supplier of uranium, South Africa's links with the global nuclear industry have been strong and can be traced back to the Manhattan project. The first research reactor was commissioned in 1965. Between 1978 and 1990 there was a secretive nuclear weapons programme. Given the considerable waste resulting from the mining of uranium, Fig notes his surprise that the first policy document on radioactive waste appeared only in 2005. The government was forced to establish a policy because of civil society litigation. Despite this, the policy was issued with almost no public input. The author posits that the industry has felt little accountability to the public, and the regulator has struggled to meet its obligations, given shortages of expert staff and budgets.

There is growing concern about current government plans for a fivefold expansion of nuclear generation capacity, given the need to ensure that the institutions governing and regulating the industry are competent and viable. This would necessarily mean a multiplication of the amount of nuclear reactor waste that has to be managed. Fig argues that it is difficult to see procurement succeeding in the short term and that Zuma's political successors are unlikely to champion this procurement very strongly. However, while vested interests still exist, it is too early to write the programme off entirely. The abiding question is whether South Africa possesses sufficient technical skills, finances and commitment to manage the resulting waste effectively and carefully regulate an expanding nuclear industry.

References

Bergmans, A.; Kos, D.; Simmons, P. and Sundqvist, G. (2015). The participatory turn in radioactive waste management: Deliberation and the social-technical divide. In: *Journal of Risk Research*, 18(3), 347-363.

Blowers, A. (1999). Nuclear waste and landscapes of risk. In: *Landscape Research*, 24(3), 241-265.

Blowers, A. (2016). *The legacy of nuclear waste*, London: Routledge.

Brans, M.; Ferraro, G. and von Estorff, U. (2015). *The OECD Nuclear Energy Agency's Forum on Stakeholder Confidence, radioactive waste management and public participation: A synthesis of its learnings and guiding principles*, Luxemburg: Publications Office of the European Union. http://publications.jrc.ec.europa.eu/repository/handle/JRC97499, last accessed 03 November 2017.

Brunnengräber, A.; Di Nucci, M. R.; Isidoro Losada, A. M.; Mez, L. and Schreurs, M. A. (Eds.) (2015). *Nuclear Waste Governance. An International Comparison*, Wiesbaden: Springer VS, Energiepolitik und Klimaschutz. Energy Policy and Climate Protection.

Cotton, M. (2017). *Nuclear Waste Politics: An Incrementalist Perspective*, New York: Routledge.

Di Nucci, M. R.; Brunnengräber, A. and Isidoro Losada, A. M. (2017). From the "right to know" to the "right to object" and "decide". A comparative perspective on participation in siting procedures for high level radioactive waste repositories. In: *Progress in Nuclear Energy*, 100, 316- 325.

Flüeler, T. (2006). *Decision Making for Complex Socio-Technical Systems: Robustness from Lessons Learned in Long-Term Radioactive Waste Governance*, Dordrecht: Springer.

IPCC (2014). Climate Change 2014: *Mitigation of Climate Change, Prepared by Working Group III contribution to the IPCC Fifth Assessment Report (WGIII AR5)*. Final draft, Genf. http://mitigation2014.org/report, last accessed 10 September 2017.

National Research Council (Panel on Social and Economic Aspects of Radioactive Waste Management, Board on RWM) (1984). *Social and Economic Aspects of Radioactive Waste Disposal: Considerations for Institutional Management*, National Academy of Science, Washington D.C.: National Academy Press.

Przeworski, A. and Teune, H. (1970). *The Logic of Comparative Social Inquiry*, Malabar, Florida: Krieger Publishing.

Schneider, M. and Froggatt, A. (2016). *The World Nuclear Industry Status Report 2016*. A Mycle Schneider Consulting Projekt, Paris, London. https://www.worldnuclearreport.org/The-World-Nuclear-Industry-Status-Report-2016-HTML.html, last accessed 12 August 2017.

Schneider, M. and Froggatt, A. (2012). *The World Nuclear Industry Status Report 2012*. http://www.worldnuclearreport.org/IMG/pdf/2012MSC-WorldNuclearReport-EN-V2.pdf, last accessed 03 November 2017.

Solomon, B. D. and Andrén, M.; Strandberg, U. (2010). Three Decades of Social Science Research on High-Level Nuclear Waste. In: *R HCPP*, 1, 13-47.

Strandberg, U. and Andrén, M. (Eds.) (2012). *Nuclear Waste Management in a Globalised World*, London: Routledge.

Verweij, M. and Thompson, M. (2011). *Clumsy solutions for a complex world: governance, politics and plural perceptions*, Basingstoke: Palgrave Macmillan.

II. “The Big 4”- China, Russia, South Korea and Japan

A Long Way Off

Nuclear Waste Governance in China

Eva Sternfeld[1]

Abstract

China has rapidly developed its nuclear power sector since the start of the 21st century. As most nuclear power plants (NPPs) only recently were put into operation, the majority of spent fuel is still stored at the reactor sites although some is transported to offsite interim storage facilities. China has expressed interest in building a geological deep repository for disposal of high level waste (HLW) from its nuclear facilities. Screening of potential sites started during the mid 1980s. R&D proceeded over the years, but received only limited funds, as attention was more focussed on China's ambitions to realise a closed nuclear fuel cycle strategy with reprocessing. So far, a site for a deep geological HLW repository in Northwest China is not confirmed and the planned construction of an underground research laboratory (URL) and a final repository have been delayed. The existing legislation regarding the treatment of nuclear waste remains fragmented and unsystematic, as China lacks a comprehensive National Atomic Law that regulates the obligation for NPP operators to treat and dispose nuclear waste. The overland transportation of spent fuel via public highways, a distance of over 4,000 km from the NPPs in the eastern provinces to the designated interim storage facilities and later final repository site, seems to be both the bottleneck and the Achilles heel of the nuclear waste strategy. Although nuclear safety issues have been discussed to some extent by the Chinese public in the aftermath of the Fukushima accident and in recent years protests against NPPs and related industries have been reported, the issue of nuclear waste has not yet been addressed in the public debate. This is expected to change as nuclear waste transports to the treatment facilities increase in the coming years.

---

1 Eva Sternfeld, Möckernstr. 112, 10963 Berlin; E-Mail: eva.sternfeld@gmx.net

A. Brunnengräber et al. (eds.), *Challenges of Nuclear Waste Governance*,
Energiepolitik und Klimaschutz. Energy Policy and Climate Protection,
https://doi.org/10.1007/978-3-658-21441-8_2

1 Introduction

China's civil nuclear power sector has a relatively short history compared with other nations. The question of spent fuel management and nuclear waste disposal is, however, becoming more urgent as the country is currently implementing the world's most ambitious nuclear power program. By July 2017, China had 36 operating NPPs with 32,6 GW installed capacity, in addition to 21 NPPs with 23 GW under construction and plans for NPPs with 43 GW already approved. It is expected that for the period of the 13th Five Year Plan (2016-2020), 6-8 nuclear power reactors will be approved per year (WNA 2017). Despite its rapid development in recent years, nuclear power only contributes a tiny share, less than two per cent, to China's electricity mix. The Chinese National Energy Admin-istration (NEA) has set targets for electricity to be produced by NPPs: 5 per cent by 2020, 9 per cent by 2030 and up to 16 per cent by 2050. Taking expected rises in total electricity production into account, this translates to an installed capacity of 200 GW by 2030, and up to 400 GW by 2050 (NEA 2012; Hu and Cheng 2015).

For many years the Chinese nuclear strategy has envisioned a "closed nuclear cycle," including reprocessing and deep geological disposal for HLW. Screening for suitable sites for a geological repository began with construction of the first commercial nuclear plants around 1985. These activities continued over the years; recently, area screening has concentrated on three granite sites in the Beishan area in Gansu province, a sparsely populated arid region in Northwest China. The program, however, is underfunded according to Chinese experts and deserves less attention than the technological advancement of reactors from energy planners (Pan and Qian 2013).

While the Fukushima accident provoked a public debate about nuclear safety in China, which caused a revision of existing nuclear legislation and policies, it did not significantly engage the question of nuclear waste management. China's awakening anti-nuclear protests in the aftermath of Fukushima have been sporadic and, as typical NIMBY protests, targeted NPPs and nuclear industries in the densely populated Eastern parts of the country. As the proposed sites for the offsite interim storage sites and the geological repository are located in sparsely populated or uninhabited areas in Northwest China, they have not been the subject of protests. Information about China's nuclear waste management program is available; scientists and officials participating in the nuclear waste management program regularly report on the technological progress of the program at international and domestic conferences and publish their findings in scientific journals. Despite its availability, this information has not sparked public debate about the financial and socio-economic aspects of this issue. However, the expected increase of nuclear waste transports via public highways across China may draw more public attention in the future.

2 Nuclear waste disposal in China

2.1 Historical background of development of nuclear power

China's civilian nuclear power industry is an offspring of the country's nuclear weapons program. The Chinese National Nuclear Corporation (CNNC), China's leading NPP operator, is a state owned company that was formed in the early 1980s out of the former Ministry of Nuclear Industry (also known as the Second Ministry of Machine Building) (CNNC 2015). Starting in the 1960s, CNNC's predecessor was put in charge of developing atomic and hydrogen bombs and nuclear submarines. In this capacity it developed a complete nuclear industry infrastructure, including uranium exploration and mining, conversion, enrichment, fuel fabrication and research reactors.

China is, however, a latecomer to the "civilian" use of nuclear power . The first commercial NPP went into operation when the nuclear energy boom in most industrialised countries was already slowing down. The first reactor, a 300 MW PWR, an indigenous Chinese design, started operation in 1991 in Qinshan, Zhejiang province, 100 km southwest of Shanghai. In 1994, a second plant, the Daya Bay NPP in Guangdong province, started operation. This project was realised with foreign technology and is equipped with two 984 MW PWR units by the French company Framatome (today's Areva). Until the early 2000s, Qinshan Unit 1 and Daya Bay remained China's only commercial NPPs. Beginning in 1997, additional projects were put on hold, when, due to the Asian economic crisis, energy demand declined and planners became concerned about excess energy production capacity and high investment costs. Construction plans were only resumed with the 10th Five-Year-Plan (FYP) (2001-2005), which for the first time explicitly incorporated the development of nuclear energy into China's energy strategy. The plan called for the construction of eight NPP reactor units (Lingao Phase II, Qinshan Units 3 and 4, Sanmen and Yangjiang). Beginning in 2003, power shortages severely affected China's main industrial areas in the eastern coastal regions and the development of nuclear power gained increasing political support. The 11th FYP (2006-2010) was even more ambitious in terms of nuclear power and included 14 NPP units in total. The 12th FYP (2011-2015) incorporates nuclear as a so-called "new energy" in China's low carbon strategy. The plan originally envisioned an expansion from 10 GW to 40 GW of installed capacity within 5 years (about 30 units), but implementation was delayed and some projects have been shelved.

China's nuclear power sector developed its own PWR reactor designs: the CNP-300. CNP-600, CNP, CNP-1000 (developed from the Areva M-310 design) and CPR-1000. In addition, it purchased units from France (M-310 3-loop PWR, EPR), Russia (AES 91), Canada (CANDU 6) and recently from the USA (AP

1000). There are also programs to promote the commercial introduction of the high-temperature gas cooled reactor (HTGR), also known as the pebble bed modular reactor (based on an earlier German design) and an experimental fast neutron reactor (based on Russian technology). So far these projects are in an early stage and it is not clear if and when they ever will replace the PWRs.

As noted above, the Fukushima accident had an impact on China's nuclear development. In March 2011, the Chinese government announced a moratorium for safety inspections and amended the country's nuclear safety provisions. For 18 months, pending approvals for nuclear power plants were suspended. The accident in Japan provoked an internal and public debate about nuclear safety issues. Experts expressed concerns about the speed of nuclear development, an inadequate regulatory system and a serious lack of educated nuclear safety experts. They further warned that the fragmented institutional responsibilities among energy, environmental, health and security authorities could lead to a dangerous muddle of competences in the case of an emergency (Sternfeld 2014: 200ff.). The moratorium officially ended in October 2012 with the adoption of the revised *"Mid- and Long-Term Plan for Nuclear Development 2011-2020"*. Compared to earlier plans, the revised version downsized the targets for 2020 to 58 GW in operation and 30 GW under construction. All proposed inland projects awaiting approval for the period of the 12th Five Year Plan were shelved. In addition, the plan announced that some older second generation PWRs will be decommissioned earlier than originally scheduled. Further, the development plan and the safety plan demanded the redesign of proposed reactors to improve safety features. They emphasized that employed technology of newly built nuclear reactors must adhere to "*the highest safety standards in the world*"; in other words, "third-generation" technology. This implies that China will eventually give up production of the second generation standard reactor CPR 1000. The 22 CPR 1000 units that were already under construction in 2011, however, have been excluded from expensive redesign (Pandza 2013: 178).

The Fukushima accident led to adjustments in China's nuclear strategy but did not initiate a phase out or retreat from nuclear power. On the contrary, China remains committed to nuclear power development, which is seen as an important pillar of its strategies for energy security, environmental protection and low carbon development. Along with wind and solar, nuclear power is being promoted to solve China's 'energy dilemma', namely to reduce China's heavy reliance on coal. Coal has thus far fuelled the economic miracle but comes at the cost of a dramatic increase in CO_2 emissions, severe air pollution, congestion of transport systems, degradation of landscapes and a high number of fatal accidents in coal mines.

Spent fuel management and nuclear waste disposal are not yet a focus of attention related to the nuclear power sector. Despite the rapid pace of recent development and the announced expansion plans, reprocessing and disposal of spent fuel remain weak links in the Chinese nuclear power strategy. Proposed projects for reprocessing, interim off-site storage and a high level waste (HLW) repository are behind schedule.

2.2 *The national inventory for spent fuel*

At present, the majority of spent fuel is stored at the reactor sites. According to the China Atomic Energy Administration (CAEA), as of 2013, 3168 tons of HLW were stored in wet storage pools at reactor sites and 394 tons were stored in interim off-site spent fuel facilities (CAEA 2014).

Table 1: Estimates for spent fuel (tons) Calculations by different institutions

<table>
<tr><th rowspan="7">Accumulated Spent fuel</th><th>1994-2005</th><th colspan="2">2006-2020</th><th colspan="2">2010-2030</th><th>2030-2040</th></tr>
<tr><td rowspan="6">1100</td><td rowspan="3">Medium & long plan 40 GWe</td><td rowspan="3">5810</td><td>Reference</td><td>14730</td><td>27200</td></tr>
<tr><td>High growth</td><td>18771</td><td>38588</td></tr>
<tr><td>Low growth</td><td>10798</td><td>16119</td></tr>
<tr><td rowspan="3">Rapid growth</td><td rowspan="3">6579</td><td>Reference</td><td>17656</td><td>29256</td></tr>
<tr><td>High</td><td>21643</td><td>40493</td></tr>
<tr><td>Low</td><td>13668</td><td>18018</td></tr>
</table>

Source: Zhou 2010a

	2010 (20 GW)	2020 (40 GW)
tHM/y	600	1000
Accumulated	3800	12300

Source: WNA China Nuclear Fuel cycle (2016b)

	2020 (58 GW)*
tHM/y	1400
Accumulated	9000

* Middle to Long-term Plan for Nuclear Energy Source: CAEA 2014

Spent fuel accumulation for different reactor types:

- Daya Bay NPP 50 tons per year (CLP Group Website)
- AP 1000 35 tons per year (Xu 2010: 187)
- Qinshan Phase III CANDU 195 tons per year

Pan and Qian (2013) and Wang (2014) calculate that by 2050 China will have accumulated 82,630 tons of spent fuel in addition to "*certain*" amounts of HLW from the defense sector (Pan and Qian 2013; Wang 2014). This includes, for example, the decommissioning of a 10 MW HWRR experimental reactor in a Beijing suburb, which had been in operation since 1958 and was permanently shut down in 2007. Decommissioning, including removal of spent fuel and dismantling of the reactor, is scheduled to be completed around 2027 (Yang n.d.).

2.3 The (interim) storage sites

Low-level waste (LLW) is stored in stainless steel casks and disposed of at regional facilities. In the case of Daya Bay NPP, a surface facility, the Beilong repository, was built about five km from the plant. The facility is designed for a total capacity of 80,000 m^3 and in its initial phase has a capacity for 8,800 m^3.[2] Another LLW facility is located in Yumen, Gansu province and has a storage capacity of 60,000 m^3. In the Yumen facility, radioactive waste is stored at a depth of 10-20 meters underground. Three more LLW repositories are planned for the east coast, northeast and central regions of China (Zhou 2013: 8).

Since 1986, a nationwide screening has been conducted to identify potential sites for HLW(Wang 2007). In 2011, the CAEA and the Ministry of Environmental Protection (MEP) approved the Beishan area in Gansu province as the most promising site. The Beijing Research Institute of Uranium Geology (BRIUG) of CNNC selected three granite sections (Jiujing, Xianyangshan Xinchang und Yemaquan) in the Beishan area for further survey and feasibility studies for an underground research laboratory (URL).[3] In addition, since 2012 the North-West Institute for Nuclear Technology (NINT) in Xian has conducted screening at a second candidate site in the Aqishan area, Xinjiang Uygur Autonomous province. This secondary site is located in the vicinity of the former Lop Nor testing site for nuclear weapons.

The design of the URL has been approved and is scheduled for construction beginning in 2020. The construction of the final repository is expected to start by 2040 and scheduled to be operational by 2050 (WNA 2016b; Wang 2007).

In 2014, the construction of a vitrification plant to melt HLW into a borosilicate glass matrix for permanent disposal began at the Guangyuan nuclear complex in northern Sichuan province. It is designed for the treatment of liquid nuclear waste from decommissioned military facilities but can also receive waste

2 For a photo of this facility, see Zhou (2013: 7).

3 The Beishan granite site is located north of the Lanzhou-Urumqi railway line, about 150 km north of Jiayuguan and 50 km northeast of Yumen City

from NPPs.[4] The facility is equipped with German technology from the Institute for Nuclear Waste Management (INE) Karlsruhe Institute of Technology (WNA 2016b; KIT 2014).

NPPs built after 2005 are equipped with spent fuel pools with a 20 year storage capacity. Two of the earlier plants, Qinshan Phase II and Lingao, use dense-pack racks to extend the storage capacity. With a maximum storage pool capacity of about 690 assemblies, equivalent to 20 years of spent fuel generation, these two NPPs are expected to run out of storage capacity around 2020 (Liu 2014; Zhou 2010a: 2).

A dry storage facility is operating at Qinshan Phase III, which runs two CANDU reactors. Construction of this facility began in 2008. All together construction of 18 dry storage modules is planned; the first two were completed in 2009. Two modules with a capacity of 24,000 bundles will be built every 5 years. As the result the reactor's storage capacity will be expanded to be sufficiently large for more than 40 years of operation (Zhou 2011: 4361). In addition, the China Guangdong Nuclear Power Corporation (CGNPC) is planning a regional demonstration program for dry storage (WNA 2016b).

In 2009, the on-site storage pool of China's first reactor, Qinshan I, was expanded and there are plans to build a second pool, which would allow for storage of spent fuel near the reactor site until 2025 (Zhou 2011: 4361). This solution could not be implemented at Daya Bay NPP, the second oldest NPP, which has used up its onsite storage capacity after years of operation. Since 2003, spent fuel has been packed in NAC-STC spent fuel transport casks with a capacity of 26 assemblies each. Twice a year, 104 assemblies are transported by two special trucks over a distance of more than 3700 km from Guangdong province to either the offsite interim storage facility at the Lanzhou Nuclear fuel complex or the pilot reprocessing facility at Diwopu 低窝铺, near Yumen city in Gansu province (Zhou 2011: 4361).[5] Since 2003, about 20 of these transports have been

---

4 The Guangyuan plutonium production complex was built in 1974 and once was China's largest plutonium production and reprocessing plant. By the late 1990s, China decided to decommission the facility and restructure the company. It is now known as the Sichuan Environmental Engineering Company attached to CNNC, which specializes in decommissioning military nuclear facilities and managing radioactive waste (NTI 2012). 800 m3 of liquid nuclear waste are reportedly stored at the facility (WNA 2016b). For the history of the Guangyuan complex also refer to Zhang (2011).

5 A video entitled "Chinese nuclear transport," available on Youtube, shows one of these special truck HLW transports taking place on public highways: https://www.youtube.com/watch?feature=player_detailpage&v=C77NRp5lK7g#t=19 (last accessed on 20.11.2016). The text at the end of the film reads: "Later we learnt that these special vehicles transport nuclear waste over a distance of 3750 km from Dayawan in Guangdong province to Diwopu (低窝铺) in Gansu province. Both locations are connected by China's only transport route for nuclear waste. Since 2003, secret vehicle convoys have travelled for several months a year back and forth. A convoy

carried out. The frequency of nuclear waste transports is expected to increase after 2020,when on-site storage facilities at Qinshan II and Ling'ao will be filled up.[6] China will need to expand its fleet for HLW transport and also develop different types of transport casks for fuel from the different reactor types. In the future, China plans to shift from road transport to sea-rail transport. Feasibility studies for these transports are currently being conducted and the sea-rail transport system is expected to be in place sometime after 2020 (CAEA 2014).

Table 2: Status of spent fuel storage in Chinese nuclear power plants

NPP name	Unit	Connected to the grid	Spent fuel storage	On-site storage capacity (years)	Year when storage expected to fill up
Qinshan 1		1991	Dense-pack wet, pool expansion	35	2025
Day Bay	Unit 1 Unit 2	1993 1994	Wet-storage	10	2003 2004
Qinshan II	Unit 1 Unit 2	2002 2004	Dense-pack wet storage	20	2024 2022
LingAo	Unit 1 Unit 2	2002 2002	Dense-pack Wet storage	20	2022 2022
Qinshan III	Unit 1 Unit 2	2002 2003	On-site wet/dry storage	40	2042 2043
Tianwan	Unit 1 Unit 2	2006 2007	Wet storage	20	2026 2027
LingAo II	Unit 1 Unit 2	2010 2011	Wet storage	20	2030 2031
Ningde	Unit 1 Unit 2	2013 2014 2015	Wet storage	20	2033 2034 2035
Hongyanhe	Unit 1	2013		20	2033

is escorted by a police car that clears the highway, followed by two special trucks transporting casks, two meters high and six meters long. The convoy covers a distance of about 200 km per day. With armed escort, three climate zones are passed and 1700 bridges are crossed. The casks carry nuclear waste, a dangerous item of which just 10 µg is a lethal concentration for humans" (author's translation).

6 CAEA (2014) calculates with more than 40 cask transports from 2020 onwards and more than 100 transports per year from 2025 onwards.

NPP name	Unit	Connected to the grid	Spent fuel storage	On-site storage capacity (years)	Year when storage expected to fill up
	Unit 2 Unit 3	2014 2015	Wet storage		2034 2035
Yangjiang	Unit 1 Unit 2	2014 2015	Wet storage	20	2034 2035
Fanjiashan	Unit 1	2014	Wet storage	20	2034
Fuqing	Unit 1&2 Unit 3	2014/5 2016	Wet storage	20	2034/5 2036
Changjiang	Unit 1 &2	2015/16	?	20	2036

Source: Zhou 2011; WNA 2016a; Bunn et al. 2016; own calculations.

A centralised offsite interim fuel storage facility with an initial storage capacity of 550 tonnes has been built in Gansu province at the Lanzhou Nuclear Fuel Complex, 25 km from Lanzhou city. As of 2014, it had already received 430 tonnes, but according to NNSA a second 760 ton pool has been completed and is awaiting approval (CAEA 2014; Bunn et al. 2016).

Nuclear waste management will become more challenging from 2022 onwards, when more NPPs are expected to run out of onsite storage facilities. By 2035, the accumulated amount of spent fuel that will need to be stored off-site is estimated to reach 3158 tons (Zhou 2011: 4365).

Table 3: Scenario for spent fuel management in interim offsite storage facilities

Offsite storage space (tons)	Estimate when storage will reach full capacity
500	2014
1000	2025
3000	2035

Source: Zhou 2013: 20; Bunn et al. 2016

2.4 *The waste management strategy (with current waste disposal plan/ concept)*

The waste management strategy and its range will depend on whether China opts for a closed nuclear fuel cycle strategy with reprocessing or a 'once through' strategy with direct disposal of spent fuel. So far, the focus is on the closed fuel cycle strategy, which would support efficient use of fuel and reduce the amount of HLW. Since the mid-1980s, the Chinese government has demonstrated interest in operating a closed nuclear fuel cycle by recovering plutonium from spent fuel through reprocessing. Given the scale of China's planned nuclear program and China's limited uranium reserves and production,[7] some Chinese energy planners support technologies that more efficiently use the uranium in order to avoid an increasing dependency on international uranium supply and world-market prices. A closed fuel cycle strategy includes MOX fuel recovery from spent fuel and fast neutron reactors, which could increase fuel utilisation by a factor of 50-60 per cent (Pandza 2013:181).

So far, a pilot reprocessing plant has been constructed at Lanzhou Nuclear Fuel Complex with an initial capacity of 50 tHM/a and an interim spent fuel storage capacity of 550 tons (500 tons for commercial reactors and 50 tons for research reactors). Operation started in 2010 and the plant supplies the China Experimental Fast Reactor (CEFR), which began operation in the same year. Demand for fuel is expected to rise if China decides to operate a demonstration fast reactor (CDFR) by 2025 and start, as originally planned, commercial operation of fast neutron reactors by 2040. There are plans for two commercial reprocessing plants with reprocessing capacities of 400 and 800-1000 tons a year based on Chinese or French technology (Deng 2010: 8). Since 2007, China has negotiated with AREVA regarding the construction of an 800 t/yr reprocessing plant in Jinta, Gansu province. The technology refers to AREVA's UP 2.800 and UP 3 plants in La Hague, which use the Purex process to recover uranium and include a facility for MOX fuel fabrication. Spent fuel storage facilities will be about 3000 tHM for the first phase and up to 6000 tHM in a second phase (AREVA 2013; WNA 2016b; Deng 2010: 8). Originally the plant was scheduled to go into operation by 2020, but because negotiations are seemingly more complicated than expected, its start has been postponed until after 2025 (Zhou 2011: 4363; Pandza 2013: 181). In April 2013, on the occasion of the visit of French president Hollande, a letter of intent between the two countries to build a used fuel treatment and reprocessing facility in China was signed. In addition, in 2010, CNNC signed a framework agreement with the Belgian companies Tractabel, Belgonucleaire and the nuclear research centre SCK-CEN to build a pilot MOX fuel fabrication plant in China (WNA 2016b).

---

7 According to recent data China's known uranium reserves account for 120,000 tU. China's seven uranium mines have an annual production capacity of 1,450 tU (with possible expansion to 2000 tU/y) sufficient to supply about 7 GW although this is by far not enough to supply the rising demand of the industry. China therefore increasingly depends on uranium imported from Kazakhstan, Uzbekistan, Australia, Namibia and Canada. In 2013 imports from the five countries amounted to 18,968 tU (WNA 2016b).

Construction of these projects has not yet started because of issues concerning price and technology transfer. There is not much transparency or public information about these issues. According to original plans, the plant was scheduled to go into operation by 2020, near Jinta in Gansu province. The schedule has been delayed and construction start is now scheduled for 2020 with the plant to be operational by 2030 (WNA 2016b). According to recent information, Jinta will be now the site for a smaller facility of 200 t/y, whereas the 800 t/y AREVA facility will be built in the east coast, where at present all Chinese nuclear power plants are located. Spent fuel shall be transported by ship, rather than on a 3000 km road or rail trip inland to Gansu. Since 2015 six possible sites for a reprocessing plant and interim storage facility with a capacity of 3000 t were under preliminary evaluation for seismic activity (Bunn et al. 2016; BJZTB 2015; Buckley 2016; CNNC 2016). A final contract between AREVA and CNNC was expected for 2017. In August 2016 a 3km^2 site near the harbor city of Lianyungang was according to a statement of Wang Yiren, deputy director of the National Defense Science and Industry Commission, chosen as the most feasible location. This news prompted thousands of cititzens to protest on the streets and there were violent clashes with the police. Days later the local Lianyungang government asked to have the city´s name removed from the list (Ma 2016; Buckley 2016). However, the delay in construction of a planned commercial reprocessing plant may be seen as a sign that the internal Chinese debate on the trajectory of nuclear waste policy is not yet decided.

At present, there are, according to Zhou (2010b, 2011), still three possible scenarios for Chinese spent fuel management:

1) Sending all spent fuel, without any reprocessing, to interim wet or dry storage facilities.
2) Need-based reprocessing: start at a small scale, reprocessing 50 tons/year and later 100 tons/year in order to manufacture MOX fuel for the CEFR and CDFR and eventually the commercial fast breeder reactors, thereby avoiding unnecessary stockpiling of plutonium. In this scenario, reprocessing of roughly 800 t/year will not need to start until 2034.
3) Capability based reprocessing: reprocessing full scale with the start of operations, meaning up to 800 tons/year. All separated plutonium will be either manufactured into MOX fuel for use in LWRs or stockpiled for later use in fast neutron-reactors. The latter will require a highly sophisticated safe storage system to prevent proliferation and safety risks (Zhou 2010b: 5; Zhou 2011: 4366).

The reprocessing options only make sense if China proceeds with the fast neutron reactor program. To use reprocessed MOX fuel in PWRs makes little economic sense, since the costs for uranium fuel account only for 5 per cent of electricity generating costs of a NPP (Pandza 2013: 187). At present, the fast neutron reactor program is already behind schedule and there are reportedly conflicting proposals for if and how the program should be continued. There currently is a 65 MW CEFR near Beijing, which started operation in 2010, but it is said to have some technical problems. For the planned CDFR, originally scheduled to start operation in 2013, it is not yet decided whether China should opt for a domestically developed 600 MW prototype or purchase the Russian BN-800 reactor. Because China and Russia did not reach an agreement over financial terms, as of the end of 2014 the planned BN-800 project in Fujian had been suspended indefinitely (WNA 2016a). Concerns about the poor safety record, especially of the Russian BN-600 and other fast breeder reactors around the world, as well as high costs might have played a role in the internal Chinese discussion and decision to suspend the project.

The commercial reprocessing program is facing similar problems. As original schedules are delayed and none of the potential commercial reprocessing projects are expected to be in operation within the next 15 years, China needs to prepare for an increased demand for additional interim storage space for HLW from 2022 onwards.

2.4.1 Strategy for HLW deep geological disposal

In 1985, CNNC proposed an R&D program with the objective of building a deep geological repository from 2040 onwards (Wang 2002: 104). Over time the technological strategy became more elaborate and the time frame and concept for the repository were adjusted. The following schedule for a three-phase building plan was published in 2006.

Table 4: Long-term plan for China's HLW repository

Activities	**Phase 1: (2001-2020)** **Site selection and site confirmation**	**Phase 2: 2021-2040** **URL construction & in situ tests**	**Phase 3: 2041-2050** **Repository construction**
Site selection and site characterization	Area selection, surface investigation, borehole drilling and testing, complete site confirmation	Supplementary work for site characterization	Monitoring on the site
Underground Research Laboratory	Feasibility study, URL	Construction of URL, *In situ* tests & demonstration of disposal technology	*In situ* tests, monitoring of repository
HLW Repository	Conceptual design	Preliminary design followed by detailed design	Construction completed around 2050
Research and Development	Studies on radionuclide migration, engineered barriers, performance assessment, methodologies	Studies on *in situ* tests, radionuclide migration, engineered barriers, construction technologies	Studies on repository closure, monitoring etc.

Source: Wang et al. 2006: 56; Pan and Qian 2013: 194

Because of the focus on the closed fuel cycle strategy, the deep geological disposal of HLW has been a lower priority in China's nuclear program in terms of funding and political support. For example, the research for deep geological HLW disposal has never been listed as a key national R&D program (Zhou 2013).

3 The legal and institutional framework

3.1 The legal framework

In recent years, China has issued a number of laws, regulations, technical guidelines and standards relevant for the treatment of nuclear waste. Many of these have been adopted from international regulations and comparable legislation in other nuclear power states. China, however, still lacks a comprehensive National Atomic Law and a National Law on Nuclear Safety, which would regulate nuclear policy including the obligation for nuclear power producers to treat and dispose of nuclear waste, transparency requirements on the origin and transport of nuclear waste, the supervision and management of HLW repository and related issues such as public participation and compensation. The existing legislation for management of nuclear waste is rather fragmented and unsystematic (Xu et al. 2014: 606).

The 2003 *Law on Prevention and Control of Radioactive Pollution* (NPC) is to date the most relevant national law concerning nuclear waste. It includes provisions concerning site selection, construction, operation, power development and decommissioning of NPPs, with the focus on preventing radioactive pollution. Chapter VI deals with the control of radioactive waste and stipulates in *§ 43* that *"High-level solid radioactive waste shall be disposed of in a centralized deep geological disposal facility."* (NPC 2003). In addition, a number of regulations (*tiaoli*) are relevant for the nuclear waste sector. These include:

- *Supervision and management regulation for civil nuclear safety* (issued by the State Council in 1986) which stipulates that the National Nuclear Safety Administration (NNSA) is the responsible institution for safety supervision.
- The *Regulations for the safe transport of radioactive materials* issued by the State Council in 2010 refer to the transport of radioactive substances in special containers (Mu et al. 2015: 167).
- The *Regulations for nuclear waste safety management* were issued by the State Council in December 2011 in the aftermath of the Fukushima accident and have been effective since March 2012. As §3 stipulates, the regulations apply to reprocessing, interim storage and final disposal of radioactive waste. Chapter 2 §10-9 deals with processing and storage of radioactive waste. Chapter 3 §20-27 deals with final disposal. §22 stipulates the procedure for HLW disposal (including screening, URL construction, site selection and construction of the final repository). §23, 3 requires that a HLW repository shall guarantee a safe disposal for at least 10,000 years. Operators of HLW repositories shall have a registered capital of no less than 100 million RMB (§23,4).
- Regulations on Nuclear Power Accident Emergency Management (under revision).

The State Council's regulations are complemented by a series of management regulations (HAF series) and technical guidelines (HAD series) issued by ministries and state commissions in charge of these issues. These include:

- Management regulation on radioactive waste supervision (HAF401) issued by NNSA.
- Classification of Radioactive Waste (GB9133-1995).

- *Measures on Management of Permits for Storage and Disposal of Radioactive Solid Waste* (issued by MEP 2014). Measures are formulated to enhance supervision and management of storage and disposal of radioactive solid waste and regulating permits for storage and disposal of radioactive waste.
- *Provisional Measures for Management of Funds for Treatment and Disposal of Used Fuel from Nuclear Power Plants* issued by the Ministry of Finance, NDRC, Ministry for Industry and Information in 2010. Document No. 58 regulates the collection and standards for collecting funds. It stipulates a levy of CNY 0,026/KWh to be used as a fund to finance reprocessing and disposal of used fuel (Ministry of Finance 2010).
- The *2006 R&D Guidelines for Geological Disposal of HLW* (issued by CAEA, Commission for Science, Technology and Industry of National Defense, Ministry of Science & Technology, Ministry of Environmental Protection, Ministry of Finance) includes the target to build a HLW repository by 2050.
- The *Guidelines for Site Preselection for a Geological HLW Repository* (issued in 2006 by the Commission of Science, Technology and Industry for National Defense) provide selection criteria for a possible site such as no permanent residents, arid climate and a granite site in a northwest region (Yuan et al 2015: 483).
- *Management Regulation for Safety Supervision of Nuclear Equipment of the National Defense Industry* issued in 1999 by the Commission of Science, Technology and Industry for National Defense); it regulates the disposal of HLW from decommissioned military equipment.
- Technical guideline for the site selection of a geological repository for radioactive waste issued in 1998 by the National Nuclear Safety Administration.
- *Regulation for management of nuclear waste* issued by the National Administration for Technical Supervision in 2002.
- *Guideline for Selection for a HLW geological disposal* HAD 401/06, issued in 2013 by the NNSA. Defines targets, procedures and principles for the site selection.

3.2 *The institutional framework*

The institutional framework for nuclear waste management in China is a complex and fragmented structure with overlapping responsibilities. Several institutions associated with different government entities are involved in the process.

Table 5: Institutions in Charge of Nuclear Waste Management

State Council						
NDRC	Ministry of Industry and Information	SASAC	MEP	National Commission for Science, Technology, Industry for National Defense	Ministry of Finance	Ministry for Public Security
NEA	CAEA	CNNC	NNSA			
ERI		BRIUG CIAE CIRP CNPE BINE		NINT		
Planning and implementation nuclear power within the energy sector	Planning of technical aspects of the nuclear fuel cycle Approval of R&D programs for spent fuel management Draft of relevant regulations and standards. Responsibility for nuclear security International cooperation (for example IAEA)	Siting design construction, commissioning, operation of spent fuel facilities and nuclear waste disposal. Screening of Beishan granite site (BRIUG) R&D radionuclide migration and waste forms (CIAE)	Supervision and administration of nuclear and radiation safety including policies, regulations. EIA	Decommissioning of military facilities Screening of Aqishan granite site	Collection and management of funds for nuclear waste management and disposal	Emergency response Safety of nuclear waste transports

The *National Energy Administration (NEA)* under the *National Development and Reform Commission (NDRC)* is in charge of planning and implementation of the nuclear power sector. NDRC and NEA, as well as the Energy Research Institute under NDRC, have been strong supporters of rapid nuclear development.

The *National Nuclear Safety Administration (NNSA)* under *MEP* is the regulatory authority in charge of approval and safety supervision of nuclear waste management. MEP/NNSA headquarter is in Beijing; in addition there are six regional offices in Shanghai, Shenzhen, Chengdu, Beijing, Lanzhou and Dalian, which are responsible for monitoring nuclear facilities within their respective regions. In 2012, in the aftermath of Fukushima, NNSA became the key nuclear safety authority which licenses, regulates and supervises civil nuclear facilities. In the same year, the staff size was expanded from 300 to 1200 employees. Relative to the gravity of its responsibilities, the authority is nevertheless still understaffed and remains financially dependent on MEP. In the past, MEP officials have criticized the speed of China's nuclear development and warned of bottlenecks such as the lack of trained personnel.

The *China Atomic Energy Agency (CAEA)* under the Ministry of Industry and Information Technology is in theory in charge of project control, fiscal approval and management of R&D programs for spent fuel management and formulation of relevant regulations and standards. In practice, CAEA's major task is the promotion of the bilateral and multilateral cooperation between the Chinese nuclear industry and international organisations (for example, the IAEA).

The key player in siting, design, construction, commissioning, operation and decommissioning of spent fuel facilities is the *China National Nuclear Corporation (CNNC),* which emerged from the former Ministry of Nuclear Industry.[8] CNNC is a state owned company under the State Owned Assets Supervision and Administration Commission (SASAC) with directors directly appointed by the State Council. It is an industrial conglomerate of more than 100 enterprises and institutes. The following CNNC subsidiaries are directly involved in nuclear waste management: the *Beijing Research Institute of Uranium Geology* (BRIUG) is in charge of site investigation, evaluation and screening of the Beishan granite site; the *China Institute of Atomic Energy* (CIAE) focuses on radionuclide migration studies and waste forms; the *China Institute for Radiation Protection* (CIRP) is responsible for safety assessment; and the *China Nuclear Power Engineering Company* (CNPE) and the *Beijing Institute of Nuclear Engineering* (BINE) work on engineering design. The *Sichuan Environmental*

8 CNNC emerged in 1989 from the former Ministry of Nuclear Industry (MIN), which had been established in 1982 as a successor to the Second Ministry for Machine Building. The restructuring marks the transition from military use to a combination of military and civilian uses and the conversion of former military facilities into facilities for the civilian nuclear industry.

Engineering Company, an entity of CNNC, which emerged from the decommissioned Guangyuan nuclear complex in Sichuan, specialises in decommissioning military nuclear facilities, nuclear waste management and related research.

In addition, the *State Administration of Science, Technology and Industry for National Defense* under the Ministry of Industry and Information oversees the safety of nuclear materials related to defense and handling of nuclear accidents. The *Northwest Institute of Nuclear Technology (NINT)* in Xian, which is attached to the National Commission for Science, Technology and Industry of National Defense, is surveying the potential HLW disposal site at Aqishan in Xinjiang province.

4 Siting procedures

Site selection for a deep geological HLW repository was divided into four stages: a nationwide screening, a regional screening, an area screening and site confirmation. To date, the first two stages have been completed and the third is in progress. A final site confirmation for HLW repository construction is still pending.

The *nationwide screening* was conducted between 1985 and 1986 and included potential sites in Northwestern China, Inner Mongolia, Eastern China and southern China (see Map Wang 2007). According to Wang et al. (2006), the screening included criteria such as socio-economic and natural factors, i.e. economic potential, biodiversity, mineral resources, land use, local public attitude, and geological and hydrological conditions as well as engineering conditions (Wang et al. 2006: 57). Wang Ju, program leader and vice-president of BRIUG, and some of his colleagues who participated in the screening process and site selection have regularly reported about the state of research on China's HLW repository project at international conferences and have published the results in Chinese and international scientific journals.

Based on the nationwide screening between 1986 and 1989, a *regional screening* has been conducted at 21 potential sites, among which the Beishan area in North-Western China has been identified as the most promising. The available reports do not provide details about progress in decision making and how, for example, attitudes of local people have been assessed.

Since 1990, the *area screening* has concentrated on the Beishan area. Three potential granite sites have been surveyed for crust stability, tectonic evolution and lithological and geophysical properties by BRIUG scientists. Since 2012, the North-West Institute for Nuclear Technology (NINT) has conducted surveys at a granite site in the Aqishan area in Xinjiang province.

4.1 Procedures and criteria for site selection

Since the early 1990s, the BRIUG has conducted feasibility studies for building a generic URL. In the early phase, two sites (Yangfang, Changping county and Shihuyu, Huairou county) in the Beijing area were selected. However, the project to build an URL in the suburbs of Beijing was suspended in 1995 (Wang 2014: 103).

In 2011, CAEA and MEP approved the Beishan area as a priority site for a generic or an area-specific URL in the near future and for construction of China's HLW repository (Wang 2014: 103).[9]

Site No. 1: The Beishan area in Gansu province

The Beishan area is located in Gansu province, Jiuquan prefecture, near Yumen city (40.6 N, 97.2 E). The rocky terrain has an elevation between 1,400 and 2,000 m above sea level. Seismic intensity is low; no earthquakes of a magnitude greater than 4.75 have been recorded in the area (Wang 2014: 103). The climate is arid with an average precipitation of 70 mm/a and average evaporation of 3000 mm/a. Year-round surface water does not exist and the area lacks groundwater resources. Due to geophysical conditions and water scarcity, the area is sparsely populated. There is no agriculture and only limited grazing activities. It is, however, easily accessible for transportation.

Since 1999 and with assistance from the IAEA technical cooperation projects (CPR/9/026, CPR/4/024 and CPR/3/008), scientists of BRIUG have conducted geological, hydrological and geophysical surveys, including studies on seismic safety, future climate changes and future geological environment changes (Wang 2010: 2). Eight granite sites in the area with different characteristics have been screened. Of these, the three sites Jiujing, (monzonitic and tonalite) Xinchang-Xiangyanshan (diorite and granite) and Yemaquan (diorite) have been identified as the highest potential for an URL and a HLW repository. Between 1999 and 2013, eleven deep boreholes and eight shallow boreholes were drilled (BS01-BS19) at the three sites and injection tests, water sampling and geo-stress tests have been conducted (Wang 2014: 103). Tests proved that the boreholes sites have extremely low permeability. Wang (2014: 104) expects that the site for the URL will be determined between 2017 and 2020. As the construction of the URL will need an additional 3-5 years, China's first URL for geological disposal of HLW may start operation between 2020 and 2025 if everything runs smoothly. In

9 The character of the facility will depend on the final decision for the location of the HLW repository.

2014, the IAEA approved the technical cooperation project CPR9045 "Supporting the Investigations on Design and Construction of the Underground Research Laboratories, and the Long Term Performance of the Buffer Material for Disposal of High Level Radioactive Waste" (IAEA 2015).

Site No. 2: The Aqishan area in Xinjiang

For many years, the Beishan area in Gansu was the only potential site discussed for a HLW deep geological disposal facility. Since 2012, however, scientists of the North-West Institute for Nuclear Technology (NINT) have conducted screening at a second candidate site in Xinjiang Uygur Autonomous Region. The survey is possibly being conducted as a reference for Beishan, since according to the *Guideline for Selection for a HLW geological repository* (2013) surveys at two alternative sites are required.

The site is located in the Aqishan area, 160 km south of Turfan city. It is located at 42 N and 90'15 E, not too far from the former nuclear testing site of Lop Nor, 41'30° N, 88'30 ° E, a 100,000 km^2 off-limits zone where, between 1969 and 1996, 22 underground tests were conducted. The Aqishan area has a flat terrain and the climate is arid, with precipitation less than 60 mm/a and an average evaporation of 2250-2900 mm/a. There are no permanent residents in the area, but it is accessible for vehicles via a Class III highway. NINT's survey focuses on the Xianshuigou granite pluton, an area of approximate 300 km^2. So far, a geological survey including remote sensing, aeromagnetic interpretation and a field survey has been conducted. The site in question is a granite batholith site with a thickness of one thousand meters. According to NINT, it is suitable for HLW disposal (Yuan et al. 2015: 483; Map of the Aqishan site: Yuan et al. 2015: 485).

Research on the technical design of the repository

The preliminary design of the repository,[10] a shaft tunnel-system in a granite site, includes natural barriers and an engineered barriers system of vitrified waste, waste canisters, buffer materials, backfill and seals. Gaomiaozi Bentonite from a large deposit in Inner Mongolia has been identified as the most suitable buffer and backfill material (Wang 2010: 8). In 2010, a mock-up facility was constructed in the BRIUG laboratory in Beijing to study the behaviour of the Gaomiaozi bentonite under coupled thermos-hydro-mechanical conditions (Wang et al. 2011;

10 See Zhou 2013: 10 for a sketch of the repository.

Liu et al. 2014). Since 2004 China has participated in international research projects concerning technical issues, including on buffer/backfilled material initiated by IAEA, as well as research between China and France. Since 2010 BRIUG is partnering with 15 European organizations in the 7th EURARTOM project PEBS (Long-term-Performance of Engineered Barrier Systems, a project to evaluate sealing and barrier performance of clay-based EBS). (People's Republic of China 2014, PEBS-EU 2014) In addition, China cooperates with the United States on underground laboratory based studies.

5 Information and participation

Technical information about R&D for potential sites is available through scientific publications and also distributed via the participation of Chinese scientists in international workshops and conferences. Legal regulations and guidelines are published on the websites of the respective government authorities.

In the aftermath of the Fukushima accident, nuclear safety issues have to some extent been discussed in the Chinese media. In April 2011, MEP/NNSA issued a *Notification on Enhancing NPP's Nuclear and Radiation Safety Information Disclosure and Regulation (trial)* and the *Information Disclosure Plan on Nuclear and Radiation Safety Regulations* According to China's Report for the Joint Convention the 2013 MEP/NNSA *Guidelines on Government Information Publicity for Environmental Impact Assessment of Construction Projects* also apply for nuclear projects. Applicants must disclose full text project information and MEP/NNSA should make the information on the environmental impact assessment statement available to the public (People's Republic of China 2014: 103). In recent years civil protests against nuclear power plants, a planned enrichment and fuel fabrication facility in Jiangmen in Guangdong province and a reprocessing facility in Lianyungang, Jiangsu province have been reported. However, there is almost no information available in the public media about issues related to reprocessing, decommissioning of nuclear facilities and a final disposal for nuclear waste. Therefore, these aspects have not played a significant role in the public debate on nuclear energy. In the relevant legal regulations, guidelines for information and public participation are missing.

6 Costs and financing

According to China's Joint Convention report the operators of nuclear facilities shall bear all costs related to nuclear safety, spent fuel management and nuclear waste management. (The People' Republic of China 2014: 44). Detailed information on expected costs and financing of nuclear waste management however is sparse. Pan (2013) provides a rough calculation for the 58 GW NPPs, which are scheduled to start operation by 2020 and will have produced 82630 tHM of spent fuel by 2050. The costs for 82630 tHM of spent fuel (not including treatment) are estimated at 134 billion RMB[11], equal to 1.25 per cent of revenue from electricity fees. Every 1 GW NPP built after 2020 will have a life cycle of 60 years and produce 1320 tHM of spent fuel. Disposal costs are calculated with 2.4 billion RMB per NPP unit. Required R&D costs for site selection of a geological HLW repository are estimated at 1.34 billion RMB. The construction of the URL will cost about 400-500 million RMB. If the construction of the URL is to be realised by 2020, an annual investment of 100 to 150 million RMB in R&D is required (Pan 2013: 195). According to Pan, however, actual funding is much lower and far from sufficient. He proposes that the government increases investment in nuclear waste treatment, especially since disposal also includes nuclear waste from decommissioned military facilities. The nuclear power industry should cover the remaining costs through the instalment of a waste management fund, which is compiled of a levy of 0,026 CNY/kWh from the electricity fee. It is suggested that of these, 0.005 RMB or 1.25 percent of the electricity fee should be used to cover the costs of deep geological disposal.

7 Conclusions

From the start of its civilian nuclear power program in the mid-1980s, China has pursued a R&D program for a final HLW repository. Since the 1990s, Chinese officials have essentially agreed that the final repository should be built in an uninhabited or sparsely populated Gobi area in North-Western China. Whereas for many years, research concentrated on the Beishan granite site in Gansu Province, since 2012 additional screening has been conducted near the former nuclear weapon test site in Xinjiang Uighur Autonomous Region. The original proposed schedule has already been delayed by more than 10 years, because of a lack of funding and a controversial debate over whether China should opt for a closed nuclear cycle based on reprocessing, MOX fuel fabrication and fast neutron reactors or a PWR based "once through" strategy. As recently plans for construction of a commercial reprocessing facility, MOX fuel fabrication and a commercial fast neutron reactor were postponed indefinitely, the management of the accumulating amounts of spent fuel is becoming an urgent issue. With onsite storage pools reaching their storage capacity, from 2020 onward China will need

11 1 RMB = 0.154 US $ or 0.128 Euro (31 December 2017).

more facilities for transport of spent fuel casks and will have to expand offsite interim storage facilities. Overland transportation of spent fuel from the NPPs in East China over a distance of roughly 4,000 km to designated interim storage facilities and final repository in North-West China seems to be the bottleneck in China's nuclear waste management strategy. In recent years, China has issued a number of laws, regulations and technical guidelines relevant for nuclear waste management; however, existing legislation remains fragmented and unsystematic as China lacks a comprehensive National Atomic Law that regulates the obligation for NPP operators to treat and dispose of nuclear waste. A similar critique can be made of the institutional framework. Although China's civil nuclear industry emerged from the military complex, it is not clear whether the civil and the defense sector cooperate in the treatment of spent fuel and nuclear waste management or if the defense sector is going to maintain separate facilities. Information on financial and socio-economic impacts of nuclear waste management is not transparent. Although the Fukushima accident recently prompted some public debate over nuclear safety, the issues of cost and the safety of nuclear waste management remain on the sidelines.

References

AREVA (2013). *Areva signs a series of strategic agreements with its Chinese partners.* http://www.areva.com/EN/news-9812/areva-signs-a-series-of-strategic-agreements-with-its-chinese-partners.html, last accessed 13.05.2015.

BJZTB (2015). *China French project discuss seismic safety evaluation for coastal sites 中法核大单走向谈判滨海启动地震安全评价招表。.* http://www.bjztb.gov.cn/news/201508/t9420808.htm, last accessed 23 August 2016.

Buckley, C. (2016). Chinese City Backs down on Proposed Nuclear Fuel Plant after protests. In: *New York Times*, 10.8. 2016. http://www.nytimes.com/2016/08/11/world/asia/china-nuclear-fuel-lianyun-gang.html, last accessed 23 August 2016.

Bunn, M.; Zhang, H. and Kang, L. (2016). *The Cost of Reprocessing in China.* Harvard Kennedy School. http://belfercenter.ksg.harvard.edu/files/The%20Cost%20of%20Reprocessing.pdf, last accessed 16 October 2016.

CAEA Nuclear Technology Support Center (2014). *Current Status of the Spent Fuel Management from Nuclear Power Reactor in China.* ppt, last accessed 15 May 2015.

CNNC (2015). *China National Nuclear Corporation.* http://www.cnnc.com.cn/tabid/643/Default.aspx, last accessed 12 December 2016.

CNNC Nuclear Fuel Reprocessing Co. (2016). *Wang Yiren about studies for Lianyungang as site for Sino-French Nuclear Recycling project* 王毅韧调研中法合作核循环项目连云港厂址. http://www.cnfr.cc/article/index/id/276/cid/4, last accessed 23 August 2016.

Deng, G. (2010). Overview over spent fuel management in China. In: *International Conference on Management of Spent Fuel from Nuclear Power Reactor*. IAEA. http://www.pub.iaea.org/MTCD/Publications/PDF/SupplementaryMaterials/P1661 CD/Session_12.pdf, last accessed 16 May 2015.

Hu, Y. and Cheng, H. (2015). Disposal capacity of spent fuel in China is not ready for the nuclear power boom. In: *Environmental Science and Technology,* (49), 2596-2597.

IAEA (2015). *Technical Cooperation Projects China*. https://www.iaea.org/technicalcooperation/documents/Projects/TCAP/China.pdf, last accessed 03 June 2015.

Karlsruher Institut für Technologie (2014). *KIT liefert Technologie für Verglasungsanlage in China*. http://www.kit.edu/kit/15762.php, last accessed 06 May 2015.

Liu, X. (2014). *Spent fuel Management in China*. NAPSNet Special Reports, August 05, 2014. http://nautilus.org/napsnet/napsnet-special-reports/spent-nuclear-fuel-management-in-china/, last accessed 15 May 2015.

Liu, Y.; Wang, J. and Cao, S. (2014). *Test plan of the China-Mock-up dismantling operation. PEBS Deliverable (D-N: DB-4).* www.pebs-eu.de/PEBS/EN/.../DB_4.pdf?__blob..., last accessed 04 June 2015.

Ma, T. (2016). *An unwelcome nuclear reprocessing plant. China dialogue.* https://www.chinadialogue.net/blog/9189-An-unwelcome-nuclear-reprocessing-plant/en, last accessed 23 August 2016.

MEP (2013). *Guidelines for site selection of HLW geological repository. Nuclear Safety Guidelines HAD 401/06-2013* 高水平放射性废物地质处置设施选址 (in Chinese). http://www.mep.gov.cn/gkml/hbb/haq/201306/W020130605487163993920.pdf, last accessed 15 May 2015.

MEP (Ministry of Environmental Protection) (2014). *Environmental Regulations and Standards Putting into Effect as of March 2014.* http://english.mep.gov.cn/standards_reports/standards/Catalogue_Standards/201403/t20140326_269719.htm, last accessed 15 May 2016.

Ministry of Finance (2010). *Provisional Measures for Management of a Fund for Treatment of used fuel from nuclear power plants* 核电站乏燃料处理处置基金征收使用管理暂行办法 (in Chinese). http://www.mof.gov.cn/zhengwuxinxi/caizhengwengao/2010nianwengao/wengao6/201009/t20100903_337280.html, last accessed 05 June 2015.

Mu, R.; Zuo, J. and Yuan, X. (2015). China's approach to nuclear safety – From the perspective of policy and institutional system. In: *Energy Policy*, 76, 161-172.

National Energy Administration (2012). New energy industry development report. *(新能源产业发展趋势研究报告)*.http://www.nea.gov.cn/2012-02/10/c_131402922.htm, last accessed 06 May 2015.

NPC (National People's Congress) (2003). *Law of the People's Republic of China on Prevention and Control of Radioactive Pollution".* http://www.npc.gov.cn/english-npc/Law/2007-12/05/content_1381967.htm, last accessed 13 May 2015.

NTI (2012). Guangyuan *Plutonium Production Reactor and Reprocessing Plant.* http://www.nti.org/facilities/730/, last accessed 06 May 2015.

Pan, Z. and Qian, Q. (2013). The Geological Disposal of High-level Radioactive Waste Strategy Research in Our Country. 潘自强，钱七虎・我国高放废物地质处置战略研究。. In: *China Nuclear Power 中国核电,* 6(3), 194-196.

Pandza, J. (2013). China's nuclear fuel cycle and proliferation risks. In: *Survival*, 55(4), 177-190.

PEBS-EU (2014). *Long-term Performance Engineered Barrier Systems*. http://www.pebs-eu.de/PEBS/EN/Home/PEBS_node_en.html, last accessed 16 October 2016.

People's Republic of China (2014). *Third National Report for the Joint Convention on the Safety of Spent Fuel Management and on the Safety of Radioactive Waste Management.*http://nnsa.mep.gov.cn/gjhz_9050/gjgybg/201512/P020151223562288196956.pdf, last accessed 16 October 2016.

State Council of the PR China (2010). *Regulation for transport of radioactive materials (放射性物品运输安全管理条例).* http://www.gov.cn/zwgk/2009-09/22/content_1423508.htm, last accessed 17 June 2015.

State Council of the PR China (2011). *Regulations for Safety Management of Nuclear Waste (放射性废物安全管理条例).* http://www.gov.cn/zwgk/2011-12/29/content_2033177.htm, last accessed 17 June 2015.

Sternfeld, E. (2014). Fukushima und die Folgen für Chinas Atomenergieprogramm. In: Mahltig, P. and Sternfeld, E. (Eds). *Harmonie und Umbruch in China,* Jahrbuch der Deutschen Vereinigung für Chinastudien 8, Wiesbaden: Harrosowitz, 189-208.

Wang, J. (2014). On area-specific underground research laboratory for geological disposal of high-level radioactive waste in China. In: *Journal of Rock Mechanics and Geotechnical Engineering,* 6, 99-104.

Wang, J.; Liu, Y. and Zhang, X. (2011). *China-Mock-up status.* Annual Reports. PEBS Project. Technical Note. PEBS Internal Deliverable DB 2-1.

Wang, J. (2010): High-level radioactive waste disposal in China: update 2010. In: *Journal of Rock Mechanics and Geotechnical Engineering*, 2(1), 1-11.

Wang, J. (2007). *Deep Geological Disposal of HLW in China: Progress and needs for Training in the Network.* PPT Presentation for technical meeting INT/9/173, May 9-11, Vienna. https://www.iaea.org/OurWork/ST/NE/NEFW/WTS-Networks/URF/documents/Status/2007/MS/China.pdf, last accessed 06 May 2015.

Wang, J.; Lin, S. and Li, C. (2006). Deep Geological Disposal of High-Level Radioactive Waste in China: A Three Step Strategy and Latest Progress". In: Witherspoon, P.A. and Bodvarsson, G.S. (Eds.). *Geological Challenges in Radioactive Waste Isolation, 4th World Wide Review*, Earth Science Division, Berkeley National Laboratory.

Wang, J. (2002). *Progress on deep geological disposal of high level radioactive waste in China: 1999-2001*. IAEA-CN.90/20. http://www.iaea.org/inis/collection/NCLCollectionStore/_Public/34/016/34016260.pdf, last accessed 28 May 2015.

World Nuclear Association (WNA) (2017). *Nuclear Power in China*. http://www.world-nuclear.org/info/Country-Profiles/Countries-A-F/China--Nuclear-Power/, last accessed 15 July 2017.

World Nuclear Association (WNA) (2016a). *Nuclear Power in China*. http://www.world-nuclear.org/info/Country-Profiles/Countries-A-F/China--Nuclear-Power/, last accessed 13 August 2016.

World Nuclear Association (WNA) (2016b). *China's Nuclear Fuel Cycle*. http://www.world-nuclear.org/info/Country-Profiles/Countries-A-F/China--Nuclear-Fuel-Cycle/, last accessed 13 August 2016.

Xu, J.; Xiong, X.; Lei, Q. and Wu, Y. (2014). Discussion on some comments on the regulatory system of high level radioactive waste disposal in China. 徐健，熊先祥，雷奇峰，武以博:我国高放废物地质处置的若干问题探讨。. In: *World Nuclear Geoscience 世界核地质科学*, 31(4), 601-606.

Xu, Y. (2010). *The Politics of Nuclear Energy in China*, London: Palgrave.

Yang, Z. (n. d.). *Research on Reactor Decommissioning in China*. ppt. http://www-ns.iaea.org/downloads/rw/projects/r2d2/workshop6/country-presentations/china.pdf, last accessed 18 June 2015.

Yuan, G.; Zhao, Z.; Chenjie, J. M.; Han, J. and Gao, W. (2015). Feasibility Study on Siting of China's High-Lecel Radioactive Waste Repository in an Area in Northwest China. In: Lollino, G. et al. (Eds). *Engineering Geology for Society and Territory – Volume 6*, 483-487.

Zhang, H. (2011). *China's Stockpile of Military Plutonium: A New Estimate*. http://belfercenter.ksg.harvard.edu/files/INMM-PU2.pdf, last accessed 06 May 2015.

Zhou, Y. (2010a). *China's current spent fuel management and future management scenarios*. http://belfercenter.ksg.harvard.edu/files/Yun-Zhou-INMM-2010.pdf, last accessed 06 May 2015.

Zhou, Y. (2010b). Why is China going nuclear? In: *Energy Policy*, 38, 3755-3762.

Zhou, Y. (2011). China's spent nuclear fuel management: Current practices and future strategies. In: *Energy Policy*, 39, 4360-4369.

Zhou, Y. (2013). *China's nuclear waste management and disposal*. ppt. http://nautilus.org/wp-content/uploads/2013/08/China_Zhou_May-28th-2013-version-2.pdf, last accessed 08 May 2015.

A Profitable Business Strategy?

Spent Nuclear Fuel and Radioactive Waste Management in Russia

Felix Jaitner[1]

Abstract

Over the past ten years, the Russian nuclear industry has undergone a profound reorganisation. In the course of this development, radioactive waste management has received substantial attention. The Russian Federation has formed a regulatory base in order to define radioactive waste and provide solutions for its disposal. The selected approach includes near-surface disposal and storage of radioactive waste in deep geological formations. Apart from the development of a necessary infrastructure, increased reprocessing of spent nuclear fuel and the development of a closed fuel cycle shall provide answers to the pressing issue of radioactive waste disposal. The guiding principle in restructuring the nuclear complex was not to increase democratic control or ecological standards, but to adjust Russia´s nuclear industry according to the demands of a market economy.

1 Felix Jaitner, University of Vienna; E-Mail: F.Jaitner@gmx.de

A. Brunnengräber et al. (eds.), *Challenges of Nuclear Waste Governance*,
Energiepolitik und Klimaschutz. Energy Policy and Climate Protection,
https://doi.org/10.1007/978-3-658-21441-8_3

1 Introduction

In 2009, the Russian government adopted the so-called Energy Strategy 2030. This paper evaluates the development of domestic energy consumption and supply through 2009. Focusing almost exclusively on fossil fuels (oil, gas and coal) and nuclear energy, Energy Strategy 2030 reinforces the dominant position of "traditional" energy sources in Russia. Special attention is paid to the development of the nuclear complex. A stated goal is to consolidate the global position of the nuclear complex and increase its significance in the domestic energy supply (Ministry of Energy of the Russian Federation 2010).

The target is defined even more precisely in the federal program, New-generation nuclear energy technologies for the period 2010-2015 and up to 2020. The explicit goal is to develop fast neutron reactors with a closed fuel cycle in order to meet the expected increase in domestic energy consumption. At the same time, the program mentions four challenges facing the Russian nuclear complex: ineffective utilisation of uranium; a decline in scientific knowledge and potential; reduced competitiveness in the production of nuclear energy on the world-market; and the high and increasing amount of spent-nuclear fuel and nuclear waste (Kurchatovskij Institut and Rosatom 2009).

The acknowledgement that the amount of spent-nuclear fuel and radioactive waste poses a relevant and increasing problem might appear to be a sign of progress. Previously, and especially in the 1990s, Russian governments tended to neglect or downplay the issue of radioactive waste. However, when considered in the broader context of the nuclear industry's growth trajectory, it is clear that this is a response to corporate rather than safety or environmental concerns. Specifically, the effort to establish the state atomic energy corporation Rosatom as a global player satisfying demand in every stage of the nuclear cycle (mining and enrichment of uranium ore, construction and maintenance of nuclear power plants, radioactive waste and spent nuclear fuel management) and demonstrate the future viability of the industry demands answers to the pressing issue of radioactive waste. A closed cycle allows reprocessing of spent nuclear fuel, which seemingly avoids the unresolved question of its eventual disposal. Such a position has been expressed by Denis Kozyrev, former CEO of FSUE, Enterprise for Radioactive Waste Management RosRAO, which is a subsidiary company of Rosatom and responsible for nuclear waste disposal in Russia: "The foundation of an effective system [...] for nuclear waste disposal is a fundamental requirement for the advancement of nuclear energy and industry" (Kozyrev 2009) [translation F.J.].

Following the definition of Brunnengräber and Mez (2016: 290), the term nuclear complex describes the shifting interlinkages among economic, political, scientific and military-industrial interests in Russia. A key feature of the nuclear complex is its non-transparent structure which impedes democratic control. The term also encompasses the reorganisation of the nuclear sector according to the demands of a market economy. With regard to radioactive waste management, newly introduced criteria such as cost-efficiency, profitability and global competitiveness indicate the valorisation of radioactive waste (tariff-system) and spent nuclear fuel (reprocessing, fuel bank). Both radioactive waste and spent nuclear fuel are transformed into commodified goods meaning that they are made exchangable through the market. The "marketization of radioactive waste management" is seen as a key solution to the unresolved question of radioactive waste disposal in Russia.

2 Nuclear waste disposal in Russia

2.1 Historical background

Despite the recent acknowledgement of the problem, nuclear waste disposal in the Russian Federation is not a new challenge. In fact, it has been an unresolved question since the foundation of the nuclear complex in the Soviet Union. Originally designed for military purposes (production of nuclear weapons, nuclear-powered ships, etc.), the nuclear industry was closely connected to the military-industrial complex and thus subject to secrecy. The subsequent expansion of nuclear industries into the civil sector (energy supply) increased the amount of nuclear waste. However, the non-transparent structure of the nuclear complex impedes valid conclusions concerning precise volumes of radioactive waste in the USSR and Russia.

Maintaining a clear priority to build-up its nuclear potential, the Soviet Union paid less attention to safe utilisation and disposal of spent nuclear fuel and radioactive waste. A clear juridical base regulating the use of atomic energy did not exist. In fact, radioactive waste was regarded as something of secondary importance (Muratov and Tichonov 2015). Moreover, the high degree of secrecy and a lack of scientific knowledge impeded research on the impact of nuclear technology on human beings and the natural environment (ibid.). Because of nuclear weapon tests and contained nuclear explosions, the nuclear industry is to be held responsible for radioactive contamination of entire areas, e.g. Novaya Zemlya, an archipelago in the Arctic Ocean in northern Russia. Accidents in nuclear power plants (Belogorsk, Chernobyl, Leningrad) contributed to the contamination of inhabited areas. Although Belarus and the Ukraine carry the heaviest burden from the reactor explosion in Chernobyl, a recent study by the Russian Academy of Science (RAN) confirmed persistent environmental damage in Russia as well (Puchkov and Bol′shkov 2016). Regarding radioactive contamination, several accidents at the Mayak Production Association (PA), which houses plutonium production reactors and a reprocessing plant, had even more devastating effects (IBRAE RAN 2006). Finally, disposal of radioactive waste in the Barents Sea and the Kara Sea is expected to lead to unpredictable long-term effects on the environment and human health. From 1959-1991, approximately 17,000 containers with radioactive waste and spent nuclear fuel were dumped into the sea offshore Novaya Zemlya. In addition, 12 nuclear-powered submarines and three nuclear-powered ice-breakers were sunk in the sea (Kronfeld-Goharani 1999: 26-27). According to a study by the Norwegian Ministry of Environment conducted in 1995, 16 nuclear reactors, which included six nuclear fuel elements, were sunk in the Kara Sea (ibid.).

With the dissolution of the Soviet Union, Russia inherited one of the biggest nuclear industries in the world along with the problem of radioactive waste disposal. The Russian government focused on the implementation of market reforms through so-called "shock-therapy". Following the dissolution of the USSR, price controls were lifted immediately, foreign trade was extensively liberalised and the development of a private financial sector was facilitated (Pomer and Klein 2001). According to the sociologist Nataliya Tichonova (2011), about one third of the Russian population became impoverished as a direct consequence of the economic transformation. The extent of the economic decline has been illustrated by the economist Jospeh Stiglitz, who observes that the Russian economy (measured in terms of GDP) suffered greater losses during this time than it did during the second world war (Stiglitz 2002). Against the backdrop of this comprehensive societal transformation, the state seemed unable to tackle the question of nuclear waste disposal. In fact, dissolving state structures and an ambiguous legal situation generated new problems. At the Kola peninsula in Northwest Russia, discharged nuclear-powered ships and submarines were not stored according to proper safety requirements (Kronfeld-Goharani 1999). Budget constraints in the military sector exacerbated the shortage of funds for necessary transport of radioactive waste and spent nuclear fuel as well as the construction of interim storage facilities (ibid.: 31). The following graph gives an overview on the accumulated amount of radioactive waste in Russia in the 1990s. The data was gathered in the course of the RADLEG[2] international research project.

---

2 The International Science and Technology Center (ISTC) in Moscow, the Russian Ministry for Atomic Energy (Minatom), the Russian Academy of Science and the Kurchatov Institute conducted extensive research on radioactive waste in Russia in cooperation with the Institute for Applied System Analysis (IIASA) in Laxenburg, Austria. The latter analysed the nuclear heritage of the Soviet Union in the course of the scientific project "Radiation Safety of the Biosphere" (Kronfeld-Goharani 2002: 9).

Table 1: Amount of radioactive waste stemming from the civil and military sector (01.01.1996)

Responsibility	Fluid Radioactive Waste		Solid Radioactive Waste		Spent Nuclear Fuel	
	m^3	Bq	m^3	Bq	Ton	Bq
Minatom (Uranium industry, nuclear fuel manufacturing, nuclear power plants, reprocessing, nuclear weapons material)	$4{,}0 \cdot 10^{8}$	$6{,}3 \cdot 10^{19}$	$2{,}2 \cdot 10^{8}$	$8{,}1 \cdot 10^{18}$	8700	$1{,}7 \cdot 10^{20}$
	m^3	Bq	m^3	Bq	Ton	Bq
Ministry of Defence (Navy) Civil nuclear-powered ships and nuclear-powered submarines	$1{,}4 \cdot 10^{4}$	$4{,}4 \cdot 10^{12}$	$1{,}3 \cdot 10^{4}$	$3{,}0 \cdot 10^{13}$	30	$5{,}6 \cdot 10^{19}$
Ministry of Economics Construction, maintenance and repair of nuclear-powered ships	$3{,}2 \cdot 10^{3}$	$1{,}8 \cdot 10^{11}$	$1{,}5 \cdot 10^{3}$	$3{,}7 \cdot 10^{12}$	-	-
Ministry of Transport Nuclear ice-breaker fleet	$4{,}4 \cdot 10^{2}$	$5{,}5 \cdot 10^{13}$	$7{,}3 \cdot 10^{2}$	$3{,}7 \cdot 10^{16}$	10	$1{,}7 \cdot 10^{18}$
Federal State Unitary Enterprise RADON Storage of radioactive waste from medicine, science, research and industry	-	-	$2{,}0 \cdot 10^{5}$	$7{,}8 \cdot 10^{16}$	-	-
Total	$4{,}1 \cdot 10^{8}$	$6{,}3 \cdot 10^{19}$	$2{,}3 \cdot 10^{8}$	$8{,}2 \cdot 10^{18}$	8740	$2{,}3 \cdot 10^{20}$

Source: Kronfeld-Goharani 2002: 8

2.2 The national inventory

2.2.1 Radioactive waste management

The federal law № 170-FZ On the Use of Atomic Energy adopted on November 21, 1995, defines radioactive waste as material and substances for which no future use is foreseen. This includes equipment and goods (including spent ionising radiation sources) containing radionuclides at concentrations greater than clearance levels established by the Russian government (Rosatom, IBRAE RAN and FBI "SEC NRS" 2014: 21). Although the aforementioned law still provides a regulatory base, further amendments have specified nuclear waste governance, e.g. the federal law № 190-FZ On Radioactive Waste Management and Amendments to Certain Legislative Acts of the Russian Federation (henceforth, On Radioactive Waste Management) (adopted in July 2011). For a more detailed discussion on the legal framework, see 3.1. The adoption of the law On Radioactive Waste Management has enabled a process of initial registration of nuclear waste, e.g. providing data on quantitative characteristics of waste. Moreover, various types of waste (removable or special) and types of (possible) storage facilities exist: temporary storage facility, long-term storage facility, facility holding special radioactive waste, conservation facility for special radioactive waste, and radioactive waste disposal facility.

There are multiple sources, which offer different estimates of the total amount of radioactive waste in Russia. In the fourth national report on the safety of spent fuel management and the safety of radioactive waste management, Rosatom and the Nuclear Safety Institute of the Russian Academy of Science (IBRAE) (2014) state the total volume of accumulated liquid radioactive waste as 489.6 million m^3 (4.2·1019 Bq) and the total volume of accumulated solid radioactive waste as 90.4 million tons (4.7·1019 Bq). The annual generation of solid radioactive waste amounts to 1.2 million m^3 with a total activity of 4.0·1018 Bq although a 13 percent reduction by volume and 2.4 times reduction in activity as compared to the 2010 inventory is emphasised. Uranium mining contributed more than any other branch to the generation of solid radioactive waste. In fact, the uranium mining company Priargunsky Industrial Mining and Chemical Union (PIMCU) accounted for more than 98 percent of total waste generation by volume (1.18 million tons) with an activity of 8.9·1018 Bq. The principle source of high-level solid radioactive waste generation represents spent nuclear fuel reprocessing. In the PA Mayak about 1250 tons of high-level, intermediate-level and low-level solid radioactive waste with a total activity of about 4·1018 Bq were generated. In the year 2013, nuclear power plants produced more than 3000 tons of solid radioactive waste with a total activity of 8.0·1015 Bq, whereas uranium treatment and enrichment and fabrication of nuclear fuel contributed 45 percent to the total volume of solid radioactive waste (except PIMCU) with a total activity of 6.3·1015 Bq. The generation of the remaining amount of solid radioactive waste with an activity of 0.5·1018 Bq is not further specified in the report. With regard to liquid radioactive waste an amount of 1.88 million m^3 with a total activity of 1.7·1018 Bq was generated. Approximately 3500 m^3 of liquid radioactive waste with a total activity of 7.5·1013 Bq was produced in nuclear power plants. Additionally, the joint stock company «State Research Centre of the Russian Federation - Research Institute of Atomic Reactors» (JSC "SRC RIAR") was responsible for 61,700 m^3 of liquid radioactive waste with a total activity of 4.2·1014 Bq (Rosatom, IBRAE RAN and FBI "SEC NRS" 2014: 28-29 and 136).

In contrast to the numbers presented by Rosatom and the Russian Academy of Science, a joint report of the Organisation for Economic Co-operation and Development (OECD) and the Nuclear Energy Agency (NEA) published in 2014 presents a different set of data. According to their survey, the total volume of accumulated liquid radioactive waste is 486 million m^3 (4.27·1019 Bq) and 87 million tons (3.59·1019 Bq). In addition, the annual generation of radioactive waste is higher than in the aforementioned report (liquid radioactive waste: 2.05 million m^3; solid radioactive waste: 1.4 million. tons) (NEA 2014: 2-3). Finally, a report published by the international NGO-network «Decommission» states the amount of liquid radioactive waste in Russia is about 500 million m^3 and the amount of solid radioactive waste is 90.4 million tons (Talevlin et al. 2015: 6). One reason explaining the different estimates on radioactive waste in Russia is insufficient access to information. Another reason might be that sources are valued differently leading to contradictory estimates on the total amount of radioactive waste.

2.2.2 Spent nuclear fuel management

Spent nuclear fuel management is conducted in compliance with the federal law, The Concept of Spent Nuclear Fuel Management of the State Atomic Energy Corporation Rosatom, which was approved by decree № 721 on December 29, 2008. In contrast to radioactive waste, spent nuclear fuel is defined by the Russian Federation as utilisable for reprocessing. Consequently, spent nuclear fuel is not considered to be radioactive waste but rather a valued resource in the nuclear fuel cycle. The core principle in the field of spent nuclear fuel management is "to ensure ecologically sound management of fission products and to recycle the recovered nuclear material into nuclear fuel cycle" (Rosatom, IBRAE RAN and FBI "SEC NRS" 2014: 16). This statement clearly emphasises the Russian nuclear complex's approach towards spent nuclear fuel, promising an ecological and safe solution to the unresolved issue of spent nuclear fuel management.

There are several sources that generate spent nuclear fuel, including nuclear-powered ships (marine nuclear propulsion units), nuclear power plants, research and foreign reactors. The transport and import of highly enriched spent nuclear fuel from Soviet-built and Russian-built reactors for reprocessing is particularly a matter of controversy and exacerbates the need to find sustainable and timely solutions for the issue of spent nuclear fuel. So far, spent nuclear fuel has been returned to Russia from Uzbekistan (2005-2006, 2012), the Czech Republic (2007, 2013), Latvia (2008), Bulgaria (2008-2009), Hungary (2008, 2013),

Kazakhstan (2009), Romania (2009), Libya (2009), Poland (2009-2010, 2012), Belarus (2010), Ukraine (2010, 2012), Serbia (2010), and Vietnam (2013) (Rosatom, IBRAE RAN and FBI "SEC NRS" 2014: 17).

As with radioactive waste, data regarding the volume of spent nuclear fuel in the Russian Federation may differ substantially. Rosatom defines the quantity of spent nuclear fuel at 21,361.714 m^3 (Rosatom, IBRAE RAN and FBI "SEC NRS" 2014: 136). Muratov and Tichonov (2015) assess the amount of spent nuclear fuel in Russia as slightly higher. According to their research, the accumulated spent nuclear fuel is 21,714 m^3 and annual production amounts to 650 tons. The environmental NGO Bellona states that the amount of accumulated spent nuclear fuel in the Russian Federation is even higher than the previous two sources, at roughly 22,000 m^3 and an annual production of 850 tons. (Nikitin et al. 2012: 57-58).

2.3 The (interim) storage sites

Previously, collection, (partial) processing and further storage of radioactive waste were tasks performed by nuclear power plants and enterprises producing nuclear fuel. In the course of the internationalisation efforts of Russia´s nuclear complex, radioactive waste management and storage practices have been reorganised and adjusted according to the Unified State System for Radioactive Waste Management (USS RWM) (see 3.1). The significance of this reform is emphasised by the state corporation Rosatom's description of the successful implementation of the USS RWM as a "priority task" in the field of radioactive waste management (Rosatom, IBRAE RAN and FBI "SEC NRS" 2014: 20). According to Rosatom, radioactive waste management shall be readjusted "in order to arrange for and ensure safe and cost-effective management of waste" (ibid.: 18).

Currently, radioactive waste is distributed among 120 enterprises located in 44 regions across the country. Combined, they own 830 storage facilities, including sites for collection and/or temporary storage and three deep well injection facilities for liquid radioactive waste disposal (ibid.: 29). Radioactive waste with a volume greater than one thousand tons for solid radioactive waste and one thousand m^3 for liquid radioactive waste is kept at 45 enterprises in 239 storage facilities (ibid.). In compliance with the law On Radioactive Waste Management specialised organisations responsible for radioactive waste management shall carry out necessary tasks, e.g. collection, segregation, processing, conditioning, transportation and storage of radioactive waste. They are also in charge of operation, decommissioning and closure of radioactive waste storage facilities. Acting as the state authority for radioactive waste management,

Rosatom has spurred a consolidation process in the field of specialised radioactive waste management under the roof of its subsidiary RosRAO. This reorganisation of radioactive waste management along new criteria, such as cost-efficiency, indicates a commercialisation or marketisation of the sector (see 3.1).

With regard to spent nuclear fuel, Rosatom pursues a different strategy. Based on the legal definition that spent nuclear fuel can be used for reprocessing, each nuclear power plant must provide storage capacity for at least three years.[3] Similar rules are applied to research reactors. Apart from interim storage at operating nuclear power plants, there have been efforts to establish a centralised spent nuclear fuel storage at two sites: the federal state unitary enterprise Mining and Chemical Combine (FSUE "MCC") and the PA Mayak.

The FSUE MCC was established in 1953 in Zheleznogorsk (previously Krasnoyarsk-26), Krasnoyarsk Territory. Until 1995, the company was involved in the production of weapons-grade plutonium. MCC´s current activities include storage of "wet" and "dry" spent nuclear fuel, including fuel from foreign reactors (Bulgaria and Ukraine) (Nikitin et al. 2012: 58). Moreover, the combine provides services for decommissioning defence facilities, as well as work related to the start-up of a commercial mixed oxide fuel (MOX) fabrication facility.

Producing plutonium and warhead components, the PA Mayak played a decisive part in the Soviet nuclear weapons program. In the 1980s, the facility operated five plutonium production reactors, five tritium production reactors, several reprocessing plants, and a plutonium metallurgy plant. According to a report by the Norwegian Radiation Protection Authority, "(s)torage at Mayak PA can be divided into storage of fissile and/or reactor-grade radioactive materials and the storage of waste products, both "historical" from weapons production and "contemporary" from reprocessing activities" (StrålevernRapport 2006: 12). A major portion of Russia´s radioactive waste and spent nuclear fuel is stored at the facility. A centralised "wet" storage facility provides interim storage for spent nuclear fuel designated for reprocessing (WWER-400, fast breeder reactor, research reactors) and in the absence of respective reprocessing technologies such fuel is currently not subjected to reprocessing (Rosatom, IBRAE RAN and FBI "SEC NRS" 2014: 26). In fact, the reprocessing plant in Mayak (RT-1) is the only one of its kind in the Russian Federation. Moreover, Mayak is one of the two principal storage sites for highly enriched uranium (HEU) and plutonium recovered from dismantled weapons (StrålevernRapport 2006: 12). A production

3 Sources differ regarding the period of temporary storage. Whereas most reports refer to a three-year time period, the environmental NGO Bellona mentions there are temporary storage facilities for at least ten years at every nuclear power plant with RBMK-reactors and the Novovoronezh nuclear power plant (Nikitin et al. 2012: 57)

facility for MOX fuel produces fuel for the BN-600 fast breeder reactor at Beloyarsk.

The open-type joint stock company Siberian Chemical Combine (JSC "SCC") was established in 1953 in Tomsk-7, now known as Seversk. The SCC is the second storage site for HEU and plutonium recovered from dismantled weapons. Seversk is one of the largest sites storing low- and intermediate-level wastes from reprocessing, with more than 30 million cubic meters stored via deep well injection.

Finally, JSC "SRC RIAR" at Dimitrovgrad employs HEU-fueled research reactors. RIAR carries out research related to the nuclear fuel cycle, particularly fuel fabrication and reprocessing technologies, as well as radioactive waste management.

2.4 The waste management strategy (with current waste disposal plan/ concept)

As the name suggests the strategic goal of the USS RWM is to establish a unified state system for radioactive waste management. The implementation process consists of several stages. During the first stage (2011-2015), the necessary regulatory and procedural framework was developed. Moreover, the aforementioned registration of radioactive waste in the Russian Federation was initialised (see 2.2.1). With regard to disposal strategies, the main efforts are targeted at disposal of low-level radioactive waste. According to Rosatom, a priority is the conservation of near-surface low-level radioactive waste storage facilities (storage facility B-9 at PA Mayak) and open storage basins (B354 at FSUE "MCC", and B-1 and B-2 at JSC "SCC"). In addition, attempts to establish safe conditions at the highly polluted Techa Resovoir Cascade o at the PA Mayak can be observed.

Another important undertaking is the commissioning of solid radioactive waste processing complexes at Smolensk and Kola nuclear power plants., Processing facilities at the PA Mayak were expanded, including the construction of a cementation complex for liquid and heterogeneous intermediate-level waste, a purification facility for water containing intermediate-level waste and of a treatment facility for low-level liquid radioactive waste. In addition, a new electric furnace EP-500/5 for high-level radioactive waste vitrification (vitrification capacity 3·1018 Bq per year) was commissioned (Rosatom, IBRAE RAN and FBI "SEC NRS" 2014: 27).

With regard to centralised storage facilities at FSUE "MCC" and PA Mayak, the plan mentions the following construction projects: a storage facility for immobilised radioactive waste with a total capacity of 28,000 m^3 in FSUE

"MCC"; a storage facility with a total capacity of 20,200 m^3 in Mayak; reconstruction activities at five radioactive waste storage facilities in in FSUE "MCC", JSC "SCC", JSC "SRC RIAR"; and the ship-repairing centre Zvezdochka. Finally, it mentions the construction of a first section of a near-surface disposal for solid radioactive waste with a total capacity of 23,500 m^3 at the Ural Electrochemical Combine (JSC "UECC") (Novouralsk, Sverdlovsk Region) (ibid.: 30).

The second stage (2015-2018) aims at establishing a disposal system for low-level and intermediate-level waste. This process involves decisions on the construction of disposal facilities, which were approved in the first stage. Apart from that, top-priority low-level and intermediate-level waste disposal facilities shall be determined in order to proceed with construction as quickly as possible.

The final stage (2018-2021) aims at developing a disposal system for high-level waste. This step involves the development of a necessary storage facility infrastructure. Special attention is paid to the development of sites holding special radioactive waste (see 4.1).

Although first steps towards the development of an appropriate infrastructure can be observed, the current waste management strategy leaves important questions unanswered. This is a particular concern regarding the management of spent nuclear fuel. Assessing Rosatom´s nuclear fuel management strategy, the environmental NGO Bellona critically concludes that there is a missing "normative-judicial base [...] regulating the responsibility of the producers of spent nuclear fuel and mechanisms for financial cooperation of the involved actor" (Nikitin et al. 2012: 68). Thus, spent nuclear fuel management is exclusively funded by the state.

3 The legal and institutional framework

3.1 The legal framework

The development of a unified system for radioactive waste and spent nuclear fuel management began after the ratification of the Joint Convention on the Safety of Spent Fuel Management and on the Safety of Radioactive Waste Management in November 2005. Another important step was the adoption of the federal target program, Nuclear and Radiation Safety in 2008 and for the period up to 2015, in 2007.

The juridical base for radioactive waste management in Russia was already established with the adoption of the federal law № 170-FZ On the Use of Atomic Energy on November 21, 1995. However, as it is limited to the civil sector of the nuclear industry and therefore exercising no regulatory power over the military complex[4], this law proved to be ineffective for tackling the issue of radioactive waste. In the course of the reorganisation process of the nuclear sector, the law № 190-FZ On Radioactive Waste Management and Amendments to Certain Legislative Acts of the Russian Federation was adopted and implemented several changes. The law stresses the need for final storage of radioactive waste. Apart from initialising a process of registration of nuclear waste (see 2.2.1) as a precondition for final storage, it defines responsibilities and priorities regarding radioactive waste, e.g. the contaminator's obligation to pay for all costs and the responsibility of the radioactive waste-producing body to guarantee safe waste management during all stages until final storage. Finally, the law revised the special status of the military complex by extending regulatory capacity to the defence sector of the nuclear industry.

The outcome of the reform is rather ambiguous. The law enhances possibilities to discuss radioactive waste management publicly because working processes, such as mining of radioactive ore, sublimation and separation productions, nuclear fuel manufacturing and spent nuclear fuel reprocessing, operation of nuclear power plants and research facilities, and the decommissioning of nuclear power facilities, are explicitly mentioned as generating radioactive waste (see chapter 5). Moreover, the law has established and defined duties and norms both of the state and Rosatom, e.g. infrastructure development. At the same time, however, the law reinforces the non-transparent structure of the nuclear complex by establishing a unitary atomic energy corporation (Rosatom). In the absence of effective supervisory authorities, both transparency and possibilities for participation of the population are restricted (see chapter 5). Another distinctive outcome of the reorganisation of the nuclear complex is that civil and military sectors are united under the umbrella of Rosatom, which grants the corporation extraordinary powers and authority in the field of nuclear power (see 3.2). Finally, these efforts to adjust the countries' regulatory base according to international standards most certainly accelerates the internationalisation of the nuclear complex.

A very sensitive issue is the import of spent nuclear fuel from foreign nuclear power plants for reprocessing. Rosatom actively promotes global commercial trade and storage of spent nuclear fuel in Russia as a part of its internationalisation strategy. In negotiations for the construction of nuclear power plants, the state corporation offers to guarantee supply of all fuel and repatriation of spent fuel for the lifetime of the plant, leading to the acceptance of bids for construction in Belarus (Schneider and Froggatt 2015: 46-47), Vietnam (ibid.: 53-54) and Hungary (ibid.: 162-163).

4 The military complex covers areas, such as development, manufacturing, testing, operation and utilisation of nuclear weapons and nuclear power units of military facilities.

According to the federal law On Environmental Protection (Art. 48 and Art. 51) and the federal law On Radioactive Waste Management (Art. 31), Russian legislation prohibits the import of radioactive waste from foreign countries for the purpose of storage, treatment, neutralisation or disposal (NEA 2014: 14). This does not include the import of spent nuclear fuel for intermediate technological storage and/or reprocessing. So far, Russia has imported spent nuclear fuel from Soviet-built and Russian-built reactors (see 2.2.2). However, it appears that the radioactive waste accumulated during reprocessing has not not returned to the respective countries from where it originated (Talevlin et al. 2015: 7). Plans to build another reprocessing plant (RT-2) in Zheleznogorsk and possible legislative initiatives on processing of radioactive waste suggest an increase in imports of spent nuclear fuel (Nikitin et al. 2012: 61-65).

The foundation of the joint stock company Angarsk International Uranium Enrichment Centre (JSC Angarsk IUEC) in September 2007 confirms Rosatom's ambitions to provide assured supplies of low-enriched uranium for power reactors to new nuclear power states. Therefore, IUEC will sell both enrichment services (SWU) and enriched uranium. In order to assure fuel supply to countries without their own fuel cycle facilities, Russia has even proposed to create an international guaranteed reserve (fuel bank) of low-enriched uranium at the IUEC at Angarsk, which would be controlled by the International Atomic Energy Agency (IAEA). Low-enriched uranium shall be available to any fully entitled IAEA member state unable to procure fuel for political reasons. The fuel will be made available to countries at market rates based on spot prices. However, foreign stakeholders will not be granted access to the enrichment technology. Although foreign companies will be allowed to acquire equity stakes, Russia will maintain majority ownership.[5]

3.2 *The institutional framework*

Following the dissolution of the Soviet Union, all civil and military nuclear activities were concentrated under the Ministry of Atomic Energy (Minatom). In 2004, the ministry was reorganised as the Federal Agency on Atomic Energy and all properties and powers transferred. Finally, in 2007, the legal status of the agency was changed into a state corporation and renamed Russian State Atomic Energy Corporation (Rosatom).

---

5 Current stakeholders are: TVEL, a Rosatom subsidary, holding 70 percent, Kazatomprom 10 percent (Kazakhstan), Nuclear Fuel 10 percent (Ukraine), Armenian NPP 10 percent (Armenia). Furthermore, negotiations on acquisition of shares with other countries have proceeded (e.g. Bulgaria, Mongolia, Jordan, Vietnam, India). TVEL aims at holding eventually only 51 percent of shares.

The legal status of Rosatom is unique. According to Russian federal law, a state corporation is officially a non-profit organisation entirely owned by the state.[6] For such a specific type of legal entity, different legislative norms are applied. With the adoption of the law On the State Atomic Energy Corporation Rosatom in December 2007, the state consolidated the entire atomic energy industry, including defence enterprises, research facilities and corresponding budget funds under Rosatom. This is confirmed by the Federal Law № 188-FZ On Amendments to the Federal Law On the State Atomic Energy Corporation Rosatom and Certain Legislative Acts of the Russian Federation adopted on July 2013. Accordingly, Rosatom

> exercises the functions of chief public funds controller, public funds recipient, chief public funds administrator, public revenue administrator and state contracting authority for long-term target programs, R&D and investment programs and projects, special environmental programs for the remediation of radioactively contaminated sites, multinational programs, the federal targeted investment program; it also places orders, concludes state contracts for goods, services and works, execution of scientific-technical, research and development, design and survey, and engineering activities for state needs, as well as other civil-law contracts according to the procedure established by the Russian legislation (Rosatom, IBRAE RAN and FBI "SEC NRS" 2014: 14).

From this, it is apparent that Rosatom has accumulated extraordinary powers. State policies, legal and normative regulations and the provision of state services are implemented by Rosatom. Moreover, the corporation manages state property in the area of atomic energy use and is responsible for the development and safe operation of atomic energy and nuclear weapons. The objective to develop the nuclear weapons industry is especially controversial since different international treaties signed by the Russian Federation (e.g. Strategic Arms Reduction Treaty, START-1) aim at nuclear disarmament. Neither the government nor the parliament exercise direct control over state corporations. Both the director general and all members of the supervisory board are directly appointed by the president. On top of that, the legal form of a state corporation ensures Rosatom's non-transparent special status. In the event of bankruptcy, the usual legislative procedure (insolvency proceedings) is not applied to state corporations. "The creditors of the corporation do not have to give their consent to their claims being transferred to a new organisation in case of corporate restructuring. Rosatom is also excluded from obligations to make public its activities, expenditures or use

6 Apart from Rosatom, several other state corporations exist, e.g. Rostec or Vnesheconombank. In the case of Rosatom, the corporation operates as a "federal nuclear organisation".

of property. The Director General also has the right to classify information as state secrets, as provided by Russian Federal Law" (Greenpeace 2014: 12).

In the absence of effective external control mechanisms, Rosatom is obliged to perform internal control functions over its subsidiary enterprises (NEA 2014: 15). According to the law On the State Atomic Energy Corporation Rosatom, the state corporation issues licenses to organisations using nuclear material and radioactive substances for defence purposes (Art. 8). Furthermore, the state corporation develops and approves safety norms and rules for the use of nuclear energy. However, this interferes with the controlling function of the Federal Environmental, Technological and Nuclear Supervision Agency (Rostekhnadzor). Serving as the state´s supervisory authority on ecological, technological and nuclear issues, Rostekhnadzor also performs normative regulatory functions in the area of atomic energy.

With regard to radioactive waste management, the national operator deserves special attention. The company is responsible for safe radioactive waste disposal and management of the selected sites, including their final closure or the construction of new sites. Furthermore, the scope of duties covers regular reports to the state and the Russian population providing details on the expected amount of radioactive waste for final disposal (Vakhrusheva 2016: 1).

3.3 Siting procedures

3.3.1 Procedures and criteria for site selection

The basic requirements for the creation of radioactive waste disposal facilities and site selection are presented in the document NP-055-04 "Radioactive waste disposal. Principles, criteria and basic safety requirements" (adopted in 2004). The law defines disposal of radioactive waste as safe emplacement of radioactive waste without intent of future retrieval (Gosnadzor 2004: 5). Furthermore, it provides the framework for radioactive waste disposal in deep geological formations located at depths of several hundred meters but also near-surface disposal in structures located at surface level and/or depths from several meters up to one hundred meters.

If a facility is fully equipped to guarantee the isolation of radioactive waste from the environment during the period of its potential hazard, it qualifies as radioactive waste disposal facility (Rosatom, IBRAE RAN and FBI "SEC NRS" 2014: 19). With regard to high-level waste, a facility holding special radioactive waste may be defined as a conservation facility. Transforming a special radioactive waste facility into a conservation facility implies a conversion of the radioactive waste storage facility status. This process includes the fulfilment of safety requirements. The decision to transform a conservation facility for special radioactive waste into a radioactive waste disposal facility is made by the government upon recommendations of the state management authority in the field of radioactive waste management, which is Rosatom (ibid.).

The law On Radioactive Waste Management has established a new classification system dividing radioactive waste into special and removable waste. Furthermore, radioactive waste defined as removable waste is divided into six classes according to its level of radioactivity. Therefore, each level is subject to different disposal requirements.

Table 2: Removable Radioactive Waste

Class	Waste Category	Disposal Requirement
1	Solid high-level radioactive waste	Final disposal in deep disposal facilities; prior storage to reduce heat generation
2	Solid high-level and intermediate-level long-lived radioactive waste containing radionuclides with half-lives > 30 years	Final disposal in deep disposal facilities; no prior storage to reduce heat generation
3	Solid intermediate-level long-lived radioactive waste containing radionuclides half-lives > 30 years	Final disposal in near-surface disposal facilities (depth of up to 100 m.)
4	Solid low-level and very low-level radioactive waste	Final disposal in near-surface disposal facilities (ground level)
5	Liquid intermediate-level and low-level radioactive waste	Final disposal in deep well injection facilities
6	Radioactive waste generated in mining and processing of uranium ore and other activities without use of nuclear energy, namely, mining and reprocessing of mineral and organic raw materials with high concentrations of naturally occurring radionuclides	Final disposal in near-surface disposal facilities

Source: Rosatom, IBRAE RAN and FBI «SEC NRS» 2014: 22-23

The order No 1/382-P, drafted in April 2013, defined a list of prospective sites in Russia for further study and site selection for creation of radioactive waste disposal facilities. So far, prospective sites in 18 regions of the country have been selected for the creation of a radioactive waste disposal facility. A selected site has to comply with the following conditions: cost-effectiveness; close proximity to places generating and accumulating radioactive waste; and appropriate geological characteristics (Talevlin et al. 2015: 7). The selection process has to fulfil certain requirements, including public hearings for the population. Without the population's formal consent and the agreement of regional administration, a disposal site cannot be opened (see also 5).

3.3.2 Compensation mechanisms and socio-economic impact

Neither the federal government nor the state corporation Rosatom provide any direct compensation to the population surrounding disposal sites. However, regional governments have the option to introduce initiatives to stimulate socio-economic development in selected regions. Due to the special risks of their occupations, employees of state companies, such as Rosatom, and workers in nuclear power plants or disposal facilities may be granted certain work-related benefits, such as early retirement, extra payment and holidays and favourable health insurance. The NGO-network, Decommission, urges that requirements for radioactive waste disposal not only focus on safety issues but also consider principles of sustainable regional development, fairness and environmental safety (Talevlin et al. 2015: 12).

4 Information and participation

The federal law, On the Use of Atomic Energy, guarantees both citizens and public associations the right to engage in the formation of certain policies in the field of nuclear energy (Articles 13 and 14). Accordingly, both citizens and public associations have the right to request and receive information concerning "the safety of planned, designed, constructed, operated and decommissioned nuclear installations, radiation sources and storage facilities" (Russian Federation 1995). However, the legislative procedure often remains vague, leading to ambiguous and sometimes even contradictory legal situations. On the one hand, the national radiation monitoring system must provide information on radiation levels in a given region (ibid.). On the other hand, a substantial share of information and functions performed by Rostekhnadzor, the supervisory body authority on ecological, technological and nuclear issues, are considered issues of national security and treated as classified. Therefore, the OECD concludes in a report on environmental policy in Russia that "complaints remain the main mechanisms of public involvement in compliance monitoring" (OECD 2006: 43).

Article 14 stresses the right "to participate in discussions of draft legislative acts and programs in the field of use of atomic energy, as well as in discussions on the issues related to siting, design, construction, operation and decommissioning of nuclear installations, radiation sources and storage facilities" (ibid.). However, the law lacks a clear formalised procedure of public participation at different stages of radioactive waste management. Apart from public hearings, participatory elements are not mentioned. In a similar vein, the fourth national report on the safety of spent fuel management and the safety of radioactive waste management predominantly focuses on safety and licensing requirements. Thus, safety issues have become a primary a concern of experts and companies (Rosatom, IBRAE RAN and FBI "SEC NRS" 2014: 96-98). Licensing of siting, construction, operation and decommissioning of spent-nuclear fuel and radioactive waste storage facilities are carried out by Rostekhnadzor. Although licensing activities shall not be granted without the formal consent of the population, public associations and/or municipal authorities as voiced during public hearings, the law does not specify sanctions in case this process is violated. The national report, however, mentions public involvement in the decision-making process only in regard to siting (ibid: 96). Moreover, the procedure for public hearings and the decision-making process are not further specified. The focus on expert knowledge and safety requirements indicates a rather technocratic approach to regulating public participation in formalised procedures. Therefore, the NGO-network Decommission demands that Russian citizens "living in the area of long-term isolation of radioactive waste in accordance with the Constitution must be able to take the final decision whether it is acceptable to place such facilities there" (Talevlin et al. 2015: 12).

5 Costs and financing

Rosatom is directly funded by the state. However, the company restricts information on the expected costs and sources of financing decommissioning of old nuclear power plants and management of radioactive waste. Despite transparency initiatives and the company's obligation to disclose financial reports listing both the results and efficiency of spending funds disbursed from the state budget and other sources, these reports remain vague regarding the industry's economic situation (Andreev 2011: 6-7). This non-transparency is further upheld by the law, On the State Corporation Rosatom., which stipulates that the federal body exercising oversight over the fulfilment of the federal budget, the Accounts Chamber of the Russian Federation, is restricted to performing external control over Rosatom's activities (ibid.: 6). Furthermore, audits of the corporation's "internal financial and economic operations, the efficiency of spending government-allocated financial resources or those held in special reserve funds" are carried out by Rosatom's own Audit Committee (ibid.).

This makes providing data on costs and the financing of decommission activities a great challenge. According to a Greenpeace report, in 2012 the company received 2.97 billion Euro from the federal budget. The lion´s share was spent on the construction of new reactors in Russia (1.44 billion Euro) and measures to improve radiation-protection (21.73 million Euro) (Greenpeace 2014: 13). This marked the first time in Soviet/Russian history, the development of a sufficient infrastructure for radioactive waste management and spent nuclear fuel including reprocessing, storage and transportation, was funded by the federal government.[7] Regional budgetary funds, Rosatom funds, such as the special reserve fund for decommissioning and radioactive waste disposal, and money received from international cooperation contribute to financing activities in the field of spent nuclear fuel and radioactive waste management and decommissioning. The special reserve fund was established to ensure nuclear safety and cover costs associated with decommissioning activities. Aside from the state, enterprises and organisations operating hazardous nuclear and radiation production and facilities are major contributors to the fund (Rosatom, IBRAE RAN and FBI "SEC NRS" 2014: 72).

A striking outcome of the reorganisation of the Russian nuclear complex is the linking of radioactive waste management to profitability standards. In addition to improving and guaranteeing safety standards, the fourth national report on the safety of spent fuel management and the safety of radioactive waste management explicitly stresses the development of "economic instruments to support decommissioning activities with due regard to social aspects and those associated with human resources" (Rosatom; IBRAE RAN; FBI "SEC NRS" 2014: 83). The main source national operators' income is suppose to be generated from payments made by the producers of radioactive waste. Producers of radioactive waste are obliged to pay for the transfer of radioactive waste for final disposal. The price is based upon tariffs determined by the Ministry of Natural Resources and Ecology of the Russian Federation (Vakhrusheva 2016: 1). The disposal tariffs, including the appropriate basic pricing principles and rules for state regulation and control, proceed according to the resolution of the Russian government № 1249 On the Procedure for the State Regulation of Radioactive Waste Disposal Tariffs (adopted in 2012). Tariffs are based on the expected expenditures of the national operator, which may include expenses on investment programs (construction of sites for disposal), the operational cost of sites or property compensation (ibid.: 2). The funds are collected in the aforementioned special reserve fund for decommissioning and radioactive waste disposal. In the future, any expenses incurred by the investment program are to be covered by the reserve fund. However, the national operator does not have full authority over these funds. Rather, investment decisions completely depend on Rosatom (ibid.: 2-3).

7 The federal target program "Nuclear Radiation Safety in 2008 and for the period up to 2015" covers most of the activities. The total budget of the program makes 2 billion Euro. About 1,03 billion Euros are allocated for resolving the nuclear legacy problem (NEA 2014: 17).

Moreover, both producers of radioactive waste and the national operator are subsidiary companies of Rosatom, which might pose conflicts of interest since waste disposal creates additional costs for the state corporation. That radioactive waste containing Cesium-137 was downgraded from second to third class by decree № 1069 in October 2012, was most likely due to cost-saving reasons and points in this direction (Talevlin et al. 2015: 7). Although the decommissioning concept stresses that disposal costs are covered by generators of radioactive waste, the final costs are covered by the state, either directly in the case that costs exceed the liability limit of the company involved or indirectly by subsidising the special reserve fund.

6 Conclusions

In the past ten years, the Russian nuclear complex has undergone a profound reorganisation involving both its legislative and institutional frameworks. Despite improvements in various areas, including infrastructural development and creation of a unified regulatory base, the outcome of the reforms has been rather ambivalent with regard to democratic control of the nuclear complex or the implementation of ecological standards. The state corporation Rosatom has assumed tasks in the fields of civil and military atomic energy previously performed by the Russian state. As Rosatom is accountable only to the President of the Russian Federation, neither the federal parliament, regional state authorities nor other public institutions can effectively control the company's actions. In fact, the reform has resulted in a new set of interlocking relationships between the state, scientific experts and the nuclear industry based on the logic of a market economy. The overarching goal of this process is to secure a significant role for nuclear energy in the domestic market and to enable Rosatom´s international expansion.

Radioactive waste management is regarded as a decisive factor in the future success of the nuclear complex. A closed nuclear cycle with reprocessing of spent nuclear fuel, deep well storage and deep geological and near-surface disposal are all factors designed to prove the superiority and sustainability of the nuclear industry over other energy sources. The foundation of an international fuel bank of low-enriched uranium to assure fuel supply to countries without their own fuel cycle facilities or repatriation of spent fuel for Russian-built nuclear power plants (Belarus, Hungary, Vietnam) point in the same direction. Guiding this reorganised approach to radioactive waste management in Russia are principles such as the valorisation of radioactive waste with a tariff-system and spent nuclear fuel through reprocessing and the creation of a fuel bank. Despite these efforts, the Russian nuclear complex has not proven able to dispose of radioactive waste safely. This reality means that major implications for human beings and the environment remain.

References

Andreev, L. (2011). *The Economics of the Russian Nuclear Power Industry*. Bellona Report, Oslo, St. Petersburg: Bellona.

Brunnengräber, A. and Mez, L. (2016). Der staatlich-industrielle Atomkomplex im Zerfall. Zur politischen Ökonomie der Endlagerung in der Bundesrepublik Deutschland. In: Brunnengräber, A. (Ed.). *Problemfalle Endlager. Gesellschaftliche Herausforderungen im Umgang mit Atommüll*, Baden-Baden: edition sigma in der Nomos Verlagsgesellschaft, 289-312.

Goskomnadzor (2014). *Disposal of Radioactive Waste. Principles, Criteria and Basic Safety Requirements*. NP-055-04. http://en.gosnadzor.ru/framework/nuclear/NP-055-2004.pdf, last accessed 10 September 2016.

Greenpeace (2014). *Rosatom Risks. Exposing the troubled history of Russia's state nuclear corporation*, Amsterdam: Greenpeace International.

IBRAE RAN [Institut problem bezopasnogo razvitiya atomnoj energetiki Rossijskoj Akademii Nauk] (2006). *Rekomendacii dlya gosudarstvennykh organov i ucherezhdenij po rabote s grazhdanami, postradavshimi v rezul'tate avarij na PO „Mayak"*. http://chernobyl-mchs.ru/upload/program_rus/program_rus_1993-2010/recom_Ural.pdf, last accessed 16 January 2017.

Kozyrev, D. B. (2009). *Plany Rossijskoj Federacii po sozdaniyu ob''ektov pripoverkhnostnogo zahoroneniya RAO. 1-15*. https://www.iaea.org/Our-Work/ST/NE/NEFW/CEG/documents/ws022009/3.%20Near%20Surface%20Repositories/3.3%20Russian%20Plans%20for%20Creation%20of%20NSR%20Rus.pdf, last accessed 10 September 2016.

Krohnfeld-Goharani, U. (1999). Ein Erbe des martimen Wettrüstens: der Atommüll der Nordmerflotte. In: *SCHIFF-Texte*, 53, Kiel.

Krohnfeld-Goharani, U. (2002). Die Umweltschäden und Entsorgungsprobleme des russischen Nuklearkomplexes. In: *SCHIFF-Texte*, 68, Kiel.

Kurchatovskij Institut and Rosatom (2009). *Yadernye energotekhnologii novogo pokoleniya na period 2010-2015 godov i na perspektivu do 2020 goda*. Rasporyazhenie Pravitel'stva Rossijskoj Federacii, No 1026-p.

Ministry of Energy of the Russian Federation (2010). *Energy Strategy of Russia for the period up to 2030*. Moscow.

Muratov, O. E. and Tichonov, M. N. (2015). *Radioaktivnye aspekty obrashcheniya s RAO i OYAT v usloviyakh innovacionnogo razvitiya yadernoj energetiki*. http://www.atomic-energy.ru/articles/2015/10/06/60277, last accessed 10 September 2016.

Nikitin, A. et al. (Eds.) (2012). *Yadernye delyashchiesya materialy (praktika obrashcheniya, obzor technologij, problemy i perspektivy)*. Ekologicheskoe ob′′edinenie «Bellona», Sankt-Peterburg, Murmansk, Oslo.

Nuclear Energy Agency [NEA] (2014). *RADIOACTIVE WASTE MANAGEMENT PROGRAMMES IN OECD/NEA MEMBER COUNTRIES*. Country Profile, 1-19. https://www.oecd-nea.org/rwm/profiles/Russian_Federation_profile_web.pdf, last accessed 10 September 2016.

OECD [Organisation For Economic Cooperation And Development] (2006). *Environmental Policy and Regulation in Russia*. https://www.oecd.org/env/outreach/38118149.pdf, last accessed 28 October 2016.

Pomer, M. and Klein, L. R. (Eds.) (2001). *The New Russia. Transition Gone Awry*, Stanford: Stanford University Press.

Puchkov, B.A. and Bol′shkov, L. A. (Eds.) (2016). *Rossijskij nacional′nyj doklad: 30 let Chernobyl′skoj avarii. Itogi i perspektivy preodoleniya ee posledstvij v Rossii 1986-2016*, IBRAE RAN, Moskva.

Rosatom, IBRAE RAN and FBI "SEC NRS" (2014). *The Fourth National Report of the Russian Federation on Compliance with the Obligations of the Joint Convention on the Safety of Spent Fuel Management and the Safety of Radioactive Waste Management*. Moscow.

Russian Federation (1995). *Federal Law On Atomic Energy Use*. http://www.vertic.org/media/National%20Legislation/Russian_Federation/RU_Federal_Law_Atomic_Energy_use.pdf, last accessed 10 September 2016.

Schneider, M. and Frogatt, A. (Eds.) (2015). *The World Nuclear Industry Status Report 2015*, Paris, London.

Stiglitz, J. (2002). *Der Schatten der Globalisierung*, Berlin: Siedler.

StrålevernRapport (2006). *Standring WJF*. Review of the current status and operations at Mayak Production Association, Norwegian Radiation Protection Authority, Østerås.

Talevlin, A. A. et al. (2015). *Radioactive Waste and Spent Nuclear Fuel Management in Russia: view of non-governmental organizations*, Sosnovy Bor - Apatity - Chelyabinsk.

Tichonova, N. (2011). Armut in Russland. In: *Russland-Analysen,* No. 222, 2 - 5. http://www.laender-analysen.de/russland/pdf/Russlandanalysen222.pdf, last accessed 16 January 2017.

Vakhrusheva, K. (2016). *Sravnenie stoimosti i metodiki rascheta platy za zakhoronenie radioaktivnykh otkhodov v Rossii i drugikh stranakh*, Sankt-Peterburg: Bellona.

Transitioning Away from Opaque Governance

Nuclear Waste Policy in South Korea

Sung-Jin Leem[1] *with Miranda Schreurs*

Abstract

Due to the fact that the temporary spent fuel storage facilities in nuclear power plants will reach full capacity in the near future, the disposal of spent nuclear waste has become a matter of great concern in the Republic of Korea (South Korea). This article discusses the current status and limits of the Korean nuclear waste storage policy and governance. Considering the ecological and social impact of radioactivity, there is a compelling need for a nuclear waste governance system based on a social consensus on nuclear waste disposal. However, the participation of citizens and stakeholders in the decision-making process and their access to nuclear related information were restricted. Successive governments introduced legal and institutional instruments to do more to engage the public on spent fuel storage after a series of failed attempts to select a nuclear waste repository site. However, anti-nuclear non-governmental organisations (NGOs) refused to participate in this government-led public discussion and criticized it for being a mere formality. There are signs of change since the election of Moon Jae-In as president in the spring of 2017. Moon's government initiated a major shift in Korea's energy policy as well as in the policy-making process. Moon's government has sought input from the public and NGOs in nuclear-related questions and is promoting more openness and transparency.

---

1 Dept. of Public Administration, Jeonju University, 560-759 South Korea; E-Mail: leemsj@gmail.com

A. Brunnengräber et al. (eds.), *Challenges of Nuclear Waste Governance*,
Energiepolitik und Klimaschutz. Energy Policy and Climate Protection,
https://doi.org/10.1007/978-3-658-21441-8_4

1 Introduction[2]

Due to South Korea's high nuclear capacity resulting from many years of nuclear expansion policy, the accumulated amount of nuclear waste has rapidly increased. Accordingly, nuclear waste has been a matter of great concern to the public, especially the issue of spent fuel is currently under the spotlight. The South Korean government is preparing a plan to select an interim storage site for spent fuel by approximately 2020, estimating that the temporary spent fuel storage facilities in nuclear power plants will reach their full capacity within some years (PECOS 2015).

Considering the ecological and social impact of radioactivity, the need for a governance system based on social consensus for nuclear waste disposal in South Korea is compelling. Moreover, the social perception of nuclear energy use has been changing since the Fukushima accident and the number of citizens and political parties requesting an alternative energy policy is steadily increasing.

Nonetheless, the government of ousted President Park Geun-Hye rushed to complete the scheduled public discussions as a mere formality and is trying to start the selection process for the storage site for spent fuel. Anti-nuclear NGOs such as the Federation for Environmental Movement, the Energy Justice Actions, and the Non-nuclear Action etc., which had already declined to participate in the discussions held by the government, protested the plan; the social conflict grew as time progressed.

This research discusses the state of Korean nuclear waste disposal policy during the period through the regime of Park Geun-hye and its limits from a governance perspective. First, Section 2 gives an overview of the status of nuclear power plants (NPP) and nuclear waste inventory in Korea. Korean nuclear waste policy can be separated into two phases, i.e., before and from 2005 to 2017, therefore, the policy of each phase will be separately analyzed in Sections 3 and 4. Section 5 presents an analysis of the legal and institutional framework surrounding nuclear waste policy, with particular attention to political transparency. Section 6 describes a survey on public participation and information policy concerning nuclear waste management examining the status and limits of the nuclear waste governance strategies. Finally, the conclusion presents lessons drawn from the research.

2 This chapters draws heavily on Leem (2015).

2 Nuclear power policy and the nuclear waste inventory

Nuclear power has been used in Korea for close to four decades, ever since the first nuclear reactor began generating power in 1978. As of July 2017 after the first shutting of the Kori 1 reactor, the first reactor to be closed down in the country's history, 24 reactors with a total installed capacity of 25,5 MW are in operation and are responsible for about one-third of the nation's total electricity production. Currently, five additional reactors (6,600 MW) are under construction. Additional reactors which were in planning stages have been cancelled by President Moon Jae In. According to the Second National Energy Master Plan up to 2035 which was released in January 2014, installed capacity of nuclear power was to account for up to 29 percent of electricity generation capacity by 2035.[3] To meet this production goal, the goal was to achieve a generation capacity of NPPs with 42,705MW, which would have required doubling the present capacity. Thus, 5-7 more reactors were scheduled to be built by 2024, in addition to the already existing and planned reactors (KHNP 2015; Yang Lee 2015; MOTIE 2014; EJA 2014; IAEA 2014). These plans have, however, been put on hold or stopped by President Moon. Table 1 and 2 show the status of nuclear power plants in South Korea.

Table 1: NPPs in Operation and Retired NPPs

Reactor	Installed Capacity (MWe)	Reactor Type	Commercial Operation (Year, month, date)
Kori 1	587		1978.04 .29 (retired on 2017.06.18)
Kori 2	640	PWR-Westinghouse	1983.07.25
Kori 3	1011	PWR-Westinghouse	1985.09.30
Kori 4	1012		1986.04.29
Shin-Kori 1	997		2011.02.28
Shin-Kori 2	997		2012.07.20
Shin-Kori 3			2016.12.16
Wolsong 1	657	PHWR-Candu 6	1983.04.22
Wolsong 2	647	PHWR-Candu	1997.07.01
Wolsong 3	651	PHWR	1998.07.01
Wolsong 4	653	PHWR	1999.10.01

3 According to the 7th Basic Plan for Long-term Electricity Supply and Demand the installed capacity of nuclear power was to be increased to 38,916MW by the year 2029 (MOTIE 2015).

Reactor	Installed Capacity (MWe)	Reactor Type	Commercial Operation (Year, month, date)
Shin-Wolsong 1	997		2012.07.31
Shin-Wolsong 2	993		2015.07.24
Hanbit 1	996	PWR-Westinghouse	1986.08.25
Hanbit 2	998	PWR-Westinghouse	1987.06.10
Hanbit 3	993	PWR (System 80)	1995.03.31
Hanbit 4	970	PWR (System 80)	1996.01.01
Hanbit 5	998		2002.05.21
Hanbit 6	993		2002.12.24
Hanul 1, Ulchin	969		1988.09.10
Hanul 2, Ulchin	965		1989.09.30
Hanul 3, Ulchin	997		1998.08.11
Hanul 4, Ulchin	998		1999.12.31
Hanul 5, Ulchin	999	OPR-1000	2004.07.28
Hanul 6, Ulchin	997	OPR-1000	

Source: KHNP 2015; MOTIE and KHNP 2014; MSIP 2014; World Nuclear Association 2017

Table 2: NPPs which were/are under Construction

Unit	Installed Capacity (MWe)	Reactor Type	Commercial Operation	
Shin Hanul 1	1400 MWe	PWR (APR1400)	2018	
Shin Hanul 2	1400 MWe	PWR (APR1400)	2019	
Shin Kori 4	1400 MWe	PWR (APR1400)	2017	
Shin Kori 5	1400 MWe	PWR (APR1400)	2019	Suspended to assess public opinon
Shin Kori 6	1400 MWe	PWR (APR1400)	2020	Suspended to assess public opinon
Shin Ulchin 3	1400 MWe	PWR (APR1400)	2021	cancelled
Shin Ulchin 4	1400 MWe	PWR (APR1400)	2022	cancelled
Shin Kori 7	1500 MWe	PWR (APR1400+)	2023	cancelled
Shin Kori 8	1500 MWe	PWR (APR1400+)	2024	cancelled

Source: KHNP 2015; MOTIE and KHNP 2014; Yonhap News Agency 2013; World Nuclear Association 2017; interviews by Schreurs with advisory staff to President Moon, Blue House, Seoul, Korea, June 19, 2017.

In South Korea, nuclear waste is categorized according to the IAEA classification system. High-level nuclear waste (HLW) is identical to spent nuclear fuel, and the rest is classified as low-and intermediate-level radioactive waste (LILW), which is subdivided into four categories: ILW, LLW, VLLW, and EW (NSSC 2013).

As of March 2015, a total of 93,680 drums (200 liters each) of low- and intermediate-level radioactive waste (LILW) had been stored on-site at NPPs. Additionally, the Korean Atomic Energy Research Institute (KAERI) and KEPCO Nuclear Fuel (KNF) had 17,922 drums in their own storage facilities (as of 2014). Starting in 2010, a portion of the nuclear waste from Hanul and Wolsong NPP, KAERI, and KNF has been transported to the nuclear waste repository in Gyeongju; in 2014 this amounted to a total of 5,032 drums. Table 3 shows the status of LILW in detail.

Table 3: Amount of solid LILW storage in each facility (March 2015)

Power Plant	Being stored (drum)	Storage Capacity (drum)
Kori	42,456	50,200
Hanbit (Yonggwang)	22,881	23,300
Hanul (Ulchin)[1]	17,497	18,929
Wolsong[1]	10,287	13,240
Shin-Kori	559	10,000
Total in NPP	93,680	115,669
KAERI, KNF[2]	17,922	
Gyeongju repository[2]	5,032	

Source: KHNP 2015, rearranged

[1] excluding waste-amount transported by KORAD to Gyeongju repository (1,000 drums in Hanul, 2,536 drums in Wolsung)

[2] in year 2014

Additionally, as of 2013 there were 3,095 drums (200 liters each) of radioisotope (RI) waste generated by industries, hospitals, universities, and research institutes that use radioisotopes. This waste is collected by the Korea Radioisotope Association and transported to a dedicated storage facility of the Korea Radioactive Waste Agency (KORAD) in Daejeon.

All LILW temporarily stored at different sites will eventually be transported to and stored in the central nuclear waste repository in Gyeongju. The first phase of construction of this underground storage facility comprising six silos was completed in 2014; it has a capacity for 100,000 drums. In the second phase, a near-surface storage facility capable of holding 125,000 drums is under construction and eventually the entire complex is expected to hold 800,000 barrels (Patel 2015)..

As for high-level nuclear waste, roughly 760 tons of spent fuel is produced annually by NPPs as of 2015 (Kang, Kim, and Lee 2015). The light water reactors in Kori, Hanbit and Hanul plants and the heavy water reactors in Wolsong produce different types of spent fuels, which are stored and managed either in wet (light water reactors) or dry (heavy water reactors) storage facilities with reinforced steel concrete structures that are specially designed for each power plant.

Spent nuclear fuels and spent test fuels from the multi-purpose research reactor HANARO in the KAERI are deposited in the storage facility within the institute. As of September 2013, the total amount of spent nuclear fuels stored within KAERI premises was about 4.1 tons (NSSC 2014a).

According to a 2008 assessment of the Korea Hydro and Nuclear Power Company (KHNP) on-site storage facilities for spent fuel would reach full capacity sometime between 2016 (Kori reactors) and 2021 (Hanbit reactors) (Kang, Kim, and Lee 2015). This served as one of the government's most compelling arguments for constructing an interim storage facility before making a concrete decision on national spent fuel management policy and subsequent procedures for permanent disposal or reprocessing.

To expand storage capacity for spent nuclear fuel in NPPs, the KHNP provides each plant with high-density storage racks and also makes use of dry and dispersed storage methods within the plants. Expansion of storage capacity is expected to prevent the interim storage facilities from reaching their saturation point until sometime between 2024 and 2029. This expansion of nuclear waste storage facilities has also met with strong opposition from residents (Hankyoreh 2015).

Table 4: Spent fuel storage status in NPP sites (as of March 2014)

Power Plant	Storage	Capacity (ton)	Stored (ton)	Share of use	Saturation Point	
					present	expanded
Kori, Shin-Kori	Wet	2,691	2,121	77%	2016	2029
Hanbit (Yonggwang)	Wet	3,318	2,202	65%	2019	2024
Hanul (Ulchin)	Wet	2,960	1,848	62%	2021	2029*-39**
Wolsong, Shin-Wo lsong	wet, dry	9,660	7,252	74%	2018	2026
Total		18,629	13,423	71%		

Source: Jang 2014; Kim 2013(*); NARS 2013(**)

3 Conflicts around nuclear waste siting: From nuclear energy's start to the designation of the Gyeongju repository[4]

Nuclear waste storage policy in South Korea began in the 1980s. The South Korean government established the Radioactive Waste Management Project Committee in 1983. The committee's first nuclear waste management plan outlined basic principles for nuclear waste management. The committee called for inland disposal of LILW and the construction of a permanent disposal facility outside of nuclear power sites. They also proposed the establishment of a government-led non-profit organization for comprehensive nuclear waste management and promoted the generator-pays-principle to cover the costs of nuclear waste management. Measures for dealing with spent fuel were not included. Based on the principles outlined by the committee, the Atomic Energy Commission approved the Radioactive Waste Management Plan on October 13, 1984, and the new Atomic Energy Act was enacted on May 12, 1986. According to this act, KAER took charge of nuclear waste management, which was ultimately handed over to KHNP in 1996.

Since 1986, there have been numerous governmental attempts to select a repository site for nuclear waste. Initially, the traditional Decide-Announce-Defend (DAD) strategy was adopted by the government. The geological suitability of potential sites was tested in secret. After a site was selected as a candidate site, this selection was announced to the residents of the impacted region. There was only perfunctory discussion with the impacted residents and stakeholders. This top-down approach aroused strong opposition from local residents and led to the failure of one attempt after the next to secure a site. The government entity conducting the selection process and strategies for selection was changed several times, but there were still nine failed attempts before Gyeongju was finally selected as a disposal site in 2005 (Kim and Yun 2014: 318).

The first candidate sites designated as possible locations for the nuclear waste repository were three small villages on the east coast. The government had to abandon the geological investigation of these sites in 1989 after being confronted with strong regional resistance.

After these failed attempts, the government changed its D-A-D strategy and tried to win public cooperation in the selection of candidate sites. In January 1994, the government implemented an Act to Promote Nuclear Power Waste Management which sanctioned the provision of considerable amounts of financial support for regional development in areas hosting a disposal site. In December 1994, the

4 This section builds on Leem (2009).

government focused attention on the small island of Gulupdo as an appropriate candidate site for nuclear waste disposal. The proposal quickly met with widespread opposition. The antinuclear movement on Gulupdo's mother island, Duckchuckdo, gained support from activists in the metropolitan city of Incheon. As an additional issue, an active fault near Gulupdo was detected in a geological survey. As a result, on November 30, 2005, the government officially halted its construction plan.

After the failure in Gulupdo, more changes were made to the government's strategy. In June 1996, the Nuclear Environment Technology Institute (NETI), a branch of KHNP, assumed responsibility for the management of the disposal facility. The Ministry of Trade, Industry and Energy (MOTIE) took over responsibilities regarding its construction and management. The government delayed announcing a new candidate site until the beginning of the 2000s, concentrating instead on advertising the benefits and safety of nuclear power usage to reduce public resistance.

In 2000, the government announced a competition to host a repository; the winning community would receive generous compensation. After the deadline was extended twice, the mayor of Buan county submitted an application on July 15th, the very last day to apply. In this way, Wydo, a small island in Buan, became a candidate site for a nuclear waste storage facility. Local citizens were outraged because the application was submitted without being discussed with the community and without their consent. Over 10,000 people gathered daily in protest. An autonomous local referendum took place in February 2004, in which 72 per cent of the eligible voters participated and 92 per cent objected to the nuclear waste disposal facility coming to their region.

A report produced by the Korean Federation of Environmental Movements at the time describes the intensity of the situation and the desperation of the local residents:

> The governor's arbitary decision to host the site instantly raised the rage of Buan residents and they started to stage demonstrations from then on, the arduous struggle against the danger, and undemocratic conduct of the government imposed on them took off. About 6,000 police from all over the country were sent to repress unarmed protestors. 150 residents so far were injured during the protests in Buan. (Wonyoung 2013).

The project had to be abandoned. After the Wydo (alternative spelled, Wido) debacle, the government once again took a new approach to site selection. The Atomic Energy Commission decided in December 2004 to build a nuclear waste repository only for LILW and not for HLW. Along with this decision, the Special Act on Assistance to the Locations of Facilities for the Disposal of Radioactive

Waste was enacted in 2005. According to the act, 300 billion KRW (239 million EUR)[5] was guaranteed to the region designated to host the repository, as a fee to accept the facility. It was also determined that the head office of KHNP would move to the region where the facility would be built.

Table 5: Process of site selection for the nuclear waste disposal facility

Period	*Candidate Sites*	Description	Results
1986-1989	Youngilgun, Ulchingun, Youngduckgun	Top-down decision on the feasible sites through technical & geological investigation	Cancellation of the plan by local resistance (March 1989)
1990	Anmyondo	Connected with the development plan of the local government	Mistrust caused by pursuing the plan secretly
1991-1993	Taeangun, Youngilgun, Ulchingun, Sanghungun, Gosunggun, Yangyanggun	Selecting the candidate sites provisionally by considering technical and social aspects, and then negotiating with local governments Voluntary application and 6 candidates identified	Aborted due to residents' resistance
1993-1994	Yongsanmyun Janghungun, Gisungmyun Ulchingun	Voluntary application Assistance programs proposed to 3 applying regions	Aborted due to residents' resistance
1994-1995	Gulupdo	Among 10 candidate regions, Gulup Island was designated as the site for the disposal facility by the government	Discovery of an active faults
2000-2001	Yeonggwanggun, Gangjingun, Jindogun, Gochanggun, Boryeng, Wandogun, Uljingun	Contest of 46 seaside municipalities Assistance programs proposed to 7 regions.	No application
2003	Youngduckgun, Ulchilgun, Yeonggwanggun, Gochanggun	Announcement of the possible candidate sites and then voluntary application from the related local governments Same priority for the applications from other municipalities else than the 4 candidate sites	No application

5 Exchange rate of 15 June 2015.

Period	*Candidate Sites*	Description	Results
2003.7 - 2004.9	Wyudo Buangun	Application from municipality of *Bu-angun*	Aborted due to residents' resistance and failure in the local referendum
2005	Gyeongju, Gunsan, Yeongdeok, Pohang	Application from 4 municipalities and local referendum conducted on 2/11/05 Division of low-/intermediate- and high-level waste facility, enactment of supporting law	Selection of Gyeongju as the nuclear waste disposal site by the government.

Source: Leem (2009), supplemented

The government made an important strategic shift accepting that local referenda be held in the affected regions. In this round of site selection, local referenda were conducted in all four candidate locations. In the end, in 2005, the government designated Gyeongju, the region which had the highest rate of acceptance, as the new construction site for the nuclear waste disposal facility.

There are some anti-nuclear groups who questioned the outcome arguing that the KHNP intervened so forcefully that they suspected vote falsification; they refused to accept the voting result. There has also been ongoing controversy over the safety of the Gyeongju facility as the bedrock in that area is very weak and the site lies on active fault lines (Leem 2009; KFEM 2014). The site has, however, in the meantime been opened.

4 The government's conception of the spent fuel repository

Based on the decision of the 249th Atomic Energy Commission of 1988, the Korean government created a plan to build a central interim storage facility for spent fuel treatment by 2016. In this facility, spent nuclear fuel would be stored until the government makes a final decision about whether to reprocess or dispose of it permanently. After several failed attempts to select a nuclear waste repository site, and especially after the debacle in Wydo, in 1994 the government of President Roh Tae Woo introduced a new political approach to gain public consensus. A public forum for discussion was created. In the policy forum, experts established basic guidelines, formualted policy advice, and developed a framework for a spent fuel policy.[6]

6 Decision of the 253rd Atomic Energy Commission in December 2004

As a result of the forum's work, the Specialized Committee on Conflict Management was formed within the Atomic Energy Commission and began public engagement in February 2007. It organized the Spent Nuclear Fuel Public Discussion Task Force in April 2007 with members of the committee and external experts, including representatives from NGOs and local citizens. The task force conducted case studies on how other countries, such as the United Kingdom and Canada, dealt with public engagement in nuclear waste management questions; the task force met over twenty times.

The Public Discussion Task Force submitted a recommendation report to the government in April 2008 and a plan for conducting public discussions in September of the same year. In this report, eight principles for public discussion were suggested: participation, responsibility, ethics, sincerity, independence, deliberation, uniqueness and transparency. It was also recommended that interim storage be given priority in discussions until a decision about final measures was made (MOTIE and KHNP 2014; Lee, H.S. 2013).

In order to provide a legal basis for the public discussion, the Radioactive Waste Control Act was introduced in 2009; it has since been amended several times. From November 2011 to August 2012, the government operated the Spent Nuclear Fuel Policy Forum, which brought together experts from various fields to collect opinions on measures for spent fuel management. The anti-nuclear NGOs, however, criticized President Lee's conservative government as being insincere and refused to participate in the forum. Nevertheless, the forum submitted a recommendation calling for an early start to public discussions and set a target to complete the construction of the interim storage facility by 2024.

Based on this forum's recommendations, a decision was made at the Second Atomic Energy Promotion Council meeting in November 2012 to create the Public Engagement Commission on Spent Nuclear Fuel Management (PECOS), a temporary advisory body of MOTIE. Established in 2013, the commission was tasked with officially starting the public discussion on spent fuel management measures. It was originally composed of 15 members, including experts in the humanities and technology as well as local stakeholders. However, the anti-nuclear NGOs refused to participate because the overwhelming majority of commission members were advocates of nuclear power. Therefore the commission started its work with only 13 members.

In February 2014, PECOS submitted a plan including guidelines for the public discussion. Since then, discussions have been conducted many times based on this plan (Hong 2014; MOTIE and KHNP 2014). The submission of the report was originally scheduled for December 2014, but after the extension of the working period, the final report and recommendation of the Commission were submitted to the government in June 2015. They were then reviewed by the Atomic Energy Promotion Commission (MOTIE and KHNP 2014; PECOS

2015). In July 2016, the government indicated it would initiate a process to narrow differences in the views of stakeholders with the goal of selecting a site for high-level nuclear waste by 2028 (Yonhap 2016).

Additionally, the government plans to set up a Radioactive Waste Management Master Plan and a local referendum will likely be held to determine a final storage site for spent nuclear waste. This process has the potential to trigger great social conflict, if more is not done to engage the public in the discussion.

5 Legal and institutional framework

The lack of transparency in nuclear waste management has been a critical issue in Korea for many years and regulatory reform has been requested by various political actors, and particularly NGOs. Under these socio-political circumstances, in April 2006 the Board of Audit and Inspection demanded the government or an institution independent of the nuclear power companies take over the direct management of nuclear waste.

The government began drafting the Radioactive Waste Control Act with the intention of making comprehensive improvements in the nuclear waste management system. This legislation, which came into effect in January 2009, consists of seven chapters, forty-five articles and eight appendices. It contains provisions for the preparation of the radioactive waste management master plan (Article VI), radioactive waste management expenses (Article XIV), charges for spent nuclear fuel management (Article XV) and the establishment and operation of a Radioactive Waste Management Fund (Articles XXVIII-XXXI).

Originally, matters concerning safety of nuclear energy and waste were integrated into regulations for the nuclear use and promotion according to the Atomic Energy Law of 1959. This law served as a basic guideline for comprehensively regulating nuclear energy use, the nuclear industry and management until its amendment in 2011. As public concerns about the safety of nuclear power plants grew in South Korea after the Fukushima nuclear accident, the Atomic Energy Law was subdivided into the Atomic Energy Promotion Act and the Atomic Energy Safety Act on July 25, 2011, with the aim of managing safety issues more systematically. The Safety Act deals with matters concerning nuclear safety, while the Promotion Act offers prescriptions for nuclear energy research, development, production and use. Through these acts, an independent system of nuclear safety management was finally created a half century after the country first started using nuclear energy for power generation.

In compliance with Article III of the Atomic Energy Promotion Act, the Atomic Energy Promotion Commission (AEPC)[7] was established on October 26, 2011. This commission, under the authority of the prime minister, was tasked with deliberation and decision making on important matters regarding the utilization of nuclear energy. At the same time, the Nuclear Safety and Security Commission (NSSC) was established in accordance with the Act on Establishment and Operation of NSSC (Article XV)[8] and the Atomic Energy Safety Act.

In contrast to European nations such as Germany, in South Korea the Ministry of Environment has no legal authority on matters of nuclear energy and waste. Instead, the ministries responsible for nuclear matters are the Ministry of Trade, Industry and Energy (MOTIE) and the Ministry of Science, ICT and Future Planning (MSIP). MOTIE oversees business and regulation of nuclear power and nuclear waste as outlined in the Electric Utility Act, while MSIP is largely responsible for nuclear energy development as prescribed by the Atomic Energy Promotion Act. Within MOTIE, the Office of Energy and Resources is responsible for general matters concerning nuclear and energy policy. It takes care of integration and coordination of nuclear power-related businesses and deals with affairs regarding nuclear waste management (MOTIE and KHNP 2014). It is responsible for:

- establishing and enforcing a basic plan for nuclear waste management and R&D of related technology,
- providing support for site selection, construction, operation and advertisement of nuclear waste management facilities, and,
- managing and operating the nuclear waste management fund.

MSIP sets goals and develops strategies for nuclear promotion policy in accordance with the Comprehensive Nuclear Energy Promotion Plan (CNEPP). In support of these goals, MSIP also comprehensively manages nuclear R&D policies and administrative management, including research on and the development of nuclear power-related facilities, nuclear materials and radioactive technologies.

AEPC is a consultative body with the Prime Minister as its head. This commission deliberates and decides on important issues concerning the utilization of nuclear energy. Preliminary decisions regarding CNEPP are also made by

7 Although the official name of the commission in Korean was changed by law, the commission continues to use the old English name, the Atomic Energy Commission, from 1959. This chapter uses the new official name of the commission as translated into English.

8 Enacted on 25 July 2011.

AEPC and then are ultimately decided on by MSIP. AEPC is also responsible for the integration and coordination of matters regarding the utilization of nuclear energy and nuclear waste, in order to control the nuclear waste in a safer and more effective way. It deliberates and decides on the general master plan for nuclear waste management, issued by MOTIE, in compliance with Article VI of the Radioactive Waste Control Act. Regarding spent fuel disposal, issues are initially addressed by AEPC, and then Ministers of MOTIE and MSIP make the final decision after consultation with the relevant authorities. In other words, AEPC, which was established with the purpose of nuclear energy promotion, has authority over the decision-making process for spent fuel management.

NSSC manages overall tasks related to nuclear power safety and security as well as nuclear non-proliferation. It is also in charge of safety regulations such as construction permits, operating licenses and the inspection of reactors including their related facilities, radioactive materials and nuclear waste disposal facilities. To get permission for operation, construction or prolonged use of nuclear waste disposal facilities and their attached structures, the operators must submit a radiological impact assessment to NSSC (Articles 63, Atomic Energy Safety Act). This commission also creates general plans for nuclear power safety. The Korea Institute of Nuclear Safety (KINS), affiliated with NSSC, carries out most of the decisions regarding safety regulations of the commission as prescribed in the KINS law (NSSC 2014a, b; NSSC 2012; MOTIE and KHNP 2014).

The establishment of NSSC was initially regarded as a step forward in the development of nuclear governance. This was because the commission, as an independent agency affiliated with the President, was expected to further strengthen the national nuclear power safety management system. However, quite contrary to this expectation, President Park Geun-hye's new government handed over command of NSSC to the Prime Minister in 2013. Consequently, the political stature of the commission was diminished while the importance of its role in rearranging safety policy on nuclear power and waste management grew.

The Korea Radioactive Waste Agency (KORAD) was established on January 1, 2009 as outlined in articles 18-27 of the Radioactive Waste Control Act. As a semi-governmental agency affiliated with MOTIE, KORAD is in charge of nuclear waste management and disposal facilities, including spent fuel management.[9] The core tasks of KORAD include:

- transportation, storage, treatment, and disposal of nuclear waste including spent fuel;

9 KHNP and MOTIE had had the responsibility for the nuclear waste management from June 1996.

- site selection, construction, operation, and post-closure management of nuclear waste disposal facilities;
- collection, survey, analysis and management of documents regarding nuclear waste management;
- public relations for nuclear waste management;
- R&D and international cooperation on nuclear waste management, and;
- operation of the nuclear waste management fund (MOTIE and KHNP 2014; Lee, K. H. 2013).

To cover the future cost of nuclear waste management and dismantling, the Electric Utility Act specifies the establishment of the Fund for Nuclear Follow-up Management. This fund covers low- and intermediate-level nuclear waste management costs, spent fuel management costs, and reserve funds for dismantling nuclear power plants. The former two costs are administered by KORAD, while KHNP is in charge of the latter.

According to the fund's annual income statement for the period of 2009-2013, total nuclear waste management costs reached 179,776 million KRW (142.9 million Euro) in 2009 and 1,136,140 million KRW (903.8 million Euro) in 2013.[10] It is encouraging that the fund is growing every year; however, fund management has been criticized because a large portion of the money has been used for asset management, while actual expenditures on nuclear waste management have been conspicuously low (CFOI 2013; MIKE 2012).

Table 6 shows important legislation and institutions in charge of nuclear energy management and security in South Korea.

Table 6: Important legislation and institutions in charge of nuclear energy

Atomic Energy Promotion Act	Covers research, development, production and use of nuclear energy, nuclear energy promotion commission, plans for promotion of nuclear energy, and nuclear energy R&D funds.	Ministry of Science, ICT and Future Planning (MSIP)
Radioactive Rays and Radioisotope Use Promotion Act	Promotes the use and R&D of radiation and radioisotopes; supports related industries.	
Nuclear Safety Act	Addresses safety management in the R&D, production and use of nuclear energy in order to prevent disasters caused by radiation.	

10 The exchange rate of 18 June 2015 was used.

Act on Establishment of the Nuclear Safety and Security Commission	Contributes to promoting public safety and environmental conservation by establishing the Nuclear Safety and Security Commission	Nuclear Safety and Security Commission
Act on Measures for the Protection of Nuclear Facilities	Establishes a system for physical prevention and protection against radioactivity and nuclear disasters.	
Nuclear Damage Compensation Act	Prescribes matters concerning compensation for nuclear damage resulting from nuclear reactor operations.	
Electric Utility Act	Establishes a basic system promoting competitiveness among electric utilities.	Ministry of Trade, Industry and Energy (MOTIE)
Electric Source Development Promotion Act	Promotes the electric source development business.	
Radioactive Waste Control Act	Improves the safe and efficient management of radioactive waste.	

Source: MSIP 2014: 319-321.

6 Siting procedures, social participation and access to information

Institutionally, commissions can be a means for non-governmental experts and anti-nuclear NGOs to take part in decision-making processes. With regard to social participation in nuclear waste issues in Korea, three Commissions – the Atomic Energy Promotion Commission (AEPC), the Nuclear Safety and Security Commission (NSSC), and the Public Engagement Commission on Spent Nuclear Fuel Management (PECOS) – play a particularly important role.

Through the amendment of the Atomic Energy Law, AEPC assumed responsibility for the functions of the former Atomic Energy Commission, except in matters related to nuclear safety. AEPC consists of a chair (the prime minister), four ex officio members (ministers) and six civil members; thus civic participation is guaranteed by law. However, as the remit of the commission has been to promote nuclear energy, there was little opportunity to discuss or question nuclear expansion policies. Commission members have typically been advocates of nuclear energy use and have tended to approach nuclear waste problems only from technological viewpoints. They have not raised questions about the fundamental direction of nuclear energy or nuclear waste management policies. The socio-economic consequences of nuclear waste management has not been given much attention. Even if anti-nuclear members were to participate, it would likely make little difference as decisions are made by majority vote (Yun 2015). If President Moon is to succeed in his plans to shift nuclear policy in new directions, than deep changes to institutions will be necessary, and resistance to change from nuclear industry supporters is likely to be fierce.

Likewise, the NSSC has nine members including the chairman. In 2015, only two members, a scientist and an NGO activist, were nominated from the opposition parties. Aside from these two members, the commission members were pro-nuclear. Although the NSSC has been more diverse than the other nuclear commissions, the majority of members have tended to support nuclear power. Thus, there has been little possibility for alternative opinions to be reflected in decision-making processes. This led opposition members to criticize the decisions made by the NSSC as being closed-door and not reflective of the views of the public.

For similar reasons, anti-nuclear NGOs declined to participate in PECOS, which they argued is biased. Due to the limited degree of NGO participation in the PECOS and low levels of public interest in their work, there have been few opportunities for the commission to gather various public opinions. Considering that the purpose of this commission is to foster public discussion, one might expect that all of the procedures and results of the meetings would be disclosed to the public, and that diverse groups of people would be involved in the discussion. In reality, this has not been the case; records of the meetings have not been made available to the public and public discussions have tended to be one-sided. Citizen participation in the public discussions has been very low. Furthermore, public discussion about spent nuclear fuel has been limited to the issue of site selection for the interim storage facility. This may now change as the government of President Moon has vowed to radically change the countries energy policy in the direction of energy saving and renewable energy and away from nuclear energy and fossil fuels, reflecting changing public perceptions of nuclear energy after the Fukushima nuclear accident and growing awareness of the threats of climate change.

If this shift in policy is to be realised, changes will be necessary in the membership and operating practices of the commissions dealing with nuclear policymaking and nuclear waste management. Transparency, starting with access to information for citizens and genuine possibilities for citizens to play a role in the decision-making process will be necessary.

In addition to the commissions, there is also a public hearing system: a legal instrument which enables citizens, stakeholders and NGO activists to gain access to information and make suggestions regarding nuclear policy-making. In the past, public hearings have often been treated as a mere formality and have been used by nuclear plant operators or government agencies as an official chance to explain their plans to the public before enacting legislation and not as a venue to response to critical or opposing positions. In some cases, public hearings have been altogether side-stepped (Yun 2015; Choi 2013). The government of Moon

Jae-In has pledged to change how energy decision making is done and made a first effort to make good on his pledge by putting the question of whether to continue with the construction of the Shin Kori 5 and 6 nuclear plants which have already been partly constructed.

7 Lessons learned

Despite democratic advances in Korea, the public has been largely excluded from the political arena surrounding nuclear power and nuclear waste management issues. The participation of citizens, NGOs and stakeholders in the policymaking process and their access to information on nuclear matters has been restricted and the policymaking process has been largely closed to anti-nuclear viewpoints, making the work of many governmental institutions controversial.

The lopsided composition of PECOS and restrictions on information release led to the nonparticipation of anti-nuclear NGOs in the commission. Consequently, public discussion on spent fuel disposal was mostly conducted by the commissioners representing or advocating the governmental plan. The Korean Federation for Environmental Movements (2015) warned that pushing into making decisions on spent fuel policy in this way, without first trying to develop a public consensus on the road forward, could lead to serious social conflicts and significant controversy over the management of spent fuel (KFEM 2015).

With temporary storage facilities for spent nuclear fuel rapidly filling up, there is an urgent need in South Korea to reach a social consensus on nuclear waste management and storage policy. To prevent further delays and opposition, nuclear governance must be based on public participation. To achieve this, the government must have an open and in-depth dialogue with various stakeholders, including civil society and local residents. Only when the policy arena is openned to the public can a transparent nuclear governance be achieved and publically acceptable decisions on nuclear waste management be made. For Moon Jae-In's government, the challenge will be to speak true to campaign pledges.

Despite the limited penetration of renewable energy to date, interest in alternative energy has been growing especially at the local level. Various local governments have linked anti-nuclear positions to renewable energy policy promotion. Particularly significant has been the 'One Less Nuclear Power Plant' campaign, conducted by the Seoul Metropolitan Government. Also noteworthy are the anti-nuclear decisions made by the local parliament and citizens of Samcheok city.

Society's perception of nuclear energy has become more negative since the Fukushima nuclear accident and public demands for alternative energy have increased. This prompted the leading opposition party, the Democratic Party, to make the phase-out policy one of its campaign promises in the 2012 presidential election. The Democraticy Party lost the 2012 election to the pro-nuclear party of Park Geun-hye. Park's controversial governing style and charges of corruption, however, led to her subsequent impeachment. In the ensuing elections, the former student activist, Moon Jae-In, was elected as president. He campaigned to end South Korea's reliance on nuclear energy and since entering office in the spring of 2017 has taken the first steps to make good on his promise by stopping plans for several new nuclear power plants.

Developing a social agreement on the path forward will require a wide-ranging and open discussion with high levels of public participation. Only in this way can a new approach to nuclear policy and waste management governance be achieved. This will be critical if a solution to the challenge of high level radioactive waste management is to be found.

References

CFOI [Center for Freedom of Information and Transportation Society] (2013). *Hoarding Radioactive Waste Management Fund, for What?*

Choi, J. H. (2013). *Legal Issues of Public Consensus on Spent Nuclear Fuel*. Public Discussion for Public Consensus on Spent Nuclear.

EJA [Energy Justice Actions] (2014). *Four Meanings of Approval of the Enforcement Plan for Shin-Kori 5, 6 Reactors*, 19 January 2014, Press Release.

Hankyoreh (2015). *13 Thousand Tons of Spent Fuel, on the Verge of Saturation*. 9 March 2015. http://www.hani.co.kr/arti/society/environment/681469.html, last accessed 03 March 2016.

Hong, D. S. (2014). *Status and Prospect of Public Consensus regarding Spent Nuclear Fuel.*

IAEA (2014). *Nuclear Power Reactors in the World*. 2014 edition.

Jang, U. S. (2014). The Status of Spent Fuel Storage and Technology. In: *KEIT, PD ISSUE REPORT*, 14(9), 39-63.

Kang, J., Kim, S.W., and Lee, B-C. (2015). "Hot Potato in South Korea: The Spent Nuclear Fuel Storage Dilemma." In: *Bulletin of the Atomic Scientists*, 71(3). http://thebulletin.org/2015/may/hot-potato-south-korea-spent--nuclear-fuel-storage-dilemma828.

KFEM (2014). *Lots of Active Defaults are detected around the Gyeonju Nuclear Waste Storage Facility*, Comment/Press Release.

KFEM (2015). Statement. *The PECOS's Recommendation is not a result of the public discussion but just a follow-up plan to expand nuclear power*. 6 November 2015.

KHNP (2015). *Nuclear Power*. http://khnp.kr/, last accessed 24 June 2015.

Kim, K. S. (2013). *Technical Development Status of Nuclear Waste Management. Proceedings*. Autumn Conference of the Korean Society for Noise and Vibration Engineering.

Kim, K. S. and Yun, S. (2014). An Analysis of Risk/Benefit Perception and Acceptability of a Nuclear Radioactive Waste Disposal Facility Placement in Buan and Gyeongju. In: *Korean Policy Studies Review*, 23(1), 313-342.

Lee, H. S. (2013). *Issues and Tasks concerning Public Consensus of Spent Fuel Management in the Point of View of NGOs and the Local Community*.

Lee, K. H. (2013). Nuclear Waste Management Policy for Wolsung Nuclear Power Plant. In: *Study of Gyeongju*, 22(2), 21-43.

Leem, S.-J. (2009). Unchanging Vision of Nuclear Energy: Nuclear Power Policy of the South Korean Government and Citizens' Challenge. In: Mez, L.; Schneider and Thomas (Eds.). *International Perspectives on Energy Policy and the Role of Nuclear Power*, UK: Multi-Science Publishing, 477-495.

Leem, S.-J. (2015). Nuclear Waste Disposal Policy and Governance in South Korea. In: *The Korean Association for Governance*, 22(3), 167-187.

MIKE [Ministry of Knowledge and Economy] (2012). *Readjust Fund for Nuclear Follow-up Management*, Press Release.

MOTIE [Ministry of Trade, Industry and Energy] (2014). *The Second Basic Energy Plan*.

MOTIE [Ministry of Trade, Industry and Energy] (2015). *The 7th Basic Plan for Long-term Electricity Supply and Demand* (2015~2029). July 2015.

MOTIE and KHNP (2014). *White Paper on Nuclear Power Generation 2014*.

MSIP [Ministry of Science, ICT and Future Planning] (2014). *White Paper on Nuclear Power 2014*.

NARS [National Assembly Research Service] (2013). *Estimation of Spent Fuel Amount by Scenarios and Tasks of Legislation and Policy*.

NSSC [Nuclear Safety and Security Commission] (2012). *Guideline on Radiation Protection*.

NSSC [Nuclear Safety and Security Commission] (2013). *Revision of Lower Regulations of Atomic Energy Safety Act regarding Nuclear Waste Classification*. 13 December 2013.

NSSC [Nuclear Safety and Security Commission] (2014a). *Nuclear Safety and Security Commission 2013 Annual Report*.

NSSC (Nuclear Safety and Security Commission) (2014b). *2013 Budget Summary*.

Patel, S. (2015). South Korea Begins Burying Nuclear Waste. In: *Power Magazine*, September 1. http://www.powermag.com/south-korea-begins-burying-nuclear-waste/, last accessed 3 November 2017.

PECOS (2015). *News Release of Pecos*. 3 June 2015.

Wonyoung, Y. (2013). "Massive Actions Against Proposed South Korean Waste Dump," Korean Federation for Environmental Movements. In: *Nuclear Monitor Issue*, 591, August 22.

World Nuclear Association (2017). *Nuclear Power in South Korea*. http://www.world-nuclear.org/information-library/country-profiles/countries-o-s/south-korea.aspx, last accessed 3 November 2017.

Yonhap News Agency (2013). *Reducing the Share of Nuclear Power. What will become of Nuclear Power Plants in Preparation?* 13 October 2013.

Yang Lee, W.-Y. (2015). *The Status and Perspective of NPP in Northeast Asia.*

Yonhap. (2016). S. Korea to Select Storage Site for High-level Nuclear Wastes by 2028. In: *Yonhap News*, 25 July,

Yun, S. (2015). Challenges and Directives of Nuclear Governance in South Korea. In: *Environmental Law and Policy*, 14, 1-48.

False Premise, False Promise

Governance and Management of Nuclear Waste in Japan

Lila Okamura[1]

Abstract

It is almost 50 years since Japan's first nuclear power plant went into operation. Prior to the Fukushima accident, the country was the world's third largest producer of nuclear energy. This huge industry and research sector was built on the premise that Japan would recycle its spent nuclear fuels. It is upon this unstable foundation that Japan attempted to establish its nuclear disposal construct. Even after the government made the decision to abandon the Monju fast breeder reactor which has been plagued by problems for years, Japan is not giving up its pursuit of a demonstration fast reactor. The nuclear fuel cycle has remained a major ambition of the government, so reprocessing cannot be excluded in any consideration of final disposal in Japan.

Nuclear waste management is always discussed separately from nuclear power policy in Japan. The power companies use nuclear power without serious consideration for nuclear waste, and even after the Fukushima nuclear accident, they tried to restart their reactors as quickly as possible.

In addition to the lack of transparency and alternatives and the confusion surrounding the management of nuclear waste in Japan, the debate has never really made it to the public arena. As a result, the site selection process in Japan has made little progress.

To change this situation, the site selection method was changed, from the system of voluntary application by local governments to a government-led selection system. The government is adopting a stronger role, and in August 2017, they issued a report which indicates that up to one-third of the Japanese territory could be considered suitable for building a disposal site. The government hopes to narrow down the candidate sites over the next 20 years.

1 Dokkyo University, 1-1 Gakuen-machi, Soka City, Saitama, 340-0042 Japan; E-Mail: lilaokamura@dokkyo.ac.jp

A. Brunnengräber et al. (eds.), *Challenges of Nuclear Waste Governance*, Energiepolitik und Klimaschutz. Energy Policy and Climate Protection, https://doi.org/10.1007/978-3-658-21441-8_5

1 Introduction

On 11 March 2011, eastern Japan was struck by a major earthquake and massive tsunami which flooded the Fukushima Daiichi nuclear power plant (NPP) knocking out its emergency cooling system. Without proper cooling, heat began to build up as the reactor went critical. The build up of heat in the reactors caused a hydrogen explosion which sent a radioactive plume into the surrounding land. Years later, the area around the plant remains off limits as efforts are made to deal with the aftermath of the world's second worst nuclear accident.

Prior to the Fukushima accident, the Japanese government had promoted an expansion of nuclear power generation. The nation's 54 nuclear reactors generated 288 TWh (288 billion kilowatt hours) in 2010, accounting for about 26 percent of Japan's total electricity consumption (WNA 2015). The government planned to build even more capacity with the goal of obtaining about 40 percent of the nation's electricity from nuclear power.

In response to the accident and both domestic and international concern, the Japanese government agreed to establish new more stringent safety standards and to test all of the country's reactors against these standards. By May 2012, all of Japan's nuclear reactors had been taken off line to await inspection under the new more stringent regime. In the meantime, TEPCO decided to retire the six reactors in the Fukushima Dai-ichi and Dai-ni nuclear power plants. In addition, several other of Japan's oldest nuclear power plants were shut down leaving 43 operational nuclear reactors as of 2017.

In September 2012 a new, independent Nuclear Regulation Authority (NRA) was established. In July the following year, it began the lengthy process of conducting safety reviews of each of the country's nuclear power stations under the new regulatory requirements. Applications for over 20 reactor restarts have been made and, as of June 2017, 12 nuclear reactors had passed the safety assessments. Five of these reactors are now operating.

The government is pushing forward with the restart of the reactors even though there is still no final disposal site for high-level radioactive waste. Currently, spent fuel (SF) is stored in cooling pools which have been built at each of the country's nuclear power stations. This fuel will continue to increase in volume as the nuclear power stations are restarted.

Discussion of nuclear energy issues in Japan has focused almost solely on the energy supply aspect. However, given the aftermath of the Fukushima accident and the issue of restarts, the time is ripe for a serious societal discussion about nuclear waste.

2 Overview of nuclear power policy and nuclear fuel cycle policy

Japan has a long history with nuclear energy and the largest nuclear energy capacity after the United States and France. Japan spent many years pursuing nuclear fuel reprocessing in the hopes of establishing a closed nuclear cycle. A brief history of Japan's nuclear programs is presented below.

2.1 The historical background of Japan's nuclear power policy

2.1.1 Nuclear energy policy

In November 1955, the Japan–US Atomic Energy Agreement was concluded. Under the agreement, the United States provided Japan with convential nuclear technology. Japan's research into the "peaceful use of nuclear energy" commenced in earnest the following month.

A decade later, in 1966, Japan's first commercial reactor, the Tokai NPP began operation. Starting in the 1970s, a succession of NPPs went into operation. The oil crisis was a significant factor behind the swift construction of nuclear power stations in Japan at this time; grants and subsidies also played a major role. Under the Three Power Source Development Laws, the national government provided tax incentives and other inducements to communities which agreed to host nuclear power plants. Many of the areas that emerged as candidate sites for nuclear power stations were experiencing depopulation, had low birth rates or an aging population as young people flocked to the cities during the era of high economic growth. Local governments in the candidate areas sought to increase employment and tax revenue by attracting NPPs (Inoue 2015:15).

In the 1980s, following the accidents at Three Mile Island and Chernobyl, there was growing anxiety in many countries about the risks of nuclear power generation. In Japan, however, the government and the nuclear industry succeeded in maintaining an image of nuclear power as an environmentally friendly form of energy. This stayed this way even after a criticality accident in the Tokai-mura uranium enrichment facility in Ibaraki in 1999 killed two workers and left another seriously injured.

Nuclear energy was promoted as a means to fuel the economy without emitting greenhouse gas emissions. Nuclear power generation was expected to play a major role in cutting domestic CO_2 emissions and enabling Japan to meet its reduction commitments under the Kyoto Protocol. However, a series of accidents at Japanese nuclear installations led to growing anti-nuclear sentiment among citizens in candidate areas. As a result, the construction of nuclear power stations did not proceed as smoothly as it had in the 1970s, or at the pace the government had projected (Sagara 2009: 50).

When the Fukushima accident occurred, the Democratic Party of Japan (DPJ) was in control of the government. The DPJ reacted to the accident by drawing up three possible energy policy scenarios for the future, including one which would see nuclear power abandoned by the 2030s. Despite the fact that it was the LDP which had promoted nuclear power, the DPJ was held responsible for many of the problems experienced in the immediate aftermath of the accident. The election of 2012 saw the Liberal Democratic Party (LDP) return as ruling party under the leadership of Abe Shinto. The Abe administration took the position that the Fukushia triple disaster (major earthquake, tsunami, and nuclear accident) was a rare occurrence and that nuclear power was itself safe and would be made even more so under the new safety standards. Furthermore, the Abe government has painted a picture of nuclear energy as being essential to the stability of the nation's electricity supply and economy. The Basic Energy Plan published in April 2014 positions nuclear power as an "important source of base load power" (ANRE 2014a: 8) and sets out a policy to restart nuclear power stations which are deemed safe. While advocating that reliance on nuclear power should be "reduced to the minimum amount necessary," the plan avoids committing to a time frame for using nuclear power.

In April 2015, the Ministry of Economy, Trade and Industry (METI) published new estimates of the cost of power generation by source of supply. The cost for nuclear power per kilowatt-hour was estimated to rise to at least 10.1 yen (ca. 7.64 cent) by 2030, making it the cheapest option among the available sources of supply when the lower end of the estimate range for each is compared. The estimates which were made for solar photovoltaics, wind power, and other forms of renewable energy were based on the purchase price under the feed-in tariff system introduced in 2012 and government research and development expenditures. With these assumptions, the report assumed that renewable energy will be substantially more expensive than nuclear power. Even in the cheapest case, domestic solar cost estimates rose from 9.9 yen (ca. 7.48 cent) under the estimate made in December 2011 to 12.5 yen (ca. 9.45 cent) in the 2015 estimate, and onshore wind power rose from 8.8 yen (ca. 6.66 cent) to 13.9 yen (ca. 10.51 cent) over the same time frame. The METI plan calls for a share of 20–22 percent nuclear power in the electricity mix in 2030.

2.1.2 The nuclear fuel cycle

The Long-Term Program for Nuclear Energy formulated in 1956 stipulated the importance of establishing a nuclear fuel cycle and commercializing fast breeder reactors. The nuclear fuel cycle basically became a "state policy" in Japan (Funabashi et al. 2012: 318).

The 1970s saw an expansion in nuclear power generation not only within Japan, but also worldwide. This gave rise to fears of uranium shortages, so proactive efforts were undertaken in Japan to develop a fast breeder reactor. The 1967 Long-Term Program for Nuclear Energy set a target for the commercialization of fast breeder reactors by the latter half of the 1980s; the Joyo experimental reactor reached its first criticality in 1977.

However, from the latter half of the 1970s into the 1980s, there was a slump in nuclear power station construction worldwide. The price of uranium resources subsequently plummeted, and as a result, developing fast breeder reactors was no longer a matter of urgency.

Nevertheless, Japan continued to promote the nuclear fuel cycle, along with nuclear power generation. The electric companies in Japan concluded a reprocessing contract in September 1977 with British Nuclear Fuels Limited (BNFL), which is now Sellafield Ltd, and in May 1978 with Cogema (now Areva). It was to be a provisional measure until a reprocessing plant in Japan could begin operations.

In 1984, the Federation of Electric Power Companies (FEPC) requested the cooperation of Aomori Prefecture with regard to the siting of three nuclear fuel cycle facilities (a uranium enrichment facility, a reprocessing facility and a low-level radioactive waste storage facility); the Governor of Aomori Prefecture granted formal approval the following year.

In 1993, in Rokkasho, Aomori Prefecture, Japan Nuclear Fuel Ltd., which had been established with contributions from the power companies, began construction of a reprocessing plant with a maximum annual processing capacity of 800 tons of uranium. Initially, it was expected to be completed in December 1997, with construction costs estimated at 760 billion yen (ca. 5.74 billion Euro), but due to various problems, the completion was postponed close to two dozen times. Japan Nuclear Fuel Limited (JNFL) estimated in October 2014 that the total construction costs had soared to 2.19 trillion yen (ca. 16.5 billion Euro), more than triple the initial estimate made in 1979.

The Nuclear Reprocessing Organization of Japan (NuRo) was established as an authorized company by METI on 3 October 2016. The organization aims to advance steady and efficient reprocessing of spent nuclear fuel, among other efforts. NuRo will collect funds and contract out reprocessing and MOX fuel fabrication to the JNFL. It shows the strong will of the government to not only to keep but also to further promote the nuclear fuel cycle.

The commercialization of fast breeder reactors fell behind schedule pushing the target period for commercialization into the 2030s. The problem was exacerbated when the Monju prototype reactor, which had reached its first critical state in 1994, suffered a sodium leak and fire in 1995, causing a further delay to commercialization plans. As a result of the Monju accident, the idea of pursuing a fast breeder reactor was abandoned. Instead, efforts to close the nuclear fuel cycle would focus - for the time being - on "pluthermal", which uses MOX fuel[2] in light water reactors.

In 2016, the government made a decision to close and decommission Monju. This does not mean that Japan has completely given up on pursuing the development of a demonstration fast breeder reactor. The government argues that by using the Joyo experimental fast-breeder reactor and collaborating on a joint project with France to develop an advanced experimental fast-breeder reactor for industrial demonstration (ASTRID) the necessary level of technological knowledge and data for development of a demonstration reactor can be obtained.

2.2 The current status of radioactive waste in Japan

In Japan, high-level radioactive liquid waste (generated by the reprocessing of spent fuel) that has been vitrified and sealed in containers is classed as high-level radioactive waste (HLW). As of 2017, Japan had 2,448 packages of waste that had undergone reprocessing and vitrification: 2,176 of these are stored at the JNFL reprocessing facility at Rokkasho-mura in Aomori Prefecture, while the remaining 272 are stored at the Japan Atomic Energy Agency (JAEA) reprocessing facility at Tokai-mura (NUMO 2017). This includes the vitrified waste packages which were reprocessed overseas and sent back to Japan. The amount of fuel reprocessed by Areva and BNGS was approximately 5,600 tons, which corresponds to about 2,200 vitrified packages. By the end of 2008 1,310 vitrified packages had been sent back from France (ATOMICA 2009). If all of the SF resulting from nuclear power generation were to be reprocessed into vitrified packages, this figure would rise to approximately 25,000 packages (NUMO 2017).

High level radioactive waste (HLW) is subject to geological disposal, as is transuranic (TRU) waste, in accordance with the 2007 revision of the Final Disposal Act. TRU is low-level radioactive waste with long half-lives containing more than a specified concentration of long-lived radionuclides, generated by the operation and dismantling of reprocessing plants and MOX fuel fabrication plants. It includes hulls and ends, emission filters, concentrated liquid and

2 The plutonium and uranium in spent fuel can be recovered through reprocessing. This plutonium and uranium is blended to produce mixed oxide (MOX) nuclear power reactor fuel. More than 30 reactors in Europe (Belgium, Switzerland, Germany and France) currently use MOX. Ten Japanese reactors are licensed to use MOX and several are currently doing so.

miscellaneous solid waste. Each type has a prescribed container, and is placed either in canisters, drums or rectangular containers, and in some cases will have undergone a solidification process (NUMO 2008). The Nuclear Waste Management Organization of Japan (NUMO) anticipates a TRU waste volume of approximately 18,100 m^3 (NUMO 2008: 5).

2.3 *The waste management strategy (reprocessing or direct disposal?)*

The nuclear fuel cycle has remained the major ambition of the government and the nuclear industry through to the present day. Reprocessing cannot be excluded in any consideration of final disposal in Japan; as such, this section provides a broad overview of reprocessing compared with direct disposal (once-through).

2.3.1 The argument made for reprocessing

Without any prospect of a fast breeder reactor in sight, there is no longer any real hope for Japan of being able to effectively utilize the industry's accumulated uranium resources, which was the primary purpose to be achieved with a closed nuclear fuel cycle. This section will concentrate on those factors which are specific to Japan.

Spent fuel is stored at each nuclear power station, but many of the storage pools are almost full (Table 1). Currently, more than 14,000 tons of SF are stored at nuclear power stations in Japan. The total capacity of all nuclear power station pools is approximately 20,000 tons, which means that almost 70 percent of storage capacity is deployed. The remaining available capacity differs between stations, but the pools in three NPPs will be full within approximately three years, as will the pool at Kyushu Electric Power Company's Genkai NPP in Saga Prefecture. Rokkasho reprocessing plant, which has a maximum capacity of 3,000 tons, had 2,964 tons as of 2016 (JNFL 2016: 6).

Table 1: Quantity of spent fuel stored at each NPP in tons (September 2017)

Electric Power Company/ NPP		Quantity of Waste Stored	Available Capacity	Remaining Operation Time (yrs) 2014 estimate
Hokkaido	Tomari	400	1,020	16.5
Tohoku	Onagawa	420	790	8.2
	Higashidohri	100	440	15.1
Tokyo	Fukushima Daiichi	2,130	2,260	n/a
	Fukushima Daini	1,120	1,360	n/a
	Kashiwazaki Kariwa	2,370	2,910	3.1
Chubu	Hamaoka	1,130	1,300	8.0
Hokuriku	Shiga	150	690	14.4
Kansai	Mihama	470	760	7.5
	Takahama	1,220	1,730	7.6
	Ohi	1,420	2,020	7.3
Chugoku	Shimane	460	680	7.0
Shikoku	Ikata	640	1,020	8.8
Kyushu	Genkai	900	1,130	3.0
	Sendai	930	1,290	10.7
JAPC	Tsuruga	630	910	9.3
	Tokai Daini	370	440	3.1
Total (rounded)		14,870	20,740	

Source: Author's compilation based on data found in ANRE (2014b: 4) and FEPC (2017)

In 1994, the municipal government of Rokkasho-mura concluded an agreement with JNFL, stating that the vitrified packages would be stored there temporarily for a period of 30-50 years, whereafter the electric power companies involved would remove the vitrified packages. The director-general of the Science and Technology Agency and the governor of Aomori Prefecture made a commitment that the vitrified packages brought to Rokkasho-mura would be removed at the end of the storage period and that Aomori Prefecture would not become a final disposal site (ANRE 2013: 4).

The proponents of interim storage claim that this agreement means that if reprocessing were to be abandoned, there would be a risk that the local government would refuse to allow spent fuel from nuclear power stations to be brought to the Rokkasho Reprocessing Plant. Moreover, the nuclear power stations might have to remove the SF that they have already taken there (Yamaguchi 2005: 7). If interim facilities for SF storage cannot be secured, the SF storage capacity at each nuclear power station will be exceeded and it will become impossible to continue nuclear power generation.

In terms of measures to deal with SF, the national government and the electric power industry revised the Reactor Regulation Act in 2000, allowing the construction of SF "interim storage facilities" outside power stations. In 2005, TEPCO and Japan Atomic Power Company (JAPC) established the Recyclable-Fuel Storage Company (RFS) and construction of an interim storage facility for SF began in the city of Mutsu, Aomori Prefecture in 2010. This facility was due to begin operating in March 2015, but its launch has been delayed until the latter half of 2018 for safety reasons (RFS 2016).

SF emitted by the nuclear power stations operated by TEPCO and JAPC will be kept in dry storage at this interim storage facility for up to 50 years. The premise of such facilities is that even after storage, the SF will be transferred to a reprocessing plant to undergo reprocessing (TEPCO 2012: 2).

2.3.2 Economic evaluation

Increasing costs of reprocessing

The Rokkasho Reprocessing Plant is not progressing as planned. According to figures released in 2003 by FEPC the total cost of building, operating for 40 years, and decommissioning the Rokkasho Reprocessing Plant is estimated at approximately 11 trillion yen (ca. 83.2 billion Euro). This figure includes the cost of reprocessing 14,000 tons of stock waste that had already accumulated by 2004, plus some 18,000 tons of flow waste that would be generated thereafter, making a total of 32,000 tons to be reprocessed over the course of 40 years. The Rokkasho Reprocessing Plant has the capacity to reprocess 800 tons annually and the estimates are based on the assumption the facility will work at full capacity for over 40 years. If operating levels were to decrease, the costs would surge even higher (ANRE 2003: 25).

There are also costs involved in storing the vitrified packages. The government expected the storage costs to be approximately 30 million yen (227,000 Euro) per package; the vitrified packages returned from overseas were estimated to cost 120 million yen (ca. 908,000 Euro) per package to store (Oshima 2010: 20). Moreover, while the packages are in storage, the nuclear fuel tax will have to be paid; this was expected to amount to approximately 730,000 yen (5,522 Euro) per package per year in the case of Aomori Prefecture, and approximately 940,000 yen (7,111 Euro) per package per year in the case of Ibaraki Prefecture. The tax cost of storing the packages for 50 years was estimated at 1 trillion yen (ca. 7.6 billion Euro).

Japan's fast breeder reactor has been shut down since December 1995, costing 50 million yen (ca. 378,000 Euro) per day in maintenance costs (Hokkaido Shinbunsha 2013: 129). Monju (see 2.1) worked for just 250 days of its 22-year lifespan and cost more than 1 trillion yen (ca. 7.6 billion Euro) (JAEA 2017).

Reprocessing is more expensive than direct disposal (once-through fuel cycle)

With no prospect of commercializing fast breeder reactors, there were fears that the cost of the nuclear fuel cycle could balloon, so the nuclear fuel cycle was put under review, with consideration being given to the option of direct disposal (once-through).

In October 2011 JAEC estimated that the cost of "re-processing" would be twice the cost of "direct disposal".

Table 2: Nuclear fuel recycling cost (Comparison in kWh)

	2011 Estimate	2004 Estimate
all reprocessing	1.98 yen (ca. 1.49 cents)*	ca. 1.6 yen (ca. 1.2 cents)
half reprocessing/half intermediate treatment[3]	1.39 yen (ca. 1.05 cents)	ca. 1.4-1.5 yen (ca. 1.05-1.13 cents)
all direct disposal[4]	1-1.02 yen (ca. 0.75-0.77 cents)	ca. 0.9-1.1 yen (ca. 0.68-0.83 cents)

*cent = Euro cent Source: Based on data from JAEC (2011: 14)

In April 2012, the JAEC re-estimated the cost of the nuclear fuel cycle. In all three comparisons, regardless of the share of nuclear power included in the electricity mix, the most cost effective method was direct disposal.

Table 3: Nuclear fuel recycling cost depending on the choice

nuclear power ratio of total electricity generation	all reprocessing	half reprocessing/ half intermediate treatment	all direct disposal
0%	----	----	8.1-8.7 trillion yen (ca. 61.2 -65.8 billion Euro)
15%	14.4 trillion yen (ca. 109 billion Euro)	14.4 trillion yen (ca. 109 billion Euro)	10.9-11.6 trillion yen (ca. 82.4-87.7 billion Euro)
20%	15.4 trillion yen (ca. 116.5 billion Euro)	15.3 trillion yen (ca. 115.7 billion Euro)	12-12.8 trillion yen (ca. 91-97 billion Euro)
30%	18.4 trillion yen (ca. 139 billion Euro)	17.3-17.4 trillion yen (ca. 130.9-131.6 billion Euro)	13.9-14.8 trillion yen (ca. 105-112 billion Euro)

Source: Based on data in JAEC (2012: 24, 66, 108).

3 Reprocessing of half of the waste, the other half is treated intermediately for 50 years.

4 All waste disposed directly 54 years after its use for power generation

This seems to suggest that direct disposal would be better in economic terms. The Draft Revision of the Basic Policy Based on the Final Disposal Act approved by the Cabinet on 22 May 2015 states that research will be conducted regarding direct disposal and other disposal methods (METI 2015: 7). The Radioactive Waste Working Group of the Advisory Committee for Natural Resources and Energy, which proposes improvements to initiatives and systems, adopted the view that "direct disposal is an alternative to the nuclear fuel cycle, not an alternative to final disposal" ANRE 2015b: 7). This in effect means that SF should be reprocessed. Moreover, even the recently revised Basic Policy stresses the importance of the nuclear fuel cycle, with the criteria for selecting final disposal sites predicated on reprocessing (ANRE 2014b: 2,3).

There is a strong tendency not to exclude the option of reprocessing, both to honor the agreement with Aomori Prefecture and to ensure access to interim storage facilities. There are also those who argue against relinquishing a technology that could potentially be used for making nuclear weapons (Hokkaido Shinbunsha 2013: 119).

3 Final disposal

3.1 The background to the selection of geological disposal

In 1962, the Japan Atomic Energy Commission (JAEC) became the first body in Japan to address the disposal of radioactive waste. In 1976, research into geological disposal got underway, with the Power Reactor and Nuclear Fuel Development Corporation (now JAEA) playing a central role in research and development. In 1992, they prepared a report setting out the technical possibilities for geological disposal and published another in 1999 detailing the technical reliability of the geological disposal of high-level radioactive waste (JAEA 2015a: 2). As a result, it was confirmed that geological disposal was technically feasible. The decision to select geological disposal rather than managing it above ground was based on the premise that areas deep underground would be the most stable locations, even in earthquake-prone Japan, provided that areas near volcanoes and fault lines were avoided (NUMO 2015b: 14-15).

In 2000, the Specified Radioactive Waste Final Disposal Act (Final Disposal Act) was enacted stipulating that: 1.) high-level radioactive waste be disposed in a stable subterranean geological formation at a depth of more than 300 meters; 2.) an implementing body be established; 3.) a three-step process be adopted for selecting disposal sites; and, 4.) a contribution-based system be used to secure the funds for final disposal.

The Nuclear Waste Management Organization of Japan (NUMO) was established in October 2000 as the implementing body. NUMO is responsible for the selection of disposal sites, the construction and management of disposal facilities, final disposal, the sealing and management of disposal facilities, and disposal operations in general.

NUMO plans to construct a facility where at least 40,000 vitrified packages and 19,000m^3 of TRU can be buried; the above-ground facility is expected to cover an area of 1-2 km^2 and the underground facility 6-10 km^2, with the cost of the disposal forecast at approximately 3.5 trillion yen, or approximately 26.5 billion Euro (NUMO 2015a:19).

3.2 Basic approach to disposal

In Japan, the 2005 Framework for Nuclear Energy Policy attached importance to radioactive waste treatment and disposal. The four principles that underlie the framework are: liability of generators, minimization of radioactive waste, rational treatment and disposal, and implementation based on mutual understanding with the people. In addition, the present generation is to assume responsibility for safe disposal of radioactive waste in order to protect future generations.

The laws regulating activities associated with radioactive waste management are the Act on the Regulation of Nuclear Source Material, Nuclear Fuel Material and Reactors (abbreviated as the Reactor Regulation Act), and the Act on the Prevention of Radiation Disease Due to Radioisotopes, (abbreviated as the Radiation Disease Prevention Act).

The Final Disposal Act was enacted in 2000 as a law specific to high-level radioactive waste. The Act on Special Measures for Handling Pollution by Radioactive Materials was enacted to address the issue of the disposal of waste polluted by radioactive materials discharged as a result of the Fukushima Daiichi NPP accident; the government has been at the center of efforts in this area.

3.3 Bodies involved in geological disposal

Ideally, an implementing body with responsibility for the disposal of high-level radioactive waste will ensure both safety and adherence to the principle of generator liability. When it met in May 2000, the JAECs Panel on the Disposal of HLW argued that the implementing body should be private rather than public, and that the government's role should be limited to establishing and enforcing safety regulations.

The NUMO, whose establishment had been authorized by the METI, was designated the implementing body for geological disposal, in accordance with the

Final Disposal Act. NUMO is responsible for selecting sites for the construction of disposal facilities; constructing the facilities; carrying out geological disposal; sealing, closing, and managing the facilities; and collecting contributions to cover costs.

The 2005 Framework for Nuclear Energy Policy states that the government, research and development institutions, and NUMO are to consistently promote research and development of geological disposal of high-level radioactive waste. NUMO was made responsible for the safe implementation of final disposal operations involving high-level radioactive waste and technical development aimed at improving economic performance and efficiency. Research and development institutions are to be led by the Japan Atomic Energy Agency (JAEA), which is described in detail in the organization chart of the Sector of Decommissioning and Radioactive Waste Management (JAEA 2015b).

4 Siting procedures and costs

4.1 The repository site selection process

The Final Disposal Act provides for the selection of a repository site in a three-stage process covering the selection of preliminary investigation areas, detailed investigation areas and a repository construction site. The opinions of the local population are to be fully respected in the selection process. At each stage of the process, reports on the investigation are to be compiled and explanatory meetings held. Without the agreement of the local population, the municipal mayors and the prefectural governors, the site selection process cannot move to the next stage.

The first stage involves a literature survey, analyzing previous earthquake activity, volcanic eruption, fault activity, uplift and erosion and other relevant tectonic phenomena. If the literature survey confirms that there is no indication of pronounced geological perturbations resulting from natural phenomena such as earthquakes, and that there will be no risk of such phenomena occurring in the future, then preliminary investigation will be carried out.

Preliminary investigations (borehole and trench) will be carried out on the surface, analyzing geological formations, rock properties, geological structures, groundwater characteristics and geodynamics. Concurrent with more detailed investigations at ground level, underground research facilities are to be constructed. This research will determine the suitability of the geological formation for a repository.

Following the identification and selection of a repository construction site, the disposal facilities are to be designed and a safety evaluation undertaken.

Construction can begin after a safety review by government experts (NUMO 2016: 5).

4.2 Geological selection criteria

Given its high volcanic activity and frequent earthquakes, are there any suitable sites in Japan for geological disposal? NUMO claims that detailed surveys of the geological environment make it possible to avoid areas affected by volcanoes or active faults, thus ensuring safety. An analysis of the distribution of volcanoes across the Japanese archipelago shows that the location of volcanoes has shown almost no change over the last several million years (NUMO 2015a: 15).

NUMO further suggests that it is possible to avoid the effects of active faults because there have been no major changes in fault movements over the last few million years; once a fault appears, it becomes the site of repeated activity, because it is a weak point. NUMO has determined that sites within 15 km of a volcano, which have experienced more than 300 m (150 m in the case of coastal areas) of uplift in the last 100,000 years, or are situated on an active fault line (with the width of the fault being given as one-hundredth of its length) are unsuitable as disposal sites.

4.3 Burden of final disposal costs

The costs involved in final disposal are premised on the nuclear fuel cycle. The largest costs are for SF reprocessing. Other costs include reactor decommissioning, geological disposal and development and siting.

Under the Spent Nuclear Fuel Reprocessing Fund Act enacted in 2005, the costs associated with reprocessing at the Rokkasho Reprocessing Plant can be passed on by adding them to electricity charges. (For further details concerning the costs associated with reprocessing, see 2.3.2) The costs for reactor decommissioning have been added to electricity charges since 1989, under the "nuclear power generation facility dismantling charge" (Oshima 2010: 21). The costs of disposing of approximately 40,000 vitrified packages, including the cost of developing the technology required for geological disposal and the costs associated with constructing, operating, and sealing the facility, were estimated at approximately 3 trillion yen (ca. 22.7 billion Euro). Under Article 11 (1) of the Final Disposal Act, power companies are obliged to make an annual contribution to NUMO, commensurate with the volume of waste resulting from nuclear power generation. Since 2000, the citizens of Japan have funded this contribution, through a portion of their electricity charges set aside for this purpose. In the financial year 2016, this contribution amounted to 10.3 billion yen (ca. 78 million

Euro), making a total of 987.1 billion yen (7.47 billion Euro) in contributions as of the end of March 2016 (NUMO 2017). In the end, it is the general public, which pays for reprocessing, decommissioning and disposal through higher electricity charges and other fees.

The selection of a final disposal site will have a huge socioeconomic impact, due to the very long time involved with geological disposal. Recognizing that it will be vital to ensure that host sites benefit from acting as hosts, METI has put together a budget for "grants for areas hosting power facilities" under the grant system based on the Three Power Source Development Laws. METI has decided to pay 1 billion yen (ca. 7.56 million Euro) per year to both the municipal and prefectural governments of the area concerned at the literature survey stage, and 2 billion yen (ca. 15.1 million Euro) per year at the preliminary investigation stage. The grant system from the detailed investigation stage onwards is to be determined in due course (Ueda and Li 2014: 7).

Once a municipality is actually confirmed as a host area for a disposal site and the site begins operating, it will receive annual income of approximately 2.7 billion yen (ca. 20.42 million Euro) in fixed asset tax for 60 years, making a total of approximately 160 billion yen (1.21 billion Euro). In addition, the area can expect to benefit from employment creation, with the economic effect of the construction and operation of the disposal site estimated at 39.8 billion yen (ca. 301 million Euro) annually (Ueda and Li 2014: 7).

5 Policy shift on site selection methods

5.1 The only district to apply for a literature survey

The Final Disposal Act prescribes a three-stage selection process with public participation and the support of local government. As described above, it also prescribes that grants are to be provided from the site survey stage for the purpose of regional development.

Despite the creation of this site selection framework, the only municipality ever to have applied for a literature survey since the application process opened in 2002 is Toyo-machi in Kochi Prefecture. Toyo-machi announced in January 2007 that it would apply for a literature survey. A town of approximately 3,000 inhabitants, Toyo-machi is located on the Pacific coast, close to the border with Tokushima Prefecture. The decision to apply was made by the town's mayor without consulting with the town council, so opinion in the town was bitterly divided (Hokkaido Shinbunsha 2013: 50). Beyond the town, the prefectural governors of both Kochi and Tokushima expressed their opposition to NUMO. The incumbent mayor resigned over the issue, calling an election in April 2007, standing again and seeking a mandate from local residents. The opposition candidate won a landslide victory and the application for the literature survey was withdrawn on 23 April (Hokkaido Shinbunsha 2013: 50). On 21 May, the town council enacted an ordinance rejecting the proposal to bring radioactive and nuclear material into Toyo-machi settling the issue once and for all (Toyo-machi 2017).

5.2 "Nationwide Map of Scientific Features for Geological Disposal"

Conducting the various surveys for selecting a candidate site could take around 20 years. The subsequent process of building a facility, burying the waste and ultimately sealing the facility could take 100 years. As not a single local government in Japan has undergone a selection survey, this means there is a long journey ahead. Japan has not yet even reached the starting point.

To rectify this situation, on 22 May 2015, the Cabinet approved a revision of the government's Basic Policy based on the Final Disposal Act. The key point in the revision is the switch from the system of voluntary application by local governments to one in which the government plays the key role in choosing the site, with the aim of speeding up the process of selecting candidate sites (ANRE 2015b). The government put together an extensive list of areas thought to be geologically suitable and released a "Map of Scientific Characteristics" at the end of July 2017, which was subsequently posted to the ANRE website (ANRE 2017). The map does not pinpoint potential nuclear waste repository sites. It illustrates in four colours (light green, dark green, orange and silver) the suitability of geological conditions throughout Japan. The areas in light green are deemed suitable as final disposal sites. If the areas lie within 20 km of the coastline they are shown in dark green. These dark green areas are also favourable from the standpoint of maritime transport. The areas in orange lie within a radius of 15 kilometres from volcanoes or are located close to active faults. These areas are therefore deemed unfavourable from the standpoint of underground stability and other factors. The areas in silver are those with reserves of oil, natural gas, coal and other minerals that could be exploited in the future. Approximately 900 municipalities, more than half of all municipalities in Japan, are classified as dark green areas. These 900 municipalities account for about 30 percent of Japan's total land area (NIKKEI 2017), the combined light green and dark green areas account for about 65 percent of Japan's total area, and are considered suitable for the construction of a disposal site (Mainichi 2017). These areas must be narrowed down, but no decision has been made yet as to how the list of candidate sites should be narrowed down.

The first stage of presenting the scientific characteristics map is finished, and Japan can now proceed to the second stage. Starting in autumn 2017, the METI began holding discussions with municipalities in dark green areas to explain the necessity for final disposal (ANRE 2017). If any municipalities show interest, research can finally go ahead. If not, the government will propose to some municipalities that they should agree to an investigation. The government hopes to narrow down the candidate sites over the next 20 years.

6 Further considerations

Why has there been so little progress made on the question of final disposal sites for radioactive waste? At the root of this non-decision are a number of uncertainties and the ambiguous definitions on which the discussion is based. The manner in which these uncertainties evolve has the potential to substantially affect future policy on final disposal.

6.1 The direction of nuclear energy policy and flow waste

It is unclear how many NPPs in Japan will actually be restarted. The government has decided that nuclear power generation will account for about 20 percent of the energy mix by 2030, but it remains unclear whether nuclear power generation will continue thereafter, or whether the nation will seek to move away from nuclear power.

Abandoning nuclear power would mean zero generation of flow waste. Consequently, the only waste which would need to be transported to final disposal sites would be the existing stock waste. Even if the country does not abandon nuclear power immediately, setting a clear deadline for abandoning it would facilitate precise calculations of the quantity of flow waste that would be emitted in the future.

If as is presently the case, the question of whether nuclear power will be sustained or abandoned remains undecided, there will continue to be uncertainty surrounding the quantity of radioactive waste that will be generated in the future. If the government were to decide to abandon nuclear power entirely, this would bring into question the need for nuclear fuel recycling, the major premise of Japan's nuclear energy policy. If the nuclear fuel cycle is not brought to fruition, the nature and quantity of flow waste will change.

6.2 Public anxiety and a low level of awareness

The method used hitherto for selecting candidate sites involved providing local citizens with a financial bonus large enough to compensate for or offset their anxiety and sense of risk regarding radioactive waste disposal. In many instances, waste disposal has been discussed solely in terms of its economic and institutional aspects, rather than following a democratic process as was the case of Toyo-machi. The mayor, who was hoping to revitalize the town, acted on his own. On 28 February 2007, when the people of Toyo-machi were still in the midst of a fierce debate that polarized the town into pro- and anti- factions, NUMO submitted an application to alter its business plan, so that it could begin the literature survey in 2007. The national government granted its approval just one month later on 28 March. Although it said that it would respect the views of local citizens concerning the selection of the candidate site, both NUMO and the national government sought to move their plans forward, without any consideration for the opinions of local people. This kind of behavior increases the sense of distrust among the general public (Nishio and Sueda 2009: 28-30).

The subsequent failure of any additional local governments to even apply for a literature survey despite the huge economic advantages involved would appear to be attributable to the fact that the feelings of anxiety and distrust among the people far outweigh any financial benefits. It is almost impossible to dispel the anxieties of the general public with economic advantages alone, so it is essential to carefully undertake a democratic process, providing accurate data and information in order to gain the understanding of the public.

According to a survey of public attitudes to HLW conducted by the Japan Atomic Energy Relations Organization (JAERO) in 2001, the proportion of respondents who reported having a good understanding of the difference between high-level radioactivity and low-level radioactivity was 6.9 percent, whereas the combined total for those who had either not much or no understanding whatsoever was around 80 percent. The proportion of those reporting that they knew nothing about geological disposal was 70 percent, demonstrating the low level of awareness among the general public (ATOMICA 2003).

Since the Fukushima accident, the topic of abandoning nuclear power has begun to crop up frequently in energy policy discussions. However, the media rarely reports anything about radioactive waste, so one certainly cannot say that public awareness of the topic has increased (JAERO 2013: 48).

6.3 *What is "high-level radioactive waste?"*

High-level radioactive waste consists of vitrified packages and TRU waste. This definition arises from the fact that the nuclear fuel cycle is the major policy premise in Japan. Spent fuel is regarded as a resource in Japan, because Japan aspires to recycle its nuclear fuel. Until now, the goal was to reprocess all SF, but given present circumstances, direct disposal of at least some of it is coming under consideration (see 2.3.2.2). Should all or any of the SF be directly disposed, the SF would also be declared as high-level radioactive waste.

From this perspective, it would seem appropriate to adopt the definition used by the Science Council of Japan: "The term 'high-level radioactive waste' does not merely refer to high-level radioactive waste emitted after the reprocessing of spent nuclear fuel; it will also be used to indicate spent nuclear fuel in the event that reprocessing of all spent nuclear fuel is halted and direct disposal is also carried out." (Science Council of Japan 2012: 4).

Discussing final disposal sites on the basis of a definition that equates high-level radioactive waste with SF would make it possible to avoid any major changes in the approach to disposal sites, even if there were a change in the current policy of reprocessing all SF. The Fukushima NPP accident is also generating high-level radioactive waste. The disposal of radioactive waste from nuclear fuel materials and reactors that have undergone core meltdowns will be much harder to dispose of than conventional high-level radioactive waste (AESJ 2015: 10). The quantity and quality of the radioactive waste which actually goes to final disposal could differ from the data on the quantity and nature of "high-level radioactive waste" as defined by law (Ueda and Li 2014: 8).

6.4 *The ambiguous principle of the liability of generators*

The costs involved in processing and disposing of high-level radioactive waste are immense and create various problems. The question of who will bear the financial burden is also a crucial point of contention.

The Framework for Nuclear Energy Policy explicitly states that the processing and disposal of high-level radioactive waste is based on the principle of the liability of generators. Electric power companies and other generators are obliged to ensure safe processing and disposal of radioactive waste, while the state must supervise them to ensure compliance.

In reality, however, the burden of ensuring the safe management of radioactive waste, responsibility for its final disposal, and compensation in the event of an accident has been effectively passed on to NUMO. The nuclear fuel cycle and radioactive waste processing cost an enormous amount of money, most of which is borne by the public in the form of electricity charges. Moreover, the national government provides vast sums in treasury funds, so it would be no exaggeration to say that it is the state that is funding the cost of the nuclear fuel cycle and final disposal operations (Ueda and Li 2014: 9). Also when it comes to the question of who shoulders the cost burden, the principle that the waste generators are liable is not functioning.

As it currently functions, the system is mitigating the liability of the electric power companies, which are also the waste generators, while increasing the authority of the state. It is unrealistic to require a single company to bear the liability for continuing to manage such waste for tens of thousands of years after disposal. Nevertheless, if mandatory generator liability is relaxed and electric power companies are relieved of their responsibilities after disposal, the only outcome will be a continuation of irresponsible power generation.

6.5 The Fukushima accident

Because of the Fukushima accident, there are three serious additional problems involved with radioactive waste which other countries do not have: dealing with the decommissioning of damaged reactors, contaminated water and decontamination in off-site areas in Fukushima.

The problem of high level radioactive waste from the Fukushima accident has yet to be addressed. Much of the high level radioactive waste is being temporarily stored at the Fukushima site itself. The immediate area around the nuclear plants remains highly radioactive and the molten fuel rods remain in the reactor core.

The decommissioning of severely damaged reactors like those at the Fukushima Daiichi NPP is unprecedented in history. It is very important to know the exact locations and the physical, chemical, and radiological forms of the corium to ensure its safe removal. The high radioactivity in the reactor means, however, that only specially designed robots can probe the unit, and this activity is still in the exploratory phase. It is estimated that it will take at least 30 years to deal with the decommissioning of the damaged nuclear reactors in the Fukushima Daiichi NPP, given the highly dangerous conditions under which the work must take place.

In August 2014, the Nuclear Damage Compensation and Decommissioning Facilitation Corporation (NDF) was set up by the government as a planning body with management support for R&D projects. The NDF supports the decommissioning work and formulates strategies to deal with the main mid-and-long term challenges associated with the decommissioning of the Fukushima Daiichi NPP, such as methods of retrieval of fuel which has melted and must then harden, as well as radioactive waste management. In May 2017, the management of the reserve fund to promote decommissioning was added to the corporation's scope of duties.

Another serious problem on site is contaminated water. The Fukushima Daiichi site contains vast amounts of highly contaminated water that were used to cool the remains of the reactor cores. According to TEPCO, groundwater enters the reactor building at a level of 300 tons per day, and becomes contaminated (TEPCO 2017). It is very important to stop the leakage of contaminated water to the outside. Since 2014, preparations have been underway for an ice wall surrounding the damaged Fukushima Daiichi, and the final section of ice is currently being frozen. However, it is uncertain whether the wall will function as envisaged. Similar technology is widely used in the construction industry, but the ice wall in Fukushima is the largest to date (WNN 2017).

Decontamination is one of the most important measures for radiation protection to remove radioactive materials from inhabited areas. Decontamination is carried out according to the Act on Special Measures Concerning the Handling of Radioactive Pollution (Act on Special Measures), which was enacted in August 2011. Areas contaminated at relatively low levels by the radioactive plume are being "decontaminated". The thousands and thousands of large garbage sized bags of contaminated soil and other debris which have been collected are being brought to temporary storage areas (often a cleared area in the mountains or fields) for low- and medium-level radioactive waste in the Fukushima area. They will remain at these locations until they can be moved into an interim storage facility. The interim storage facility is to be established in Fukushima prefecture where huge amounts of soil and waste were generated by decontamination activities. The interim storage will operate for 30 years, and then final disposal of the waste will be carried out.

In 2016, the Ministry of Environment announced that material with less than 8000 becquerels per kilogram of caesium would no longer be specified as contaminated waste, or subject to restrictions on disposal. Where radioactivity is less than 8000 Bq/kg, contaminated soil may be used for embankments. Use is unrestricted, if the radioactivity level is less than 100 Bq/kg. Most of the stored waste in Fukushima has decayed to below the 8000 Bq/kg level. According to the Ministry of Environment, the current status of specific (designated) waste (over 8,000 Bq/kg) is 194,080.8 tons (MoE 2017).

7 Conclusions

In Japan, the discussion on final disposal began at more or less the same time as the nation began developing and using nuclear power. As a result of research and development over many years, the discussion of the technical aspects of vitrifing waste and disposal methods has advanced. In addition, the government has established financial support to help deal with the enormous costs that are expected to be incurred between the selection of the candidate sites and actual final disposal. Despite the large financial incentive, however, not a single local government is willing to apply for the investigation stage, the financial benefits cannot offset the fears and skepticism of the people.

Although the variables in the equation have changed, such as the Fukushima accident and the Monju shutdown, the outcome has been left unchanged. SF reprocessing is a constant in Japan's nuclear and waste management policy, a default answer which has stifled discussion and hindered planning.

Moreover, several uncertainties make decisions in Japan more difficult. Future nuclear policy will determine the type and volume of waste which will be emitted. Discussing the restart of Japan's nuclear power stations must take into account the direction of the nation's nuclear energy policy and its disposal challenges.

The "high-level radioactive waste" accounted for in present legislation are the canisters of vitrified waste. This legal framework is the basis for researching, financing, storing and disposing of this waste, but the framework is too tight and rigid. The waste from the Fukushima accident and other types of waste are subject to specific legislation; they differ greatly from the vitrified packages in form, quantity and required disposal methods, and they are discussed separately.

The uncertainties will remain until the government decides on the direction it wishes to take, and this in turn renders the disposal site issue even more intricate and formidable. It is necessary to have clearly defined criteria, accurate data and information, in order to have open democratic discussions. In Japan, the repository search process has been restarted, but the dialogue with the municipalities is based on vague criteria. It will prove enormously difficult to gain the trust and understanding of citizens. Japan must finally cease to discuss its nuclear policy and nuclear waste policy independently if the country is to develop a realistic solution to the dilemma of final disposal.

References

AESJ (Atomic Energy Society of Japan) (2015). *Fukushima Daiichi Gensiryoku Hatsudensho Jiko ni yori Hassei suru Houshasei Haikibutsu no Shori Shobun* [Processing and Disposal of Nuclear waste from the accident of Fukushima Daiichi Nuclear Power Plant]. http://www.aesj.or.jp/special/report/2013/s_haikibutsusyori-syobun20140331.pdf, last accessed 26 June 2015.

ANRE (Agency for Natural Resources and Energy) (2003). *Sougou Shigen Energy Chosakai Denki Jigyou Bunakai Kost tou Kentou Shou Iinkai (Dai 4 Kai) Heisei 15 nen 11.11.* [Advisory Committee on Energy and Natural Resources Electrical Industry Committee Electrical Industry Committee (4th Session) 11 November 2003]. http://www.enecho.meti.go.jp/committee/council/electric_power_industry_subcommittee/006_004/pdf/004_006.pdf, last accessed 18 September 2015.

ANRE (2013). *Koureberu Houshasei Haikibutushobun ni tuite* [Disposal of high level radioactive waste]. METI Symposium "Geological Disposal", Tokyo 23 May 2015. Material.

ANRE (2014a). *Energy Kihon Keikaku* [The Basic Energy Plan]. http://www.enecho.meti.go.jp/category/others/basic_plan/pdf/140411.pdf, last accessed 29 June 2015.

ANRE (2014b). *Kakunenryo Saikuru/Saishu Shobun ni kansuru Genjyo to Kadai* [Current status and issues about fuel cycle and final disposal]. http://www.meti.go.jp/committee/sougouenergy/denkijigyou/genshiryoku/pdf/006_03_00.pdf, last accessed 26 June 2015.

ANRE (2015a). *Koureberu Houshasei Haikibutsu no Saishushobun ni muketa aratana Torikumi* [New effort for final disposal of high level radioactive waste]. METI Symposium "Geological Disposal", Tokyo, 23 May 2015. Material.

ANRE (2015b). *Saishushobunhou ni motozuku Kihonnhoushin no Kaiteian ni tuite* [Revised draft of the Basic Policy based on Final Disposal Act]. http://www.meti.go.jp/committee/sougouenergy/denryoku_gas/genshiryoku/houshasei_haikibutsu_wg/pdf/017_01_00.pdf, last accessed 05. October 2017.

ANRE (2017). *Kagakuteki Tokuseimap Kouhyouyou Saito* [Website for the Map of Scientific Characteristics]. http://www.enecho.meti.go.jp/category/electricity_and_gas/nuclear/rw/kagakutekitokuseimap/, last accessed 05 October 2017

ATOMICA (RIST Research Organization for Information Science and Technology) (2003). *Koureberu Houshasei Haikibutsu ni kansuru Kokumin no Ishikichosa* [Survey of public attitudes to high-level radioactive waste]. http://www.rist.or.jp/atomica/data/dat_detail.php?Title_Key=10-05-01-15, last accessed 29 June 2015.

ATOMICA (2009). *Kaigai Itaku Saishori kara Henkan sareru Houshasei Haikibutsu* [The returning of nuclear waste from Overseas]. http://www.rist.or.jp/atomica/ (05-01-04-05), last accessed 26 June 2015.

FEPC (The Federation of Electric Power Companies of Japan) (2017) *Shiyouzumi Nenryo no Chozou Jokyo to Taisaku* [Storage Condition and Measures of Spent Fuel]. http://www.fepc.or.jp/resource_sw/chozo.pdf, last accessed 25 November 2017.

Funabashi, H., Hasegawa, K. and Iijima, N. (2012). *Kakunenryo Risaikuru-shisetsu no Shakaigaku – Aomoriken Rokkashomura* [Sociology on Nuclear Fuel Cycle Facilities in Rokkasho Village], Tokyo: Yuhikaku.

Hokkaido Shinbunsha (2013). *Genshiryoku Fu no Isan* [Nuclear power Negative Legacy]. Hokkaido Shinbunsha.

Inoue, T. (2015). *Genshiryoku Hatsuden to Chiho Zaisei* [Nuclear power generation and Local government finance], Kyoto: Koyoshobo.

JAEA (Japan Atomic Energy Agency) (2015a). *Chiso Shobun Gijyutsu ni kansuru Kenkyu Kaihatsu* [Geological Disposal Research and Development], Ibaraki: JAEA (Brochure).

JAEA (2015b). *Geological Isolation Research and Development.* (n.d.). Retrieved, 24 September 2015. http://www.jaea.go.jp/04/tisou/english/organization_chart/organization_chart.html, last accessed 20 September 2015.

JAEA (2017). *Monjyu nituite Okotaeshimasu.* [FAQs about Monju]. https://www.jaea.go.jp/04/turuga/anncer/page/kaitou/kaitou2-2.html, last accessed 24. November 2017.

JAEC (Japan Atomic Energy Commission) (2011). *Kaku Nenryou Saikuru Kosuto no Shisan* [Estimates of the nuclear fuel cycle cost]. http://www.aec.go.jp/jicst/NC/tyoki/hatukaku/siryo/siryo3/siryo1-1.pdf, last accessed 20 September 2015.

JAEC (2012). *Kaku Nenryou Seisaku no Sentakushi ni kansuru Kentoukekka ni tsuite Sankoushiryou 3* [Study results on the choice of the nuclear fuel cycle policy, Appendix 3]. http://www.aec.go.jp/jicst/NC/iinkai/teirei/siryo2012/siryo22/siryo1-2.pdf, last accessed 20 September 2015.

JAERO (Japan Atomic Energy Relations Organisation) (2013). *Genshiryoku Riyou ni kansuru Yoronchousa no Kekka ni tuite* [Survey results of a Poll about Utilization of nuclear energy]. The 23rd JAEC Material No.1 (2013). http://www.aec.go.jp/jicst/NC/iinkai/teirei/siryo2013/siryo23/siryo1.pdf, last accessed 26 June 2015. http://www.jcp.or.jp/web_policy/2014/11/post-635.html, last accessed 29 June 2015.

JNFL (Japan Nuclear Fuel Limited) (2016). *Saishorijigyo no Gaiyo* [Outline of Reprocessing]. http://www.nuro.or.jp/pdf/20161125_03.pdf, last accessed 24 November 2017.

Mainichi (2017). *Kakugomi map Tekichiha Kokudo no 65%* [Map for nuclear waste. Suitable areas are about 65% of Japan's total area.]. https://mainichi.jp/articles/20170728/k00/00e/040/286000c, last accessed 05 October 2017.

METI (Ministry of Economy, Trade and Industry) (2015). *Tokutei Houshasei Haikibutsu no Saishuu Shobun ni kansuru Kihonkeikaku* [Basic Policy on the Final Disposal of Designated Radioactive Wastes]. http://www.meti.go.jp/press/2015/05/20150522003/20150522003-1.pdf, last accessed 29 June 2015.

MoE (Ministry of the Environment) (2017). *Sitei Hikibutsu ni tuite* [About the off-site Waste Contaminated with Radioactive Materials]. http://shiteihaiki.env.go.jp/radiological_contaminated_waste/designated_waste/, last accessed 05 October 2017.

Nishio, B. and Sueda, H. (2009). *Genpatsu Gomi ha Fu no Isan* [Nuclear waste is a Negative legacy], Tokyo: Sousisha.

NIKKEI (2017). *Kaku no Gomi Saishuushobunjyou "Tekichi" 900 Jichitai ni* ["Suitable areas" for final disposal of nuclear waste number about 900 municipalities]. https://www.nikkei.com/article/DGXLASGG28H1D_Y7A720C1000000/, last accessed 05 October 2017.

NRA (Nuclear Regulatory Authority) (2014). *Gensiryokusisetsu ni kakawaru Heisei 25 Nendo Houshcasen Kanritou Houkoku* [Radioactive management at Nuclear Facilities in 2013]. https://www.nsr.go.jp/data/000048084.pdf, last accessed 29 June 2015.

NUMO (Nuclear Waste Management Organization of Japan) (2008). *Chiso Shobun sono Anzensei* [Safety of Geological Disposal], Tokyo: NUMO.

NUMO (2015a). *Chiso Shobun Jigyo no Gaiyo* [Summary of Geological Disposal Project]. METI Symposium "Geological Disposal", Tokyo on 23 May 2015. Material.

NUMO (2015b). *Sittehoshii ima Chisoushobun* [Explaining Geological Disposal], Tokyo: NUMO (Brochure).

NUMO (2016). *"Geological Disposal Through Global Collaboration"*. http://www.numo.or.jp/en/publications/pdf/NUMO_En_Nov_2016.pdf, last accessed 24 November 2017.

NUMO (2017). NUMO Top>*Yokuaru Goshitumon* [NUMO Top> FAQs]. http://www.numo.or.jp/q_and_a/, last accessed 24 November 2017.

Oshima, K. (2010). *Gensiryokuseisaku Taikou Minaoshi no Hituyousei ni tuite* [The Need for a Review Nuclear Energy Policy Framework]. The 48th JAEC Material 1-1 (2010). http://www.aec.go.jp/jicst/NC/iinkai/teirei/siryo2010/siryo48/siryo1-1.pdf, last accessed 26 June 2015.

RFS (Recyclable Fuel Storage Company) (2016). *risaikuru nennryoubichikusennta-*siyouzuminennryo *chozoujigyouhennkou kyokasinnseisho no ichibu hoseitou ni tsuite* [Recyclable Fuel Storage Center- about partial change of business change permission application]. http://www.rfsco.co.jp/press/h28/p02.pdf, last accessed 22 June 2017.

Sagara, N. (2009). *Nihon no Genshiryoku Seisaku no Hensen to Kokusaiseisaku Kyouchou ni kansuru Rekisiteki Kousatsu* [Nuclear Policy and International cooperation in Japan from historical consideration]. RIETI (Research Institute of Economy, Trade & Industry, IAA), Policy Discussion Paper 09-P-002. http://www.rieti.go.jp/jp/publications/pdp/09p002.pdf, last accessed 26 June 2015.

Science Council of Japan (2012). *Koureberu Houshasei Haikibutu no Shobun ni tuite* [Disposal of high level radioactive waste]. http://www.scj.go.jp/ja/info/kohyo/pdf/kohyo-22-k159-1.pdf, last accessed 16 September 2015.

TEPCO (2012). *Chukanshori Shisetsu ni tsuite* [About the interim storage facility]. http://www.tepco.co.jp/company/corp-com/annai/shiryou/report/bknumber/0512/pdf/ts051205-j.pdf, last accessed 20 September 2015.

TEPCO (2017). *Osensui Taisaku no omona Torikui* [Main mesures for contaminated water]. http://www.tepco.co.jp/decommision/planaction/waterprocessing/index-j.html, last accessed 05 October 2017.

Toyo-machi (2007). *Toyomachi Houshasei Kakubusshitu no mochikomi Kyohi ni kannsuru Jyourei* [Toyo-machi by-law rejects radioactive waste]. http://web.town.toyo.kochi.jp/joho/reiki_int/reiki_honbun/g900RG00000341.html, last accessed 26 June 2015.

Ueda, K. and Li, H. (2014). Sekinin to Hiyouhutan kara mita Nihon no Koureberu Houshasei Haikibutsumonndai [Who is responsible and who should pay for Nuclear Waste Management in Japan]. Hokkaido University. In: *Economic Studies,* 63(2), 1-11. http://eprints.lib.hokudai.ac.jp/dspace/bitstream/2115/54572/1/ES_63 (2)_1.pdf, last accessed 26 June 2015.

WNA (World Nuclear Association) (2015). *"Nuclear Power in Japan"*. http://www.world-nuclear.org/info/Country-Profiles/Countries-G-N/Japan/, last accessed 28 June 2015.

WNN (World Nuclear News) (2017). *"Last part of Fukushima ice wall being frozen"*. http://www.world-nuclear-news.org/RS-Last-part-of-Fukushima-ice-wall-being-frozen-2308171.html, last accessed 05 October 2017

Yamaguchi, S. (2005). Kakunenryou Saikuru wo meguru Giron [Discussion about Nuclear Reprocessing]. NDL (National Diet Library). In: *Chosa to Jyohou*, 473(2005). http://www.ndl.go.jp/jp/diet/publication/issue/0473.pdf, last accessed 26 June 2015.

III. A Missed Opportunity: The Case of the Eastern European Countries

Postponed Policy

Nuclear Waste Governance in Hungary

Zsuzsanna Koritár[1]

Abstract

The main source of nuclear waste in Hungary is the four VVER-440 reactors at the Paks site, which began producing commercial electricity between 1982 and 1987. At the time of plant construction, in the 1970s, an agreement was made between Hungary and the Soviet Union that spent fuel should be transported to the Soviet Union for reprocessing. This was the practice for more than a decade; the last shipment took place in 1998, with a total of 2,331 spent fuel cartridges transported to Russia. Although this final shipment took place 18 years ago, there has been no final decision regarding Hungary's nuclear waste management strategy. The reference scenario foresees domestic deep geological disposal, however the option of shipping spent fuel to Russia for reprocessing still remains open. Spent fuel is currently stored in an interim storage facility next to the Paks NPP.

A complex screening procedure for a potential waste disposal site (hosting either high level waste from decommissioning or spent fuel as well as high level waste) started in 1999. As a result, six geological formations were "recommended for further research". The most favoured is the Boda Siltstone Formation in the Western Mecsek Mountain, in the southern part of Hungary. According to plans, the site will start to operate in 2064.

The Hungarian national programme, based on the Council Directive 2011/70/EURATOM was prepared by the government and submitted to the European Commission in August 2015. The document was made public in June 2016, and its final version was adopted by the government in August 2016.

---

1 Energiaklub Climate Policy Institute; 1056 Budapest, Hungary, Szerb 17-19; E-Mail: koritar@energiaklub.hu

A. Brunnengräber et al. (eds.), *Challenges of Nuclear Waste Governance*, Energiepolitik und Klimaschutz. Energy Policy and Climate Protection, https://doi.org/10.1007/978-3-658-21441-8_6

1 Introduction

The main source of nuclear waste in Hungary is the four VVER-440 reactors at the Paks site, which began producing commercial electricity between 1982 and 1987. In 2014, over 50 percent of electricity production originated from this nuclear power plant (NPP) in Hungary. The NPP is 100 percent state-owned.

The original lifetime of the reactors in the Paks NPP is 30 years. The license was valid for Unit 1 until 2012, for Unit 2 until 2014, for Unit 3 until 2016 and for Unit 4 until 2017. A decision to prolong operation was made by the company, which was supported by political consensus; the lifetime extension procedure started in the early 2000s. Unit 1, 2 and 3 have already received their renewed licenses, while the process is ongoing in case of Unit 4.

On January 14, 2014, Hungary and Russia concluded an agreement on atomic energy cooperation, in which the countries agreed, inter alia, to construct two new nuclear reactors of 1,200 MW capacity each with a Russian loan of 10 billion Euros. Their operation is planned to start in the mid-2020s. The agreement also makes possible the shipment of spent fuel to Russia for reprocessing or for technological storage.

2 Nuclear waste disposal in Hungary

2.1 Historical background

The 1996 Act on Atomic Energy was the first law to regulate radioactive waste management in Hungary. It was also the first official document to mention disposal of spent nuclear fuel (SF), based on the international standards of national disposal (OECD 1995). According to the Act, spent fuel is not defined as radioactive waste until the decision about its management is finalized; in other words, it only becomes waste when its final disposal is decided on.

At the time of NPP construction, in the 1970s, an agreement was made between Hungary and the Soviet Union which stated that after three years of cooling in a water pool, SF was to be transported to the Soviet Union for reprocessing. The service was free of charge, based on the original agreement. However, the agreement was amended unilaterally by the Soviet Union several times; these amendments increased the cooling time from three years to five years, as well as introduced a charge for shipping and reprocessing that increased over time. It soon became clear, from the Hungarian perspective, that this practice could not be maintained for long. Thus at the beginning of the 1990s, a decision was made to prepare for an alternative, namely building a national interim storage next to the Paks site.

Nevertheless, until 1998, Paks NPP transported spent fuel for reprocessing to Russia without the obligation to take back radioactive by-products. This occurred despite the adoption of the Russian Environmental Protection Law in 1993, which ordered the return of reprocessed radioactive wastes to the country of origin. According to special agreements between Hungary and Russia, Hungarian shipments could ignore this obligation. In 1997, the first three modules of the interim storage facility were built and transport to Russia was no longer necessary. The last shipment of spent fuel took place in 1998, with a total of 2,331 spent fuel cartridges transported to Russia.

The Hungarian Atomic Energy Authority (HAEA) was assigned by the government to establish a public authority responsible for collection, storage and decommissioning of radioactive waste. In 1998, the Public Agency for Radioactive Waste Management (PURAM) was founded for this purpose. Since then, PURAM has been managing and supervising three facilities: the Radioactive Waste Treatment and Disposal Facility (RWTDF) at Püspökszilágy for the disposal of radioactive wastes of institutional origin, the National Radioactive Waste Repository (NRWR) at Bátaapáti for the final disposal of low- and intermediate-level wastes of NPP origin and the Interim Spent Fuel Storage Facility (ISFSF) at Paks for the interim storage of spent fuel from the NPP. A fourth facility will be added to this list, when the final disposal site for high-level radioactive waste (HLW) is selected (Fig. 1.). Screening for this site started in 1999 and as a result, six geological formations were recommended for further research. Site selection is a longer procedure, with a planned conclusion in 2030.

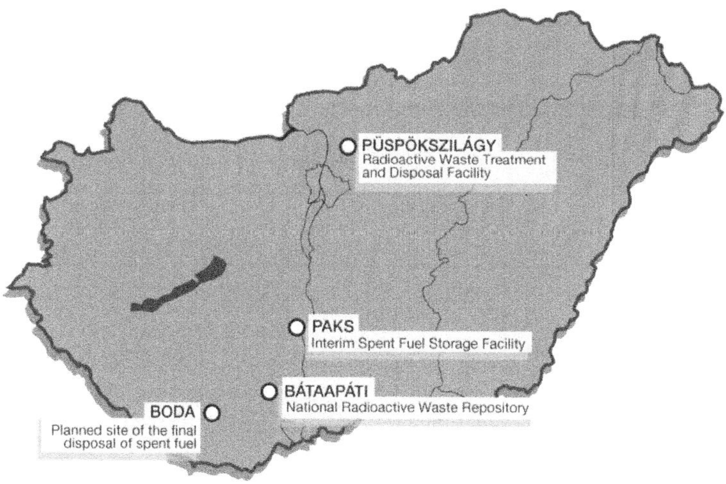

Figure 1: Facilities managed by PURAM (by Dénes Fellegi).

There is an important event in the history of spent fuel management which needs to be noted. A serious incident happened at the Paks NPP Unit 2 reactor in April 2003, when 30 fuel assemblies were damaged during a cleaning process. The incident was provisionally registered as Level 2 on the International Nuclear Event Scale (INES). The incident has been re-evaluated and given a final rating of INES 3. The incident occurred in the fuel cleaning system, which was located under 10 metres of water in a cleaning tank and next to the SF cooling pond as well as the reactor in the reactor hall. An investigation by the HAEA concluded that the incident was caused by inadequate cooling of the fuel elements. In July 2003, an International Atomic Energy Agency (IAEA) mission concluded that neither the HAEA nor Paks NPP used conservative decision making in their safety assessments, given to this unproven fuel cleaning system (IAEA 2003). The damaged fuel assemblies stayed in the tank for three and a half years, then they were removed, put in cases and inserted into the cooling pond next to the reactor. The remediation process was finished by the end of 2006 and the fuel elements were transported to the Russian Mayak facility in August 2014.

2.2 The national inventory

The largest portion of HLW originates from the four reactors of the Paks NPP. The fuel is uranium dioxide, of which some 42 tonnes are loaded in each reactor. One hundred SF assemblies per unit are generated every year. After its removal from the reactor, SF is placed in a water pool (decay pool) for 3 years and then is transferred to the ISFSF for an additional 50 years. As of 1 January 2014, the Paks NPP had produced 1,095.2 tons of heavy metal with 1,751 SF assemblies in the decay pools and 7,687 assemblies in the ISFSF (see Table 1.). An additional 8,290 SF assemblies are expected by the end of operation (PURAM 2014).

Table 1: Spent fuel inventory of the Paks NPP as of 1 January, 2014.

	Storage capacity		Stored amount	
	Number	t_{HM}	Number	t_{HM}
Decay pools	2,600	308.4*	1,751	207.7
ISFSF	9,308	1,104.1*	7,687	887.5

* The heavy metal equivalent of spent fuel was calculated with 118.62kgU per assembly. Source: PURAM 2014

Note that these figures do not include the fuels damaged in 2003 during the INES 3 incident.

A small amount of SF is generated in reactors that do not produce commercial electricity, such as the educational (Hungarian Technical University) and research (Hungarian Academy of Science) reactors in Budapest. The fuels used here differ in every parameter from those used by the Paks NPP (see Table 2).

Table 2: Spent fuel generated by educational and research reactors.

	SF generated until 1 January 2014		SF to be generated during whole operation	
	Number	kg_{HM}	Number	kg_{HM}
Research reactor (fuel type: VVR-M2(20)	39	8.58	810	178.2
Educational reactor (type: EK-10)	0	0	56	68.91

Source: PURAM 2014

SF generated by the research reactor is stored in two onsite water pools. An average of 70 fuel assemblies are replaced annually; on 1 January 2014, 39 SF assemblies with a heavy metal content of 8.58 kgU were stored in the water pools and all others were transported to Russia. The educational reactor has a facility to store SF but it has never been used, as the original load made in 1971 is still in use (PURAM 2014).

HLW other than SF is generated in the Paks NPP, at a rate of five m^3 each year. It is stored temporarily onsite in vertical storage tubes. As of 1 January 2014, 99.3 m^3 of HLW had been generated; a further 120 m^3 is expected by the end of operation. An additional 496.2 m^3 (gross amount) HLW is estimated as a result of decommissioning.

2.3 The (interim) storage site

In the 1990s and following changes in the political and economic situation in Central and Eastern Europe, it became clear that the practice of shipping SF to Russia couldn't be maintained for long. Therefore the management of the Paks NPP made a decision to build an interim storage for SF as an alternative to the transports. Experts from the NPP selected the Modular Vault Dry Storage model for this facility. Planning, licensing and construction of the first modules took place from 1992 to 1996. The ISFSF was put into operation in 1997 and has been managed by PURAM since 1998.

Each module of the facility has a license for 50 years. Operation and extension happens in parallel to one another; currently the third extension phase is underway with the construction of modules 21-24. For the four operating blocks (and excluding the planned two new reactors), a total of 36 modules are planned.

2.4 *The waste management strategy (with current waste disposal plan/concept)*

In Hungary, HLW has been generated since the 1960s. Due to lack of prohibiting legislation, it was transported to the Radioactive Waste Treatment and Disposal Facility (RWTDF) at Püspökszilágy, where institutional radioactive wastes are disposed. This practice was stopped after the Paks NPP was put into operation.

Nevertheless, there has been no final decision regarding a strategy for nuclear waste management. The reference scenario foresees domestic deep geological disposal (DGD), however the option of shipping SF to Russia for reprocessing remains open. This latter option is theoretically ensured for the operating reactors by a protocol between Hungary and Russia, which was signed in 2004; and for the two planned reactors by the international agreement signed in 2014. Taking official calculations into consideration, direct disposal seems to be the more practical and economical option (PURAM 2010). Despite this, the Hungarian national policy adopted by the parliament in May 2015 states that no final decision has been reached regarding SF management; therefore, the selection of a DGD site has to be conducted in addition to current interim storage procedures.

A complex screening procedure for a potential waste disposal site started in 1999 and as a result, six geological formations were "recommended for further research" (Konrád et al. 2006). The most favoured is the Boda Siltstone Formation in the Western Mecsek Mountain, in the southern part of Hungary. At this site, a second phase of geological field investigations started in 2016. According to plans, the disposal site will begin operation in 2064 (PURAM 2014).

The legal framework on radioactive waste disposal only mentions the ensuring of retrievability during operation of the disposal site. No reversibility or retrievability options are discussed in the strategies or regulations for after site closure.

The Hungarian national programme (NP 2016), based on the Council Directive 2011/70/EURATOM, was prepared by the government and submitted to the European Commission in August 2015. After the Strategic Environmental Assessment (SEA 2016) and related public involvement process was carried out, the national programme became public in June 2016. Although the document was open for public evaluation, none of the concerns raised have been taken into account, evidenced by the government's approval of the final version without any changes in August 2016.

The national programme contains the same information as the earlier plans and strategies, described above, nothing specific or new emerged with its adaptation. Questions on management options, such as domestic DGD or shipments to Russia, for SF remain open, as do questions on financing and schedule.

3 The legal and institutional framework

3.1 The legal framework

In 1996, the Hungarian Parliament adopted the Atomic Energy Law, which entered into force on 1 January 1997. It is based on the rules and guidelines of the IAEA, incorporating EU directives and Organisation for Economic Co-operation and Development/ Nuclear Energy Agency (OECD / NEA) recommendations. This law sets the framework for the application of nuclear energy, including SF management. Accordingly, the NPP operator shall ensure that the smallest amount of radioactive waste is generated and must obey the polluters pay principle, while the management of radioactive waste is the responsibility of the state. Therefore, the Atomic Energy Law regulates the establishment of an agency, PURAM, with the mission to manage interim storage and final disposal of nuclear waste, as well as the establishment of the Central Nuclear Financial Fund (CNFF), where all these activities are to be financed from (see section 6.)

A number of lower regulations exist to describe in detail the operation of PURAM (215/2013 Government Decree), as well as the safe storage and disposal of radioactive waste (155/2014 Government Decree). The latter includes provisions on siting, building, operating, closure and monitoring of disposal sites. The so-called active institutional time interval (the time of active monitoring of the site after closure) regarding all radioactive waste disposal sites is determined to be least 50 years, which can be extended by the atomic energy authority as needed. During building and operation of disposal sites, PURAM is required to submit a yearly report on management and safety to the atomic energy authority. This authority is required to conduct an intermittent inspection every 10 years regarding safety of the disposal site. The appendix of this decree contains detailed safety regulations on planning, establishment, operation, emergency preparedness and response, maintenance and supervision, radioprotection[2] and management of accidents.

Interestingly enough, there was a ministry regulation on site selection (geological requirements as well as details of licensing procedure), however, after approximately one year of being in force, it was abolished.

2 Radioprotection means the protection against radioactive radiation and the protection against the leaking of radioactive substances. This word appears in connection with the safety of radioactive waste disposal sites.

3.2 The institutional framework

For nuclear facilities defined in the Atomic Energy Act, including facilities for SF management, the responsible authority is the HAEA. It is a regulatory body operating under government control, but with independent responsibilities defined in a specific law. The minister in charge of supervision is appointed by the Prime Minister.

The fundamental objective of regulatory supervision is to prevent nuclear incidents and accidents. The HAEA's competence includes the establishment of safety requirements, nuclear safety licensing and control, approval of nuclear emergency preparedness and response plans of nuclear facilities and registration, transport and control of radioactive substances. The HAEA's work is supported by a scientific council, with a maximum of 12 experts; its members and president are appointed by the minister supervising the HAEA. Their tasks include advising on important strategic questions, as well as on research and development regarding nuclear safety.

The HAEA regularly reviews and assesses the operation of the licensees and the safety and security performance of the facilities. The essential part of the review and assessment activity is the investigation of events occurring at nuclear facilities.

Licensing of the management of SF and the establishment of a disposal site is a complex procedure with a number of authorities involved. According to the regulations, the main authority is the National Public Health and Medical Officer Service, which is responsible for licensing radioactive waste management facilities. Other distributions of authority include:

- radioactive and nuclear substance registry and control, as well as certain safety elements of installations and radioprotection, are under the supervision of the HAEA;
- environmental and water quality protection aspects are enforced by the regional environmental authority (as the establishment of a nuclear waste disposal site requires an Environmental Impact Assessment, which is to be conducted by the environmental authority);
- geological safety is under the supervision of the mining authority.

4 Siting procedures

For deep geological disposal of SF, no particular site has been selected in Hungary. A national screening procedure was conducted in the early 2000s, which resulted in the selection of six geological formations (with different lithology, such as clay, siltstone and granite). The most advantageous seems to be the Lower Permian Boda Siltstone Formation, with an area of 150 km^2 and thickness of nearly 1,000 m. It is mainly homogeneous and shows fine impermeability (Konrád and Hámos 2006). In 2004, field investigations started but only lasted a few years; they had to be stopped because research capacity was concentrated on the LILW final repository, Bátaapáti, which has since been built and began operation in 2012.

4.1 Procedures and criteria for site selection

The screening was a combined, two-level procedure. Within this process, all registered Hungarian formations were considered.

In the first phase (negative screening phase), evidently unsuitable formations were dropped; e.g. limestones or formations with thickness less than 300 m. By the end of this phase, 20 geological formations remained in 32 territorial units. These were subject to a deep investigation in the next phase.

The second phase (detailed assessment phase) was a qualification with uniform evaluation aspects. A total of 62 properties were analysed and each was assigned a weight number reflecting its importance. Evaluation covered qualities such as thickness, homogeneity, porosity, permeability, stability, phenomena of tectonic origin, hydrogeology, cover and underlayers, seismicity and volcanic activities.

After evaluation, each formation was put in one of the following categories: (1) recommended for further research, (2) could be considered, (3) not recommended for further research. Six formations with nine territorial units were classified in the first category, with the Boda Siltstone being the most ambitious.

5 Information and participation

Research reveals that there has been a turnaround in the social acceptance of radioactive waste management facilities over the past decade (Vári and Ferencz 2006). This means that - along the lines of the developed Western countries - the purely technically-oriented, hierarchical approach has been replaced by a more democratic one. The most important element of this approach is to ensure concerned municipalities a generous compensation and some degree of control. However, Vári and Ferencz's analysis also shows that important components of developed Western countries' radioactive waste policy, e.g. transparency of decisions, a broad social debate, crystallized national energy policy and an agreed-upon strategy for radioactive waste management, have not been included in the Hungarian experience.

Information distribution and public engagement is mainly conducted through the municipal associations. Their main task is participation in safety oversight of nuclear facilities and provision of up-to-date information on important events in course of the siting, planning, construction, and operation of the facility to local populations.

The national programme has brought no change in this status quo. Although Article 12 point (j) of the 2011/70 Euratom Directive prescribes the development of a transparency policy to ensure access to information and effective public participation in the decision-making process, the national programme neglects to include this policy. It also lacks a targeted strategy to engage the general public regarding transparency; the national programme only discusses interaction with the municipal associations.

5.1 Compensation mechanisms and socio-economic impact

The Atomic Energy Law provides for support to associations of local governments regarding their access to information and control activities. In this vein, government decree (214/2013), which was put into force on 1 January 2014, regulates the distribution of supporting funds, their transfer and their accounting. The municipal associations can be established around nuclear facilities (or future facilities) and are supported annually from the Central Nuclear Financial Fund. Currently, four municipal associations exist and represent 34 local governments: the Isotope Information Association (ITT) around the Püspökszilágy RWDTF facility (for non-NPP LILW); the Social Association for Control and Information (TETT) around Bátaapáti (final repository for NPP LILW); the Social Control and Information Association (TEIT) around the ISFSF; and the Western-Mecsek Social Information Association (NYMTIT) around the possible future disposal site for HLW and SF. Their combined support exceeds 1 billion HUF (3.3 million EUR) per year, which is distributed according to a particular algorithm described in the government decree.

By law, this financial support was originally intended to cover costs of the municipalities' dissemination activities regarding the nuclear facility in their vicinity. However, local governments have also used this money for development within their municipalities; at the time, this was clearly against regulations. A

2005 official investigation of the State Audit Office of Hungary revealed this practice and recommended stopping it (Állami Számvevőszék 2005). However, the reaction of the Hungarian Parliament was quite different: they amended the law and allowed funds to be spent on regional development as well.

Thus, since this amendment, the financial support acts as a kind of compensation for municipalities, which accept the operation of a radioactive waste facility in their vicinity. However, this may be questionable in the case of NYMTIT, as there are not yet operating facilities in the vicinity. The local governments within NYMTIT get "compensation" for accepting research activities. In the long-term, the local government of Boda (a member municipality of the NYMTIT) was promised that they may hold a referendum to decide whether or not to accommodate the planned underground research laboratory and subsequently – if tests prove the suitability of the rock formation – the repository itself. This is only a promise, however, and not a law (Vári and Ferencz 2012).

6 Costs and financing

The Central Nuclear Financial Fund (CNFF) was established in 1998 and based on the Atomic Energy Law. The CNFF is a treasury fund, which means that its budget, including annually planned expenditures and income, is part of the state budget. Until 2014, the manager of the Fund was HAEA; after 2014 it has been managed by the Ministry of National Development.

The CNFF is dedicated to financing the final disposal of low, intermediate and high level radioactive waste (including spent fuel), as well as the interim storage of SF and the decommissioning of the NPP. It also covers the costs of operation of PURAM. The CNFF's income is provided by the users of nuclear energy. The biggest contributor to the fund is the Paks NPP itself, which provides around 90 percent of the fund's annual income.

It is worth mentioning that there was no money allocated for waste management in the first 16 years of operation of the Paks NPP. For the total calculated expenditures of the CNFF, see table 3.

The plan – which does not include the planned new units at the Paks NPP – shows, that slightly less than half of the CNFF will finance the expenditures concerning spent nuclear fuel and nearly a quarter will be used for the decommissioning of the operating four units of the Paks NPP. As the Bátaapáti LLW-ILW final repository has only been in operation since 2012, most of its costs do not appear in the plan.

Table 3: Total expenditures for the activities to be financed from the CNFF for the four operating blocks, until 2084

Total expenditures until 2084		
In millions of EUR, including VAT		
Bátaapáti LLW-ILW final repository		**258.36**
Construction, expansion	68.77	
Operation	100.07	
Conditioning	23.75	
Shut down, monitoring	65.76	
HLW and SF final repository		**2,443.53**
Preparation	190.59	
Construction	962.51	
Operation	1,069.39	
Conditioning	-	
Transportation	-	
Shut down, monitoring	221.03	
SF interim storage		**395.86**
Construction, expansion	197.65	
Refurbishment	1.7	
Operation	196.51	
Püspökszilágy LLW-ILW final repository (no NPP waste)		**152.62**
Safety enhancement	7.43	
Shut down, monitoring	45.20	
Operation	99.98	
Decommissioning Paks NPP and SNF interim storage		**1,267.77**
Other costs		**893.0**
Compensation for municipalities	322.08	
Fee of manager of the Fund	32.55	
Fee for supervision	197.66	
PURAM operation costs	340.70	
Total		**5,411.15**

Source: PURAM 2014

In 2014, the income of the fund was 32,226 million HUF (~104 million Euro), while the expenditures were 12,493 million HUF (~ 40 million Euro). The fund thus increased by 19,733 millions of HUF (~ 64 million Euro). The total amount in the CNFF was 242,975 million HUF (784 million Euro) at the beginning of 2015.

The methodology of the calculations regarding expenses is based on an algorithm which was determined in 2000 by the Expert Committee of the then-existing National Atomic Energy Committee (OAB 2000). By law, the state annually sets an interest rate on the reserves of the fund, which is connected to the prime rate of the central bank. Thus, inflation is taken into account.

It is worth mentioning that plant lifetime extension (PLEX) was already taken into account in calculating the annual payments, although none of the units were licensed for PLEX at the time (and to date, the first, second and third unit have received their PLEX licenses). While the total income of the CNFF, calculated on basis of the total operating time of the Paks NPP, significantly increased, the necessary annual payments slightly decreased. The high annual payments for 2012-2017 calculated in earlier plans were no longer necessary, due to the longer operating time.

The tasks that are to be financed from the CNFF fall within the state's responsibility. If the amount in the CNFF proves to be insufficient, the state is obliged to pay. According to the recent status of the CNFF, the satisfactory operation of the fund is guaranteed only as long as there is a major contributor to the fund, namely, a NPP. It is unclear whether the state budget could cover the expenses in the event that there is no contributor to the fund (as the NPP is shut down); it is clear that high expenditures will continue to appear year by year.

7 Conclusions

The issue of nuclear waste management within the radioactive waste discussion is relatively young in Hungary. Establishing a national HLW disposal site is a general aim; however, the question of SF seems to remain tied to Russia. Shipment of SF to Russia (for reprocessing) happened in the past and is still considered an option, even though economic analyses show that it is a costly alternative.

The national programme for radioactive waste management leaves this question open, with a wait-and-see approach. Subjects such as reversibility and retrievability do not appear in the discussion framework; these subjects seem to be steps ahead of where the issue of SF is right now. Currently, interim storage of SF is applied, while the first phases of geological research are now being carried out in the area found to be most advantageous for a final repository according to the screening process.

Financing of radioactive waste management happens through a budgetary fund (CNFF). It is fairly transparent, and its incomes and expenses can be followed through PURAM's activities and reports, as well as through the balance-sheets of the Hungarian State Treasury. However, as the 2005 report of the State Audit Office of Hungary found, money in the CNFF is in practice not allocated on a separate bank account, and its reserves are continuously used by the government for other purposes. This means that the significant costs of HLW final disposal will be a heavy financial burden for future generations.

The Council Directive 2011/70/EURATOM prescribes for Member States the preparation of a national policy and a national programme. The Hungarian national policy, which was adopted by the Parliament in May 2015, states that no final decision has been reached yet regarding SF management, and therefore besides interim storage, the selection of a DGD site has to be conducted. The Hungarian national programme was prepared by the government and submitted to the European Commission in August 2015. Its final version was adapted by the government in August 2016.

Since the actual building of the disposal site will happen in decades, public participation is not a general practice regarding this issue. It is represented by the municipal associations comprised of local governments, whose primary current task is providing information to the public on the research activities.

References

Állami Számvevőszék (2005). *Jelentés a Központi Nukleáris Pénzügyi Alap működésének ellenőrzéséről.* 0509. [State Audit Office of Hungary: Report on the monitoring of the operation of the Central Nuclear Financial Fund]. https://asz.hu/storage/files/files/%C3%96sszes%20jelent%C3%A9s/2005/0509j000.pdf?ctid=758, last accessed 22 February2017.

HAEA (2014). *Nemzeti jelentés. Ötödik jelentés, készült a kiégett fűtőelemek kezelésének biztonságáról és a radioaktív hulladékok kezelésének biztonságáról szóló közös egyezmény keretében.* [National Report. 5th Report based on the convention on the safety of radioactive waste and spent fuel management]. http://www.haea.gov.hu/web/v3/OAHPortal.nsf/77443B9F6801CC1AC1257E59003ABB7D/$FILE/jc_nj_5_OAH_v12.pdf, last accessed 22 February 2017.

HAEA (2015). *HAEA organigram.* http://www.oah.hu/web/v3/HAEAportal.nsf/C3EE3EDD33E78576C1257C6F004CFA71/$File/HAEA%20organigram20150610.pdf, last accessed 22 February 2017.

IAEA (2003). *Report of the Expert Mission 'To Assess the Results of the Hungarian Atomic Energy Authorities Investigation of the 10 April 2003 Fuel Cleaning Incident at Paks NPP'*, 16 - 25 June 2003, Paks. http://www.oah.hu/web/v3/HAEAportal.nsf/8CC2F7526765592BC1257C5C003815CC/$FILE/iaeaexpertmission2003.pdf, last accessed 22 February 2017.

Konrád, G. and Hámos, G. (2006). A magyarországi nagy aktivitású radioaktív hulladéktároló telephely kijelölésének földtani szempontjai és az eddigi kutatások eredményei. [Geological aspects of determining high activity radioactive waste depository sites in Hungary and the results of the recent research]. In: *Acta GGM Debrecina*, 1(issue), 33-38.

NP (2016). *Magyarország nemzeti programja a kiégett üzemanyag és radioaktív hulladékok kezelésére.* http://www.kormany.hu/hu/dok?page=16&source=9&year=2016#!DocumentBrowse, last accessed 22 February 2017.

OAB (2000). *A Központi Nukleáris Pénzügyi Alapból finanszírozandó tevékenységek hosszú távú terveinek és a vonatkozó költségbecslés kialakításának szabályai.* [Hungarian Atomic Energy Committee: Rules of developing long term plans and cost estimation related to activities financed from the Central Nuclear Financial Fund].

OECD (1995). *Environmental and ethical basis of geological disposal of long lived radioactive wastes.* Paris. https://www.oecd-nea.org/rwm/reports/1995/nea120-environmental-ethical-aspects.pdf, last accessed 22 February 2017.

PURAM (2010). *Megalapozás a hazai radioaktív hulladékok és kiégett fűtőelemek kezelésének és elhelyezésének új programjához.* [Foundations to the new program of domestic radioactive waste and spent fuel management and disposal].

PURAM (2014). *Az RHK Kft. tizennegyedik közép- és hosszú távú terve.* [14th middle and long term plan of PURAM]. http://www.rhk.hu/docs/14_KHTT_m.pdf, last accessed 22 February 2017.

SEA (2016). *National Program of Hungary on the Management of Spent Fuel and Radioactive Waste, Strategic Environmental Assessment.* http://www.kormany.hu/download/5/93/a0000/Nat_Progr_rad_waste_SEA_EnvRep_HU.pdf, last accessed 22 February 2017.

Vári A and Ferencz, Z. (2012). *Identifying remaining socio-technical challenges at the national level: Hungary.* Insotec project. https://docs.google.com/viewer?a=v&pid=sites&srcid=aW5zb3RlYy5ldXxpbnNvdGVjfGd4OjFiOGNmODc2ZTRmNzEyMzQ, last accessed 3 June 2015.

Vári, A. and Ferencz, Z. (2006). Fordulatok és kérdőjelek a radioaktív hulladékok hazai kezelésében. In: *Társadalomkutatás*, 24(4), 469-492.

Progress on Nuclear Waste Management

Nuclear Waste Governance in Lithuania

Povilas Poskas[1]

Abstract

There is only one nuclear power plant in Lithuania – the Ignalina NPP. Both of its units are permanently shut down and under decommissioning. Unlike many other nations with nuclear power plants, Lithuania never operated a research reactor. During the Ignalina NPP's operation from 1983 until 2010, roughly 21,500 spent nuclear fuel (SNF) assemblies were accumulated. A portion of this fuel has already been transferred to a facility for dry interim storage in casks for 50 years and the remaining portion will be transferred to the new facility that is currently under construction.

Since 2003, two near surface repositories (NSRs) have been under implementation near the Ignalina NPP: a landfill facility for 60,000 m^3 of Very Low Level Radioactive Waste and a NSR facility with engineering barriers for 100,000 m^3 of Low and Intermediate Level Radioactive Wastes. In the site selection process for the NSR facilities, Lithuanian authorities concluded that public acceptance and good relations with neighbouring countries are as important as geological criteria.

Possibilities for disposal of SNF in a geological repository were analysed in 2001–2004 with support from Swedish experts. Because of these investigations, crystalline rocks and clay formations were declared as prospective for SNF disposal. In parallel, a repository in crystalline rocks was proposed and a related generic safety assessment of this repository concept was performed. Thus, at the moment the Lithuanian SNF disposal program is in the initial site investigation and preliminary facility design stage. It should be emphasized that existing knowledge has largely been gained from international cooperation.

Regarding EC Directive 2011/70/EURATOM, a new national program on the management of spent nuclear fuel and radioactive waste has been prepared and approved by the government in 2015. As initial activities, more detailed planning and preparation of a research program for implementation of the geological repository have been proposed. The intention is to continue investigation of clay and rock formations with support from safety assessments.

---

1 Lithuanian Energy Institute, Breslaujos g. 3, LT-44403 Kaunas; E-Mail: povilas.poskas@lei.lt

A. Brunnengräber et al. (eds.), *Challenges of Nuclear Waste Governance*,
Energiepolitik und Klimaschutz. Energy Policy and Climate Protection,
https://doi.org/10.1007/978-3-658-21441-8_7

1 Introduction

There is only one nuclear power plant (NPP) in Lithuania—the Ignalina NPP. It is situated nearby the town of Visaginas in northeast Lithuania, near the borders of Latvia and Belarus (Figure 1) and on the bank of the largest Lithuanian body of water, Druksiai Lake (Figure 4). Lithuania never operated a research reactor.

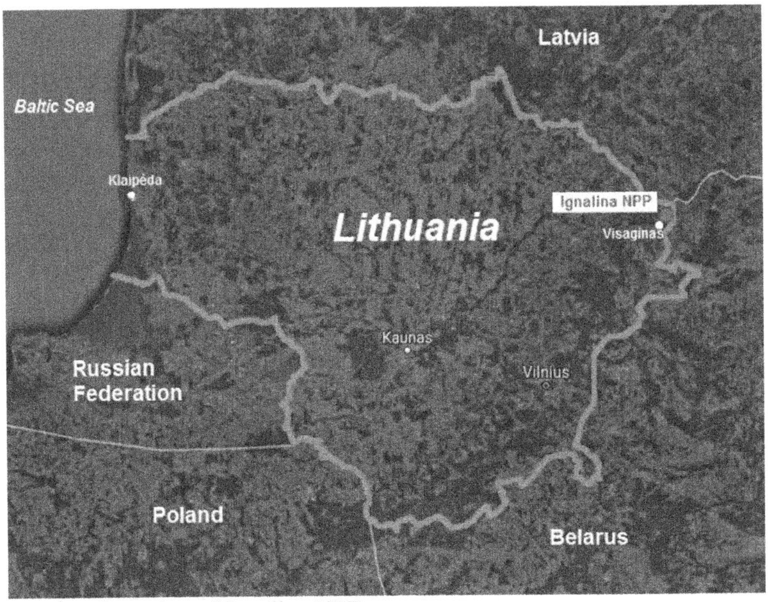

Figure 1: Location of Ignalina NPP.

The Ignalina NPP possesses two similar light water cooled graphite moderated reactors, also known as RBMKs. Using light water for cooling and graphite for moderation enables the use of low-enriched uranium for fuel. The RBMK-1500 is the latest and most advanced version of this type reactor design. The first grid connection for Unit 1 was achieved 31 December 1983, and for Unit 2 at 20 August 1987. Original design lifetime for these reactors was projected to end in 2010–2015.

The Ignalina NPP during operation was a vital component in Lithuania's energy balance; it generated more than 70 percent of the total electricity in Lithuania (including energy exports). There were a variety of reasons for this high percentage, but the main one was that nuclear power had a significantly lower production cost than other forms of power under the economic and technical conditions in Lithuania. After the Chernobyl nuclear accident, safety systems at the Ignalina NPP were re-evaluated and authorities decided to decrease the maximum thermal power of the units from 4,800 to 4,200 MW. This limited the maximum electric power to about 1,250 MW per unit. On 5 October 1999, the *Seimas* (the Parliament of Lithuania) approved the National Energy Strategy, in which it was stated that the Ignalina NPP's first unit would be shut down before 2005. This decision took into consideration substantial long-term financial assistance from the European Union (EU), G7 and other states as well as international institutions. On 10 October 2002, the *Seimas* approved an updated National Energy Strategy in which it was stated that the first unit would be shut down by 2005 and the second unit by 2009 if funding for decommissioning was available from the EU and other donors. The units were shut down on 31 December 2004 and 31 December 2009, respectively.

The following sections present a historical background and review progress in the disposal of nuclear waste and activities related to the implementation of EC Directive 2011/70/EURATOM in Lithuania.

2 Nuclear waste disposal in Lithuania

2.1 Old Soviet time facilities

A storage facility exists at the Ignalina NPP for bituminized evaporator bottoms of liquid radioactive waste generated there. This is an aboveground, two-storey assembled concrete monolith building (Building 158) consisting of 12 inside steel-lined vaults for loading the bitumen compound. A preliminary long-term safety assessment of this facility performed in 1998 concluded that this radioactive waste facility could be converted into a disposal facility if a multilayer earth cover was used, but a more detailed analysis is necessary. At present, this facility is licensed as a storage facility, based on the Safety Analysis Report (SAR) prepared in 1999–2000.

There is also an old "Radon" type facility for institutional waste at Maisiagala (40 km from the capital Vilnius). This repository, built in 1964, was designed for institutional waste and is a typical of the "Radon" type facilities constructed in the early 1960s and used throughout the former Soviet Union. It was closed in 1989. Waste at this site was disposed of in a reinforced concrete vault with internal dimensions 14.75 × 4.75 × 3 m (200 m^3 in volume). In addition, sealed sources were deposited in two stainless steel containers, each with a volume of 10 litres. Medical sources were deposited with a biological shielding. At the end of the operation, the residual volume was filled with concrete and sand.

Based on updated information, in 2005-2006 a SAR for the existing disposal facility was prepared and proposals for safety improvements were made. After review of the SAR, by regulatory authorities, it was decided to upgrade and license the Maisiagala facility as a radioactive waste storage facility because safety indicators for a disposal facility could not be met. The state nuclear power safety inspectorate (VATESI) licence was issued in 2006. Additionally, there are plans to retrieve waste from this facility and dispose of it in the NSR currently under design at the Sabatiske site (close to the Ignalina NPP). The retrieved sealed spent sources will be transferred to the Ignalina NPP site for interim storage.

2.2 New near-surface disposal facilities

Lithuanian legislation states that very low level short-lived waste could be disposed of in a simple near surface repository of landfill type (VATESI 2010). Here, treated and untreated radioactive waste that met acceptance criteria could be placed. Activities related to the set up of a landfill repository for very low level waste at the Ignalina NPP site began in 2003, with preparation of a reference design and site selection. The design and construction contract for buffer storage facility and for landfill repository design was signed at the end of 2007. Implementation of the project is ongoing. The total volume of the landfill repository will be 60,000 m^3. The waste will be disposed of by campaigns (4,000 m^3 of Very Low Level Radioactive Waste per campaign). Between campaigns, the waste will be stored in the buffer storage facility. Operation of the landfill repository is planned for 2018–2038. After closure, active passive control will be performed until 2068 and 2138, respectively.

The activities related to the implementation of a near-surface repository (NSR) for low and intermediate level (short-lived) waste in Lithuania commenced in 2002 with preparation of the reference design (2002) and siting of the repository (2003–2007). Finally, the Sabatiske site for the NSR was approved by the Lithuanian Government (11 November 2007). The design project has been under development since 2009 by the consortium: AREVA (France), ANDRA (France), the Lithuanian Energy Institute (Lithuania), Specialus Montazas-NTP (Lithuania), Pramprojektas (Lithuania). The total volume of the near surface repository will be 100,000 m^3. Operation of the near surface repository is planned for 2022–2038. After closure, active and passive control is foreseen until 2138 and 2338, respectively.

2.3 Interim storage of SNF and long-lived waste

Lithuania's legislation forbids reprocessing of SNF in Lithuanian territory, but does not disallow reprocessing SNF in other countries and returning the secondary waste back to Lithuania. However, SNF reprocessing has no relevance in the current conditions because the use of reprocessed SNF is limited and secondary waste must be treated similarly to SNF (i.e. disposed of in a deep geological repository).

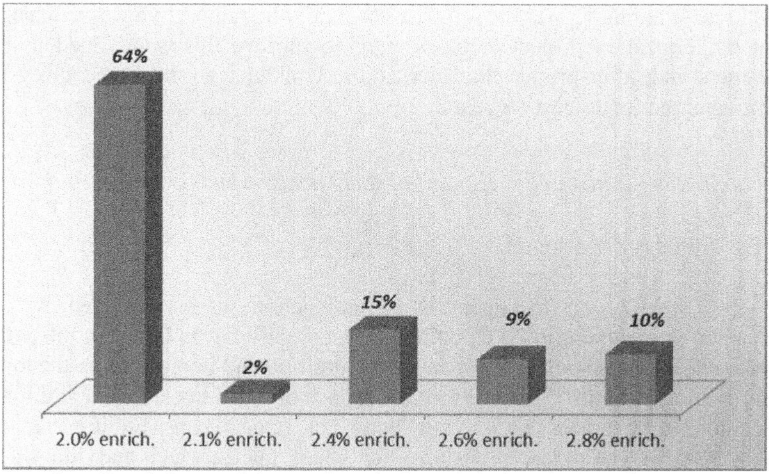

Figure 2: Breakdown of the nuclear fuel used during the Ignalina NPP's lifetime

Nuclear fuel with different U-235 initial enrichment, from 2 percent to 2.8 percent and with and without erbium absorber was used during the NPP's operation (Figure 2). By 2010, 21,571 SNF assemblies that contain about 2,400 tons of uranium (110–112 kg per assembly) had been accumulated.

In 1992, the decision was made to build an interim dry SNF storage facility at the Ignalina NPP site. After the tendering process, GNB Gesellschaft für Nuklear Behälter mbH casks were chosen. Twenty of these casks are ductile cast iron CASTOR RBMK-1500 casks; the remaining one hundred are metal-concrete CONSTOR RBMK-1500 casks. The facility design operation lifetime is until 2050.

As indicated above, 21,571 SNF assemblies had been accumulated at the Ignalina NPP as of 2010; 6,016 of these were accommodated within the existing SNF dry-storage facility at the Ignalina NPP. To accommodate the remaining 15,555 SNF assemblies, a new interim storage facility was needed. After the tendering process in January 2005, the contract for this new facility was awarded to Consortia GNS Gesellschaft für Nuklear-Service mbH and RWE NUKEM GmbH, Germany. The Lithuanian Energy Institute (LEI), as the local sub-contractor, is supporting the Consortia in the environmental impact assessment program and report, and also preliminary and final safety analysis reports preparation and licensing. Within this facility, the SNF will be loaded in the higher-capacity CONSTOR RBMK-1500/M2 casks.

The operational long-lived ILW is stored in large vaults within concrete structures (Building 157) at the Ignalina site. The solid radioactive waste management modernization project includes plans to retrieve this waste, load it into containers and, after proper characterization, transfer it to the new facility for interim storage for at least 50 years.

2.4 Activities related to the disposal of spent nuclear fuel and long-lived waste

2.4.1 Strategy for disposal

The first strategy on management of radioactive waste approved by the Lithuanian government in 2002, defined several activities to be taken related to spent-nuclear-fuel disposal. These included drafting and implementing the long-term research program, "Possibilities to dispose of spent nuclear fuel and long-lived radioactive waste in Lithuania" amd analysing the possibilities for: a.) having a deep geological repository for spent nuclear fuel and long-lived radioactive waste, b.) creating a regional repository through joint efforts with neighbouring countries, c.) disposing of spent nuclear fuel in other countries, and determining the cost and justification for such disposal, and d.) prolonging the storage period for interim storage facilities by 100 years or more.

The Research Program for Assessment of Possibilities for Disposal of Spent Nuclear Fuel and Long-lived Radioactive Waste for the Years 2003–2007 was prepared and approved in 2003. In parallel, the Swedish Ministry of Foreign Affairs allocated special funding to support activities in Lithuania related to the closure of the Ignalina NPP. One of the subject areas identified for support was the development of national competence on issues related to disposal of SNF.
In 2008, the Strategy for Radioactive Waste Management was revised and subsequently, the national research program for 2008–2012 was developed and approved (Government Resolution No. 860).

Regarding EC Directive 2011/70/EURATOM, a new National Program for the Management of Spent Fuel and Radioactive Waste was prepared and approved by in 2015 (Government Resolution No. 1427). The ultimate strategic goal of the program is safe management of all radioactive waste and spent nuclear fuel available in Lithuania, which includes protection of people and the environment from the harmful effects of ionizing radiation and avoiding the imposition of undue burden on future generations. Storage of spent fuel and radioactive waste, including long-term storage, is an interim solution and thus cannot be an alternative to disposal.

The first objective of the program is to reduce generation of radioactive waste. Legislation requires reducing the volume of radioactive waste as practically and reasonably possible. Minimization of waste generation should be achieved via waste clearance (reuse of materials, devices and equipment that have been contaminated with radionuclides or disposal as non-radioactive waste).

The second objective of the program is to achieve a high level of nuclear and radiation safety and environmental protection for spent nuclear fuel and radioactive waste. The actions to be taken concerning short-lived low- and intermediate-level radioactive waste have been described above. Long-lived low- and intermediate-level radioactive waste and spent sealed radiation sources will be separated from short-lived wastes and loaded into appropriate containers. The containers holding long-lived radioactive waste will be stored in a long-lived waste storage facility. Graphite from the dismantled reactors will be moved to the storage facility between 2022 and 2038. The waste will be stored up to 2066 and then is to be disposed of in a geological repository. The design lifetime of the existing dry storage facility for spent nuclear fuel is until 2050 and that of the new facility is until 2067. After the storage period, the spent nuclear fuel should be disposed of in a geological repository. Since the operation period of the existing spent nuclear fuel storage ends earlier than the planned commissioning of the geological repository, the possibility of extending the storage period of the dry storage for spent nuclear fuel will be analyzed. This analysis program is expected to conclude in 2025.

The third objective of the program is to ensure sustainable management and long-term safety of spent fuel and long-lived radioactive waste. Laws prohibit the processing of spent nuclear fuel in Lithuania. Spent nuclear fuel can be recycled abroad, with the resulting secondary waste returning to Lithuania. However, this solution is not economical and the resulting secondary long-lived high-level radioactive waste has to be managed the same way as spent fuel is. As the storage of spent nuclear fuel and radioactive waste is only a temporary solution, the spent fuel and long-lived radioactive waste eventually has to be disposed of in a geological repository. The geological repository will be needed before the end of the spent nuclear fuel and long-lived radioactive waste storage period (2050 - 2067). Usually the installation programs for geological repositories (research and development, site selection, construction) last approximately 30 years. The National Programme for the Management of Spent Fuel and Radioactive Waste foresees that the geological repository in a suitable geological formation in Lithuania will start operation in 2067. Therefore, in 2015 – 2016, a repository development project was established.

2.4.2 Development of the disposal concept and generic safety assessment studies

During 2001-2004, the Lithuanian Energy Institute and supporting Swedish experts were developing the repository concept (LEI 2005a) and the generic safety assessment (LEI 2005b) of a repository in crystalline rock.

A detailed repository design is highly specific to waste type and geological conditions. For all waste types, construction of access and emplacement shafts and tunnels involves the excavation of a substantial underground facility. This includes the removal of several hundreds of thousands of cubic meters of rock, to possibly millions of cubic meters for larger waste disposal programs. Geological repositories currently being considered have underground dimensions varying from a few square kilometres to roughly twenty square kilometres, depending on the inventory of waste, its thermal output, and the repository design.

The proposed repository concept for Lithuania is based on the KBS-3 concept developed by SKB for disposal of SNF in Sweden. The KBS-3H design with horizontal canister emplacement has been proposed as the reference design for Lithuania. The scheme of the repository is shown in Figure 3.

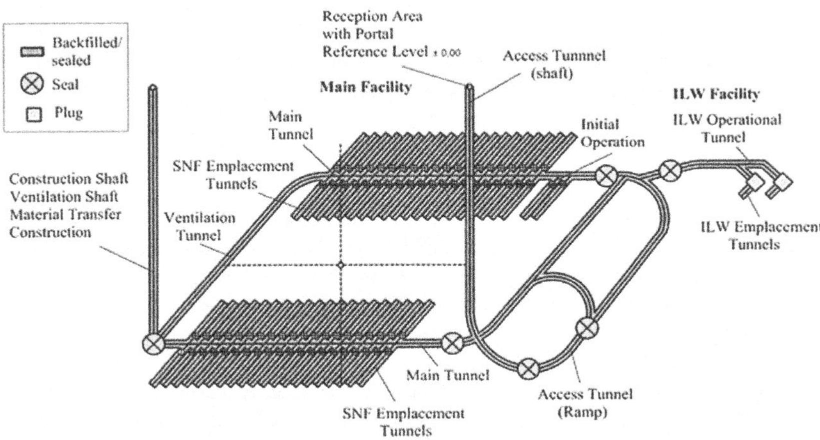

Figure 3: The scheme of the repository after final sealing and closure of the facility (LEI 2005a)

The repository concept is described at the level of detail needed to perform a generic safety assessment and cost analysis. This generic repository concept for SNF disposal in crystalline rock in Lithuania envisions waste disposal in horizontal emplacement tunnels with a diameter of 1.85 m and length of 250 m. The repository would be constructed in the crystalline basement at a depth of 300-500 m. Disposal canisters with RBMK-1500 SNF would be emplaced with 1.2 m between the canisters. A copper canister is assumed for the disposal of SNF in the basement rock. The proposed canister for SNF is designed with two components: an outer corrosion protection of copper and a cast iron insert with channels for the RBMK-1500 SNF half-assemblies. This is to improve the mechanical strength of the canister. The copper canister has a wall thickness of 50 mm and will be made of oxygen-free copper with low phosphorus content. The canister insert will be made of cast iron with a minimum wall thickness of 50 mm. Preliminary data for the RBMK-1500 SNF reference canister propose a 1,050-mm diameter and a 4,070-mm length. One disposal canister can hold 32 RBMK-1500 fuel half-assemblies, and for Lithuanian SNF disposal, about 1,400 canisters would be necessary. Thus, the required area for the repository construction would be about 0.4 km^2.

A generic safety assessment of the repository is presented in Volume 3 of Investigations of Possibilities (2005). Because of assumed similarities in the repository environment and repository concept, the selection of scenarios was based on the safety assessment performed in Sweden. For this stage of generic safety assessment, only two scenarios were chosen: a base scenario and a canister defect scenario. In performing this safety assessment, Lithuanian parameter values were used to the extent possible.

To analyse system evolution under the above-mentioned scenarios, a thermal evolution, criticality, and dose assessment of the copper canister loaded with RBMK-1500 SNF were performed. Analysis of thermal-evolution effects at the top of the crystalline rock showed that over a period of one million years after deposition of the spent fuel, the heat released from the canister with RMBK-1500 SNF will have only a marginal impact on thermal conditions at the top of crystalline rock.

An essential component of the safety assessment was to calculate radionuclide release and doses to critical group members. The computer codes COMPULINK7 and CHAN3D, provided by SKB, were used to calculate radionuclide release through the near- and far-field regions from a canister with an initial defect (canister defect scenario). For dose assessment, the computer code AMBER 4.4 (UK) was also used. These calculations showed that the majority of the analysed radionuclides identified as safety concerns will be effectively retarded in the near-field (by the bentonite buffer). Release rates of Cs-135, I-129, Tc-99, and Ra-226 are the most significant. The near-field release rates are dominated by Cs-135 and I-129. Over an extended period, the release rate is dominated by Ra-226, which is formed by an in-growth from the chain decay of U-238. Modelling of the transport of radionuclides through the far-field was performed for the radionuclides most dominant in the near-field. The values of parameters that influence on radionuclide transport were selected from the study area in Lithuania. When they were not available, values from the Beberg site (in Sweden) were used. Results of total dose behaviour demonstrated that the dose constraint of 0.2 mSv/y will not be exceeded in a period of a million years and, in fact, will be lower than this constraint by two orders of magnitude. The total dose rate for the initial period is dominated by non-sorbing radionuclide I-129; later, it is dominated by Ra-226.

Subsequently, modelling studies were performed to gain a better understanding of these processes in the repository and in the surrounding environment (Justinavicius et al. 2014; Narkuniene et al. 2015; Poskas et al. 2014).

International cooperation was critical in this process. An effective catalyst of competence development was a multiyear cooperation project with the Swedish International Project for Nuclear Safety (SIP) ("Competence development in the area of SNF disposal", 2001-2004). Later, to ensure effective transfer of knowledge and practices from other countries, the Lithuanian Energy Institute participated in various research projects coordinated by the International Atomic Energy Agency (IAEA) and took advantage of IAEA fellowships, training courses and workshops. To participate in the European research community on nuclear issues, the Lithuanian Energy Institute took part in the EC 7th Framework Program and in HORIZON 2020 projects.

3 The legal and institutional framework

3.1 The legal framework

Lithuania has established an appropriate legislative and regulatory framework to govern the safety of spent fuel and radioactive waste management. All legal acts concerning spent fuel and radioactive waste management are prepared according to the best in-country and international practices, including IAEA recommendations. This legislation covers all areas of spent fuel and radioactive waste predisposal management and disposal of very low level waste and low and intermediate level wastes (VATESI 2014).

A list of the primary legal acts regulating the management of spent nuclear fuel and radioactive waste in Lithuania include the laws on: Management of Radioactive Waste (1999, last amended 2014), Nuclear Energy (1996, last amended 2014), Radiation Protection (1999, last amended 2014), and Ratification of the Joint Convention on the Safety of Spent Fuel Management and on the Safety of Radioactive Waste Management (2003). Various regulations were developed addressing nuclear safety and disposal of wastes, including:

- Nuclear Safety Requirements BSR-3.1.1-2010, the General Requirements for Dry Type Storage for Spent Nuclear Fuel (2010) and Nuclear Safety Requirements BSR-3.1.2-2010, Regulation on the Pre-disposal Management of Radioactive Waste at the Nuclear Facilities (2010).
- Regulation on Disposal of Low and Intermediate Level Short Lived Radioactive Waste P-2002-2 (2002).
- Regulation on Disposal of Very Low Level Radioactive Waste P-2003-02 (2003);
- General Radioactive Waste Acceptance Criteria for Disposal in Near Surface Disposal Facilities (2009).
- Order of the Minister of Health and the Head of the State Nuclear Power Safety Inspectorate No. V-1271/22.3-139 On the Rules of Radioactive Substances, Radioactive Waste and Spent Nuclear Fuel Import, Export, Transportation in Transit and inside the Republic of Lithuania (2008, amended 2012).

The basic provisions for the management of spent nuclear fuel and radioactive waste are established in the Law on the Management of Radioactive Waste. This law defines principles of radioactive waste management, the competence of authorities, the duties and responsibilities of the waste generator as well as of the radioactive waste manager and provisions for licensing.

The basic radiation protection and safety requirements, corresponding to IAEA recommendations and EU legal requirements, as well as the allocation of responsibilities to bodies involved in the different steps of spent fuel and radioactive waste management are established in the Law on Nuclear Energy and the Law on Radiation Protection.

The Law on Nuclear Energy and the Law on Nuclear Safety, in concert with regulations made under other laws, establish the licensing system for activities related to nuclear materials and nuclear cycle materials (their transportation, acquisition, etc.), as well as for nuclear facilities in the following life-stages: site evaluation, design, construction, commissioning, operation, and decommissioning. Supervision of the closed radioactive waste repository and the acquisition, storage, use and transport of nuclear or nuclear fuel cycle materials is also executed according to the afore mentioned laws.

A licence for the construction of a nuclear facility may be granted only if the parliament (in the case of the NPP) or the government (for other facilities) has adopted a legal act on the facility in question.

3.2 The institutional framework

The state nuclear power safety inspectorate (VATESI) is the competent authority for the licensing of activities involving nuclear materials and nuclear fuel cycle materials, as well as activities carried out in nuclear facilities within their legally defined lifetimes.

During site evaluation, VATESI is responsible for reviewing and assessing the site evaluation report. Positive conclusions in respect of this report must be made also by the Ministry of Health, the Civil Aviation Administration, the Lithuanian Geological Survey, the Lithuanian Hydro Meteorological Service and the Fire Prevention and Rescue Department. Before the design activities start, technical specification for design must be approved by VATESI. Design of a nuclear facility has to be performed and assessed according to the requirements established by the competent institutions, including VATESI, the Ministry of Environment (MOE), the Ministry of Health, the Ministry of Interior and other institutions involved according to the laws on Construction and on Nuclear Energy and their corresponding regulations.

Following the provisions set out in the Law on Radiation Protection, VATESI issues licences and permits for the nuclear energy area activities involving the sources of ionising radiation. These primarily involve a licence or a temporary permit to carry out activities under ionising radiation at a nuclear facility and a licence or a temporary permit to store, maintain and use sources of ionising radiation at a nuclear facility.

The Radiation Protection Centre (RPC) under the Ministry of Health is responsible for issuing licences for the transportation of radioactive waste and for small producers (waste producers excepting the operator of a nuclear plant) to manage institutional waste excluding disposal; to collect and sort radioactive waste; undertake its pre-treatment, treatment, and conditioning; and to store, recover and decontaminate it. In order to carry out the one-time transport of radioactive waste from small producers, a single permit is needed in addition to a licence issued by the RPC.

According to the Law on the Management of Radioactive Waste, sealed sources can be imported into Lithuania if there is a plan to return them to the supplier after their useful life. Additionally, the recipient must cooperate with the Radioactive Waste Management Agency (RATA) for the management of radioactive sources if, due to unforeseen circumstances, it is not possible to return them to the supplier. The recipient is thus obliged to insure the source for value of RATA services. For small producers, cooperation with RATA and insuring source is required before the use licence will be granted.

The MOE coordinates and methodically manages the environmental impact assessment for proposed national and transboundary economic activities. It decides whether the proposed economic activities are to be allowed in the selected site (since 2010, this function has been delegated to the Environmental Protection Agency (EPA) under the MOE) and takes part in state supervision and control of nuclear facility design and construction. Following the procedure prescribed by legislation, the MOE participates in the issuing of licences for radioactive waste management activities.

The EPA organises, coordinates and performs state environmental monitoring, and controls environmental monitoring of economic entities. It also exchanges monitoring information with other countries.

There are additional ministries and institutions that are involved in regulating some specific issues in radioactive waste management according to their competence; however, these institutions are not regulatory bodies as defined in this Convention. These institutions are the Ministry of Energy, the Ministry of Social Security and Labour, the Ministry of Transport and Communications, the Ministry of National Defence, the Ministry of the Interior, the State Security Department and the Governmental Emergencies Commission. The competence of these institutions is defined in the Law on Nuclear Energy.

4 Siting procedures and criteria for site selection

4.1 Experience from the NSR site selection

Activities related to site selection were started in 2003. At the first stage of the desktop studies (based on the assessment of geological, tectonic and hydrogeological conditions as well as principal selection criteria), three areas were selected near the Ignalina NPP. These were the Dysnai, Zarasai, and Visaginas areas. The principal criteria for selection were as follows:

- Legal (compliance with the legislation of Lithuania and its international agreements and treaties);
- Environmental (the repository must have minimal impact on the environment);
- Public acceptance (the project must be acceptable to local, national and international communities);
- Technical and safety aspects (the site must be geologically stable, and the natural phenomena and processes must have little effect on the engineered barriers and the waste);

- Location (the site must be located as close to the Ignalina NPP as possible in order to minimise the probability of accidents and costs related to transportation of waste).

At the second stage of the desktop studies, nine sites within these areas were selected for more comprehensive analysis. The terrain, topography, geomorphology, and infrastructure as well as the geological and hydrological structures of the sites were analysed and assessed. Main site selection criteria included:

- No possibility for flooding of the foundation, sufficient surface inclination so that water can drain away into a surface water body, with preference being for a large hill.
- High resistance to erosion, meaning a relatively smooth site with shallow water flow speed v below the critical speed v_{cr}
- Slope stability of friction material: 1.) Compressibility, compression strength, shear strength, and internal friction angle and stiffness (E-modulus) of the bottom bed shall comply with requirements for massive construction projects; 2.) Low pore water pressure; 3.) A maximum seismic intensity on the MSK scale <=6; 4.) Good constructability. Feasibility of excavation, homogeneous ground; 5.) Low hydraulic conductivity with a filtration coefficient k less than 10^{-7} m/s or even 10^{-9}m/s; 6.) A low and steady groundwater level at least 3 m below the bottom barrier to mitigate the risk of flooding, and; 7.) Short distance to the Ignalina NPP, and no transportation of waste through large settlements and protected or recreational territories.
- Favourable infrastructure and logistics.

The sites in the Dysnai and Zarasai areas were found to be less suitable because of their greater distance from the Ignalina NPP, higher population density, and relatively unfavourable public opinion. Ultimately, Apvardai and Galilauke in the Visaginas area (Ignalina District) near the Ignalina NPP were selected as the two most promising sites.

In 2004, an environmental impact assessment (EIA) study for a potential NSR to be constructed at both the Apvardai and Galilauke sites was developed. The EIA report was subject to approval by regulatory authorities and the local municipalities. The EIA law also requires that the comments and concerns of the local public are considered. The population of the Ignalina District was informed and consulted in many ways, including articles in local newspapers, broadcasting programs on local TV, and seminars and meetings. In 2005, all relevant institutions approved the study. However, the Municipality of Ignalina District approved the EIA study on the condition that measures would be taken to mitigate the negative social and economic impacts of the NSR facility. After analysing the municipality's demands, the Ministry of Economy recommended looking for an alternative site outside the boundaries of the Ignalina District, with preference given to Visaginas Municipality and the territory belonging to the Ignalina NPP. In 2005, a new search for a site in the territory of Visaginas Municipality was conducted and the Stabatiske site (Figure 4) was deemed to be the most suitable. The EIA report was updated to include the studies related to this new site. The majority of the local population in Visaginas district is familiar with nuclear activities and thus accepted the proposal to construct an NSR there.

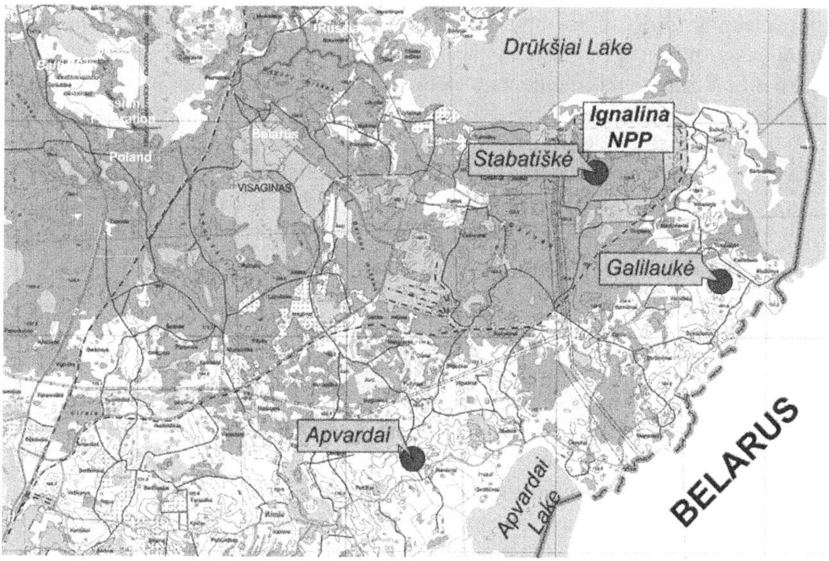

Figure 4: Location of the potential NSR sites. Source: Prepared by the author

Consultations with Latvia and Belarus began in late 2005 and continued through 2007. These neighbouring countries were informed about the site selection and the EIA in accordance with the provisions of the Joint Convention on the Safety of Spent Fuel Management and on the Safety of Radioactive Waste Management and the Espoo Convention on the Environmental Impact Assessment in a Transboundary Context. The Ministry of Environment in Latvia and the Ministry of Natural Resources and Environmental Protection in Belarus submitted their comments and proposals regarding the EIA report in 2006. The comments and proposals were discussed at bi- and trilateral meetings. Belarus was especially concerned about the Galilauke site due to its proximity to their border and, as a result, this site was wholly rejected. This issue became a relevant topic in the bilateral diplomatic relations and was discussed not only at expert levels, but also among prime ministers. Opposition to the Galilauke site was raised by Belarus in various international forums, including the IAEA. A special inter-governmental working group was established and met twice a year to discuss nuclear energy development and radioactive waste management issues. The Latvian Government and the general public did not show as much interest and concern as the Belarusian Government. The proposed sites are located some 10–20 km from the Latvian border.

In December 2005, an IAEA experts mission was organised to review the site selection process for the NSR in Lithuania. The experts made some comments, but in general emphasised that the site selection program met international safety requirements and that the proposed technology with waste packages being housed in concrete vaults sealed with clay was very reliable.

In June 2007, the MOE approved the EIA report and indicated that it agreed with the establishment of the NSR at either the Galilauke or Sabatiske sites, with priority given to the latter site. Based on this conclusion, the Lithuanian government approved the Stabatiske site for construction of the NSR in November 2007.

4.2 Selection of the possible geological formations for disposal of SNF

Based on prior investigations, several geological formations were identified as potentially suitable for a deep SNF repository (Kanopiene and Marcinkevicius 2000): rocks of the crystalline basement, Lower Cambrian clay, Permian sulphate deposits (anhydrite), Permian rock-salt, and Lower Triassic clay.

In 2001-2004, further investigations were performed by the Geological Survey of Lithuania and the Institute of Geology and Geography, with support of Swedish experts, to define suitable geological formations. The initial phase of the investigation, from 2001-2002, consisted of a general overview of the region's geological structure and the composition of the sedimentary cover and crystalline basement based on the most important geological parameters describing the suitability of formations (such as simple tectonic structure, absence of intraformation aquifers, low neotectonic and seismic activity, good isolation, and favourable mechanical properties of the rock). Following this, the aerial distribution of these favourable formations was defined. The screening was also based on an evaluation of the general geological parameters – such as lithological homogeneity of the prospective layer, thickness, depth and lateral extent, and tectonic structure – of the candidate formation.

After these investigations, experts concluded that rock salt did not have a high potential for SNF disposal. In contrast, the crystalline basement rocks were considered among the best candidates for a geological repository. The best prospects for the crystalline basement appeared to be located in the south-eastern part of Lithuania, where these rocks have only a 200–300 m thick sedimentary cover. The prospects for clayey formations were reduced to the Lower Cambrian Baltija Formation and the Lower Triassic, as they best meet the requirements for depth, thickness, lithological composition, and homogeneity of sediments.

In 2003, more detailed investigations of these geological media were made, as desktop studies alone are not a sufficient basis for making a prioritisation. The direct observation and characterisation of cores from reference wells were carried out for detailed evaluation of candidate formations, analysis of similarities, and differences between the cores of the different zones at different depths. At the same time, a sampling for a preliminary analysis of mechanical properties was carried out leading to the selection of the Lower Triassic formation as the first and the Cambrian Baltija Formation as the second priority for investigations of their clayey formations.

It was determined that an area of 100 km^2 of crystaline rock could be found between major fracture zones at an acceptable depth and with a normal fracture content to fulfil desired requirements, both with respect to tightness and stability. As there is also broad experience and competence concerning SNF disposal in crystalline rocks in Sweden, crystalline rocks were selected for further characterization in 2004.

Laboratory investigations were performed on crystalline rocks to gather general information about the strength parameters of these rocks as a function of different petrological compositions. Rock samples were taken from old cores collected in boreholes drilled in southern Lithuania during deep geological mapping at a scale of 1:200,000. Depths of the samples varied from 427 to 606 m and were selected as these are optimal values for deep repository construction. The main characteristics of the crystalline rocks, such as density, porosity, unconfined compression strength, and rock temperature, were collected from different literature sources. All the data were filed systematically and a simple statistical analysis was performed. Tectonics and hydrology of the crystalline basement were analysed using existing archival information.

The four-year long investigation concluded that the Lower Triassic formation was most suitable.

5 Information and participation

Information and participation in the environmental impact assessment process is based on application of several national and international legal documents (Ragaisis et al. 2014).

The Law on Environmental Impact Assessment is supported by dozens of regulations, which specify requirements for preparation of the EIA program and report, on informing the public and public participation in the EIA, on review of EIA documents at the MOE and subordinate institutions, and other miscellaneous topics, e.g., guidance on quality control for the EIA.

Annex 1 to the law defines that the EIA is obligatory for activities identified as construction and decommissioning or dismantling of nuclear power plants, storage or disposal of spent nuclear fuel, and construction and decommissioning of radioactive waste management, storage or disposal facilities. The EIA process consists of two main stages: the scoping stage (deciding on the coverage of the EIA) and the EIA study stage. Each stage includes preparation of relevant EIA documents (i.e., the program during the scoping stage and the report during the study stage), informing the public, and coordinating the review of the EIA document with relevant parties and the competent authority. At each step, the EIA document must be updated in accordance with the comments and proposals received (Figure 5).

The MOE participates as a stakeholder in the strategic environmental assessment of national level plans and programs: it examines strategic environmental assessment documents, drafts plans and programs, and provides conclusions concerning them; organizes and coordinates EIAs in the transboundary context; takes part in state supervision and control of design and construction of nuclear facilities; and, following the procedure prescribed by legislation and other legal acts, takes part in the issuing of licences for radioactive waste management activities.

The EPA participates as a stakeholder in the strategic environmental assessment of plans and programs at the local level (municipality lever or smaller): it examines strategic environmental assessment documents draft plans and programs and provides conclusions concerning them; coordinates the EIA process for proposed economic activities; and makes decisions whether these activities are to be allowed in the selected site.

The relevant parties (state and municipal authorities) analyse programs and reports and draw conclusions on programs, and reports about the feasibility of the proposed economic activity within the scope of their respective competencies. The relevant parties are the MOE, the Ministry of Health, the Radiation Protection Centre (RSC), the Fire and Rescue Department under the Ministry of the Interior, the Department of Cultural Heritage under the Ministry of Culture, county administrations, municipal administrations, the Lithuanian Geological Survey under the MOE, and the State Nuclear Power Safety Inspectorate (VATESI), among others.

The public can submit proposals concerning the EIA of the proposed economic activity and the likely effect of this activity on the environment. The public concerned can be entitled by relevant laws to apply to the Court for a public interest defence contesting the material or procedural legality of decisions, actions or inaction as they related to the EIA.

A decision made by the competent authority on the feasibility of the proposed activity is valid for five years. Recent amendments to the law allow for an extension of five additional years upon motivated request from the organiser of the proposed activity.

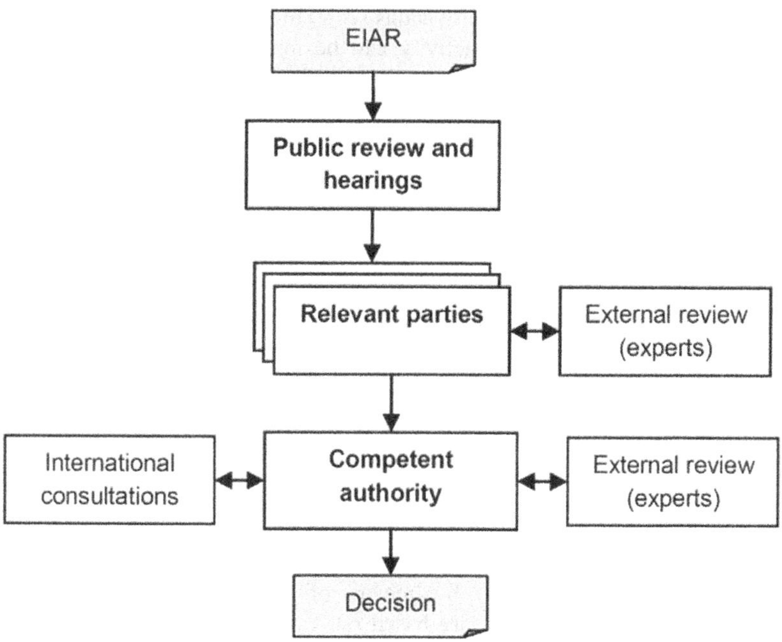

Figure 5: Process of review and coordination of the EIA report (Ragaisis et al. 2014)

Time allocated for public participation in the EIA and the duration of EIA document review is defined by the Law on Environmental Impact Assessment and its supporting regulations. Based on the most recent requirements, public informing, coordination and approval of an EIA program takes approximately three months. A similar process for an EIA report should take approximately six to nine months (this timeframe is influenced by the relevant party or the competent authority's decision regarding the necessity of external support). These estimations are based on a rather optimistic assumption that the need for updated EIA documents will be minimal and only one review is sufficient to lead to a decision from relevant parties and the competent authority. If the EIA report changes significantly in the course of review, the competent authority has the right to request that the review process start over.

As a party to the Convention on Environmental Impact Assessment in a Transboundary Context (Convention 1991), Lithuania is obliged to implement the EIA process in compliance with provisions set by the Convention. A decision on the feasibility of the proposed activity can be made only after international consultations with relevant foreign states are finished.

As of this writing, there are no specific regulations for an EIA or compensation mechanisms for socio-economic impact in the case of the geological repository.

6 Costs and financing

Cost estimates for the disposal of SNF and long-lived intermediate level waste in the crystalline basement in Lithuania is presented in the Concept of Repository in Crystalline Rocks, in Investigations of Possibilities (LEI 2005a). This preliminary assessment was based on the experience accumulated during development of the Swedish KBS-3 concept, as applied to the Lithuanian case. The method used in the analysis is based on the application of a concept known as the "successive principle," which has been used particularly as a tool for managing uncertainties that may develop due to unforeseen events in the future. The input data for the calculations are obtained from the "most likely" costs, or the so-called reference costs, by means of conventional (deterministic) calculations. These calculations are based on a functional description of each facility, resulting in layout drawings, equipment lists, personnel forecasts, etc., under established fixed conditions but without allowances for variations and uncertainties. To provide guarantees to cover losses resulting from future unforeseen events, reasonable additional costs (cost variations) are included in the probabilistic calculations. The influence of a specific variation on the costs is evaluated, and the result provides a mean value of the future costs and the standard deviation of the cost for a desired degree of confidence. If the 50 percent degree of confidence is accepted, the planned total expenditures would be about 2,600 million Euros.

7 Conclusions

The Lithuanian SNF disposal program remains in an initial site investigation and preliminary facility design stage. Research and international cooperation activities undertaken in the field of SNF disposal in Lithuania starting in 2000 have led to the following outcomes:

Prioritisation of potential geological formations for SNF disposal in Lithuania.

- Proposal of a preliminary repository concept (design of the repository and canister).
- Preliminary calculations of temperature distribution, criticality, and dose rate from disposal canisters, modelling of radionuclide and gas migration from the repository.
- Development of expert competence in SNF disposal.

According to the new national program on management of spent nuclear fuel and radioactive waste approved by the government in 2015, the near future activities related to SNF disposal will be oriented towards planning for implementation of the geological repository, investigations and characterization of prospective regions, further development of the repository concept and general safety assessment, as well as building competence through international cooperation.

References

Convention on Environmental Impact Assessment in a Transboundary Context, done at Espoo, Finland, on 25 February 1991.

Government Resolution No. 860 on Approval of the Strategy of Radioactive Waste Management (2008) (amended 2009).

Government Resolution No. 1427 on Approval of the National Programme for the Management of Spent Fuel and Radioactive Waste (2015).

Justinavicius, D.; Sirvydas, A. and Poskas, P. (2014). Thermal analysis of reference repository for RBMK-1500 spent nuclear fuel in crystalline rocks. In: *Journal of Thermal Analysis and Calorimetry*, 118(2), 767-773.

Kanopiene, R. and Marcinkevicius, V. (2000). *Possibilities for developing deep radioactive waste repository in Lithuania*. Conference Proceedings, Budapest, Hungary, 251–261.

Lithuanian Energy Institute (LEI) (2005a). *Concept of Repository in Crystalline Rocks. Investigations of Possibilities to Dispose of Spent Nuclear Fuel in Lithuania: A Model Case*, Report, Vol. 2.

Lithuanian Energy Institute (LEI) (2005b). *Generic Safety Assessment of Repository in Crystalline Rocks. Investigations of Possibilities To Dispose of Spent Nuclear Fuel in Lithuania: A Model Case*, Report, Vol. 3.

Lithuanian Geology Survey (LGT) (2005). *Suitability of Geological Environment in Lithuania for Disposal of Spent Nuclear Fuel. Investigations of Possibilities To Dispose of Spent Nuclear Fuel in Lithuania: A Model Case*, Report, Vol. 1.

Narkuniene, A.; Poskas, P.; Kilda, R. and Bartkus, G. (2015). Uncertainty and sensitivity analysis of radionuclide migration through the engineered barriers of deep geological repository: Case of RBMK-1500 SNF. In: *Reliability Engineering and System Safety*, 136, 8–16.

Poskas, P.; Narkuniene, A.; Grigaliuniene, D. and Finsterle, S. (2014). Comparison of radionuclide releases from a conceptual geological repository for RBMK-1500 and BWR Spent Nuclear Fuel. In: *Nuclear Technology*, 185(3), 322-335.

Ragaisis, V.; Poskas, P.; Šimonis, V.; Smaizys, A.; Kilda, R. and Grigaliuniene, D. (2014). The environmental impact assessment process for nuclear facilities: A review of the Lithuanian practice and experience. In: *Progress in Nuclear Energy*, 73, 129-139.

Republic of Lithuania (2015). *Lithuanian National Report on Implementation of Council Directive 2011/70/EURATOM of 19 July 2011*. Establishing a Community Framework for the Responsible and Safe Management of Spent Fuel and Radioactive Waste. Vilnius.

VATESI (2014). *Lithuanian National Report under the Joint Convention*, ISSN 2351-6445.

VATESI (2010). Requirements on Pre-Disposal Management of Radioactive Waste at Nuclear Energy Objects BSR-3.1.2-2010. Approved by the Order No. 22.3-120 of the Head of VATESI dated December 31, 2010. In: *State Journal*, No. 3-121.

Delays in Finding a Solution

The Governance of Nuclear Waste Disposal in Slovakia

Peter Mihók[1]

Abstract

After its founding in 1993, Slovakia faced two main challenges in the governance of nuclear waste disposal, mainly with regards to a final solution for spent nuclear fuel (SNF): 1.) the SNF generated in the territory of the former Czechoslovakia was to be disposed on the territory of the Czech Republic due to its more suitable geological conditions, and 2.) help was expected from Russia, as the former Czechoslovakia cooperated closely with the former Soviet Union. Despite the fact that a national agency for nuclear waste was envisaged in the Slovak legislation, its establishment was delayed, and relevant roles were 'entrusted' to the state-owned company JAVYS only in 2011. This contributed to the delayed restart of the national SNF geological repository development programme, which was suspended in 2004. Despite the government's request to restart this programme as soon as possible, as affirmed in the Nuclear Back-end Strategy of 2008, only some desk research activities have been carried out since 2013. The plan for a final solution for SNF was most recently reported on in the joint National Policy and National Programme document (prepared in order to comply with the Council Directive-2011/70/Euratom), approved in 2015.

1 Matej Bel University, Cesta na amfiteáter 1, 974 01 Banská Bystrica, Slovak Republic; E-Mail: peter.mihok@umb.sk

A. Brunnengräber et al. (eds.), *Challenges of Nuclear Waste Governance*,
Energiepolitik und Klimaschutz. Energy Policy and Climate Protection,
https://doi.org/10.1007/978-3-658-21441-8_8

1 Introduction

When Slovakia was established in 1993, it had the fourth-largest share of electricity generation from nuclear (about 54 percent) of all the nations in the world (Bennett 1994: 12), and has maintained a more or less stable share since then. In 2015, Slovakia ranked third, after France and Ukraine, with a share of about 55.9 percent (IAEA 2017a).

As will be shown in part 2.1, Slovakia faced significant challenges in the governance of nuclear waste disposal, especially with regards to a spent nuclear fuel (SNF) final solution. In addition to discussing how Slovakia dealt with these challenges, this chapter provides an overview and update on relevant information as reported to the European Commission in the joint National Policy and National Programme in 2015, and selected results of the research carried out within the IPPA project (2011 – 2013), which concerned transparency, participation and compensation related to nuclear facilities.

2 Nuclear waste disposal in Slovakia

2.1 Historical background

Construction of all existing nuclear reactors on the territory of Slovakia began during the communist regime (1948 to 1989), during which Czechoslovakia cooperated closely with the Soviet Union (see Table 1). This cooperation influenced the management of SNF from the reactors. SNF from an experimental nuclear power plant (NPP) A-1, which was shut down as a result of the INES-4 accident during refuelling in 1977, was exported to the Soviet Union without any obligations concerning the reimport of radioactive wastes (RAW) from reprocessing. In contrast, SNF and high-level radioactive wastes (HLW) from the other NPPs located in the territory of the former Czechoslovakia, which was ineligible for export back to the Soviet Union and later Russia, was to be disposed of in the Czech Republic "because of the much more suitable geological conditions there" (Mihók and Mršková 2013: 8). Only after a political decision to divide Czechoslovakia into two countries was made in 1992 did the Slovak Government address the intention to implement the SNF final solution by means of construction of a deep geological repository (DGR) in Slovakia (Resolution of the Slovak Government No. 930/1992).

Table 1: Overview of nuclear reactors in Slovakia

Locality, name, Unit no.	Unit type, Net capacity	Construction start	Operation since	Shut down
Bohunice, A-1 (exper.)	HWGCR KS 150, 93 MW_e	1958	1972	1977
Bohunice, V1, Unit 1	VVER 440-230, 408 MW_e	1972	1978	2006
Bohunice, V1, Unit 2	VVER 440-230, 408 MW_e	1972	1980	2008
Bohunice, V2, Unit 1	VVER 440-213, 471 MW_e	1976	1984	-
Bohunice, V2, Unit 2	VVER 440-213, 471 MW_e	1976	1985	-
Mochovce, Unit 1	VVER 440-213, 436 MW_e	1983	1998	-
Mochovce, Unit 2	VVER 440-213, 436 MW_e	1983	1999	-
Mochovce, Unit 3	VVER 440-213, 440 MW_e	Construction started 1987, suspended 1993 and restarted 2009.		
Mochovce, Unit 4	VVER 440-213, 440 MW_e			

Source: IAEA 2017b.

The first Conception of RAW disposal in Slovakia was approved in 1994 (Resolution of the Government No. 190/1994). It set general principles and tasks for state authorities in the newly established country related to the development of a legal and institutional framework and a final solution for SNF inclusive of a financing scheme (see section 6). The main amendment of this first conception, approved in 2008 as the 'Strategy of nuclear back end in Slovakia' (Ministry of Economy (MoE) 2008), is addressed in section 2.4.

Until 2006, all Slovak commercial nuclear facilities were owned by the former state monopoly, Slovenské elektrárne (Slovak Electric hereafter referred to as the SE company or just SE). The monopoly managed operations and planning in power plant operations (both non-nuclear and nuclear), operation of SNF and RAW facilities, and electricity transmission. While separation of the transmission grid was requested by the government, the restructuring of the nuclear facilities resulted from negotiations held in tandem with the sale of 66 percent of SE's shares. This sale was motivated mainly by SE's significant debts. The Italian company Enel, chosen in 2004 as a preferred bidder in this sale process, was not interested in holding a share in SE's divisions for decommissioning of nuclear units and SNF/RAW disposal, or in the Bohunice V1 NPP. In 2005, SE therefore established a daughter company, GovCo (from 'the Government company'), which acquired all the property not wanted by Enel on 1 April 2006. In 2007, GovCo changed its name to 'Nuclear and Decommissioning Company, plc. ('Jadrová a vyraďovacia spoločnosť' (JAVYS)).

JAVYS's mandate and activities concerning SNF and RAW are further dealt with in section 3.2.

Apart from NPP decommissioning and SNF and RAW activities, the state-owned JAVYS also holds a 51 percent share in the company 'Jadrová energetická spoločnosť Slovenska' (JESS), which is a joint venture with the Czech power group 'ČEZ' that holds the remaining 49 percent of shares. JESS was established in 2009 as a direct response to the Slovak government's decision to construct a completely new NPP at the Bohunice location, next to the shutdown NPP V1 owned by JAVYS (Resolutions of the Government 275/2008 and 948/2008). The Environmental Impact Assessment (EIA) process, concluded in 2016, recommended that a Generation III+ pressurized water reactor with a maximum gross generation capacity of 1,700 MW_e be built. Rosatom and China General Nuclear Power Corporation negotiated investment options for this project with the Slovak government and the Ministry of Economy (MoE); there are no publicly available proceedings of the negotiations.

2.2 *The national inventory*

Information concerning the inventory of SNF and RAW in Slovakia is available in "The national programme for handling of SNF and RAW in the Slovak Republic", or simply, The National Programme (National nuclear fund 2015a). An overview of key information is provided in Tables 2, 3, and 4.

Table 2: The current RAW inventory (NPPs in operation)

<table>
<tr><th>Type of waste (activity in GBq, NPP V2 / NPP Mochovce 1&2)</th><th>Bohunice NPP V2</th><th>Mochovce (Units 1&2)</th><th>Total</th><th>Total</th></tr>
<tr><td>concentrate (167,5/447,1 GBq)</td><td>1 560 m³</td><td>983 m³</td><td>2 543 m³</td><td rowspan="2">2 785 m³</td></tr>
<tr><td>sorbents (5,4/3,0 GBq)</td><td>156 m³</td><td>86 m³</td><td>242 m³</td></tr>
<tr><td>compressible (354,5/7,2 GBq)</td><td>80 tonnes</td><td>15 tonnes</td><td>95 tonnes</td><td rowspan="3">244 tonnes</td></tr>
<tr><td>combustible (354,5/7,2 GBq)</td><td>106 tonnes</td><td>19 tonnes</td><td>125 tonnes</td></tr>
<tr><td>metal (354,5/7,2 GBq)</td><td>19 tonnes</td><td>5 tonnes</td><td>24 tonnes</td></tr>
</table>

Source: National nuclear fund (2015a), Table 2, pp. 31-32; calculations by the author.

Table 3: Assessment of future RAW generation (assuming 40-year lifetime of NPPs)

Type of waste (activity in GBq, V2 / Mochovce 1&2 / Mo. 3&4)	Bohunice V2	Mochovce (Units 1 & 2)	Mochovce (Units 3 & 4)	Total
concentrate (114/887/1480 GBq)	1 060 m^3	1 950 m^3	3 253 m^3	6 263 m^3
sorbents (2,8/7,9/11,6 GBq)	80 m^3	228 m^3	330 m^3	638 m^3
compressible (396/92/132 GBq)	95 tonnes	207 tonnes	296 tonnes	598 tonnes
combustible (396/92/132 GBq)	102 tonnes	268 tonnes	380 tonnes	750 tonnes
metal (396/92/132 GBq)	32 tonnes	23 tonnes	40 tonnes	95 tonnes

Source: National nuclear fund (2015a), Table 2, pp. 31-32; calculations by the author.

Table 4: Assessment of future RAW generation (assuming 60-year lifetime of NPPs)

Type of waste (activity in GBq, V2 / Mochovce 1&2 / Mo. 3&4)	Bohunice V2	Mochovce (Units 1 & 2)	Mochovce (Units 3 & 4)	Total
concentrate (226/1369/1962 GBq)	2 112 m^3	3 010 m^3	4 313 m^3	9 435 m^3
sorbents (0/13/16,6 GBq)	224 m^3	372 m^3	474 m^3	1 070 m^3
compressible (tonnes)	205 tonnes	337 tonnes	426 tonnes	968 tonnes
combustible (tonnes)	216 tonnes	436 tonnes	548 tonnes	1 200 tonnes
metal (tonnes)	62 tonnes	33 tonnes	50 tonnes	145 tonnes

Source: National nuclear fund (2015a), Table 2, pp. 31-32; calculations by the author.

Slovak authorities assume that "together approximately 620 casks of wastes from Slovak power plants will not be disposable in the National RAW repository" (ibid.: 32), without providing any information about possible destinations for these wastes.

The National Programme document provides information about the weight (in kg, in scientific notation) and activity (in Bq, in scientific notation) of wastes expected to arise from the decommissioning of the Slovak NPPs along the categories: a) activated or contaminated components, b) activated or contaminated construction objects, c) activated or contaminated facilities, and d) contaminated soil in the case of the decommissioned experimental NPP A-1.

These wastes total 44,000 tonnes for the experimental A-1 NPP and 243,000 tonnes for the V1 NPP (ibid.: 32). The contaminated parts of the NPPs still in operation were assessed at 20,400 tonnes for the V2 NPP and 20,200 tonnes for the Mochovce NPP Units 1 and 2 (ibid.), and 25,700 tonnes for Units 3 and 4 of the Mochovce NPP, which are under construction (ibid.: 33).

Furthermore, the document provides information and data concerning institutional and captured RAW. The concluding part of the document, which is dedicated to the national inventory, contains an assessment of SNF unit production (SNF assemblies), provided in graph form only for the variant assuming a 60-year lifetime for NPPs. According to the forecast, the number of SNF units (about 11,000) should increase to 32,658 units.

2.3 The SNF interim storage facilities

Currently, there is one SNF interim storage facility in operation located near the Bohunice NPPs. It is owned and operated by JAVYS. Its construction (1983 – 1987) followed the Soviet Union's decision to extend SNF interim storage time in Slovakia from 3 to 10 years (MoE 2014: 50 – 52). This wet type facility (with the operating temperature 31–40°C, limit value max. 50°C) has been in operation since 1986. Its current capacity of 14,112 SNF assemblies is estimated to be fully utilized by around 2022. In order to respond to future SNF production, both expansion of an existing facility in Bohunice and construction of a completely new SNF interim storage facility at the Mochovce NPP are planned.

In February 2016, an EIA procedure for an expansion of the facility in Bohunice was concluded with a recommendation for JAVYS to opt for dry type storage in a building connected with original facility's main building (Ministry of Environment 2016: 2). Increases in storage capacity are planned for two phases – in the first phase, by 10,100 SNF assemblies, and in the second, by 8,500 assemblies (ibid: 1). These additional capacities are to be constructed between March 2018 and March 2020 and to operate from 2021 to 2121 (ibid.).

The new SNF interim storage facility in Mochovce is to be constructed by the nearby NPP owner and operator, the SE company. Because the original EIA procedure concluded in 2003 was not valid under the new laws resulting from Slovakia's accession into the EU, a new EIA procedure started in 2013 (MoE 2014: 55; VÚJE 2013). The procedure passed scoping with a decision issued in February 2014, but as of February 2017, it had not yet been concluded (Ministry of Environment 2017). The facility is planned to have a storage capacity of 21,200 SNF assemblies, and its operation period is planned for 60 years with a possibility of extension to 100 years (VÚJE 2013: 15).

2.4 *The waste management strategy*

The current Slovak SNF and RAW management strategy was (re)approved by the government on 8 July 2015 in the joint National Policy and National Programme document, prepared in order to comply with the Council Directive 2011/70/ Euratom establishing a community framework for the responsible and safe management of SNF and RAW (MoE 2015). From the official name of this document in its Slovak original, it is explicitly clear that it is an update of 'The Strategy for the Final Stage of Peaceful Utilization of the Nuclear Energy in the Slovak Republic' approved by the Government on 15 January 2014 (MoE 2014). Apart from cost estimates valorisation, the joint National Policy and Programme document was prepared essentially by restructuring the content of the strategy approved in 2014 into the structure adopted in the aforementioned Euratom Directive. The only additional content prepared specifically for the National Programme concerned the "specification of radioactive wastes quantities and detailed specification of their characteristics and flows" (MoE and Ministry of Environment 2015: 4); in other words, reporting national inventory as required by this Directive.

The vast majority of the content of the National Policy and Programme had already been prepared and consulted on in the mid-2000s in procedures concerning 'The Strategy of nuclear back end in Slovakia' (MoE 2008); this strategy was approved on 21 May 2008 (Government Resolution 328/2008). With regards to the SNF final solution scenarios under consideration, all of the Slovak strategic documents are limited to open fuel cycle. They do not consider an option for SNF reprocessing because VVER-440 reactors are not licensed to use MOX type nuclear fuel in Slovakia (National nuclear fund 2015b: 6).

An option to export SNF to Russia was included in the list of three potential scenarios for SNF final solution in the original 'Strategy document' approved in 2008 (MoE 2008: 15). However, it was clearly stated that under current legislation, this option was not legally implementable without an obligation to bring RAW resulting from reprocessing back to Slovakia. Moreover, this scenario has been assessed as the economically most difficult[2] (ibid.). The updated strategy document from 2014 provided more detailed information concerning the export to Russia scenario than the original strategy approved in 2008. The updated strategy highlighted that with exception of the contract signed in 1958 concerning SNF from experimental A-1 NPP, the signed contracts did not imply an obligation by Russia, as a supplier of fresh nuclear fuel, to accept SNF permanently (MoE 2014: 50). For this reason, both the updated strategy document from 2014 and the

2 This assessment is not publicly available, as it was submitted to the government in 2001 in a classified document.

joint National Policy and Programme document from 2015 do not consider the SNF export to Russia scenario at all. In both of these documents, the authorities are explicit that the final SNF solution in Slovakia follows a double path scenario, i.e. preparation for the development of a deep geological repository of SNF and HLW in Slovakia, and participation in activities which would lead to international deep repository (MoE 2015: 6f). According to the National Programme, the Slovak authorities should make a decision concerning the continuation of the 'scenario of a double path' in 2020 (ibid.: 16). The National Policy states that Slovakia will deal with the retrievability of disposed SNF within this double path scenario framework (ibid.: 7).

With regards to RAW disposal, the National Programme set objectives to construct and commission a facility for handling institutional and captured RAW (by 2016, met on schedule), to construct facilities for remelting of metal radioactive wastes and integral storage of RAW in Jaslovské Bohunice, and to construct a repository of very low-level wastes and a new repository structure after filling the second double row of the National RAW repository, all by 2018 (ibid.: 15f).

3 The legal and institutional framework

3.1 The legal framework

Slovakia has not yet adopted any specific enforceable national-level legislation concerning technical or social aspects of the SNF final solution. The Council Directive 2011/70/Euratom was transposed into Slovak legislation by amendment of the Atomic Act (541/2004 Coll.) and the Act on the National nuclear fund (238/2006 Coll.) with the new Act No. 143/2013 Coll. (MoE and Ministry of Environment 2015: 4; NNF 2015a: 12). These amendments, together with the Atomic Act, are considered to compose the primary nuclear legislation of Slovakia; while the Act on the National Nuclear Fund (NNF) and related Decrees of the Government concerning financing of nuclear back-end are referred to as the third pillar (NNF 2015a: 114f).

The Atomic Act contains detailed provisions for regulating safety and reliability of operations as well as closure and decommissioning of all nuclear installations. It covers all SNF/RAW facilities and areas, such as generation of nuclear energy/electricity, SNF/RAW management, SNF/RAW disposal, and management of nuclear material in quantities greater than one effective kilogram (subject to some legal exceptions); additionally, uranium enrichment and production of nuclear fuel fall under the jurisdiction of this act. The main principles which were incorporated into this act are briefly outlined in section 3.2.

The key principle, which concerns the roles and rights of the nuclear regulatory authority, provides legal ground to consider the authority's decrees on nuclear safety and regulations on radiation protection as a secondary nuclear legislation of Slovakia (see: National nuclear fund 2015a[3]: 112).

None of the amendments to the Atomic Act, which were approved after finalisation of the reports sent to the EC under the Council Directive 2011/70/Euratom in 2015, concerned SNF/RAW management strategy or safety measures. Rather, they concerned amendment of the EIA Act (314/2014 Coll.), financial provisions concerning nuclear accidents (54/2015 Coll.), criminal liability of legal persons (91/2016 Coll.) and Civil Dispute Code, Civil Non-dispute Code and Administrative Procedure Code (125/2016 Coll.). Therefore, all the parts of the National Programme and report to the European Commission from 2015 (Nuclear regulatory authority 2015) and also an overview of the Slovak Atomic Act and related regulatory framework from 2013 (OECD 2013) were still valid at the time of finalisation of this chapter.

The National Nuclear Fund Act (238/2006 Coll.) contains provisions concerning the financing of decommissioning of NPPs, SNF/RAW storage outside of NPP premises, SNF/RAW disposal, and provisions concerning preparations and updates of relevant strategies and reports, mainly those required by the Council Directive 2011/70/Euratom. Some of this act is briefly outlined in section 6. It is, however, worthwhile emphasizing that a major amendment of this act was planned for 2016 according to the National Programme document approved in 2015 (MoE 2015: 15). In January 2016, the MoE announced that it planned to publicly disclose the first draft of this amendment in May or June 2016 (MoE 2016), but this was postponed indefinitely (MoE 2017a). The deadline to submit a proposal to amend the Act on NNF to the government was set for 30 September 2017 (ibid.), but this was changed on 7 September 2017 to the new deadline of 30 September 2018 (MoE 2017b).

3.2 The institutional framework

The general principle imposed by the Slovak Atomic Act is that any use or handling of nuclear energy sources requires an authorisation (construction permit, operation license etc.) issued by the Slovak Nuclear Regulatory Authority (Úrad jadrového dozoru (ÚJD)). This principle applies to nuclear energy generation, SNF/RAW facilities operation, and nuclear facilities decommissioning. The provisions of the Atomic Act apply in parallel with provisions of general laws

3 The abbreviation ÚJD SR [used in the names of decrees in the quoted document] stands for the Nuclear Regulatory Authority, from the Slovak 'Úrad jadrového dozoru'.

such as the Construction Code and laws on impact assessments. ÚJD is an independent body liable to the Government.

From 2006 to 2011, the Atomic Act contained a provision according to which disposal of SNF/RAW could be realised only by a legal entity independent from SNF/RAW producers, founded or established by the MoE of the Slovak Republic. However, the amendment of the Atomic Act in September 2011 (350/2011 Coll.) added a new option for entrusting an existing legal body (independent of SNF/RAW production) with tasks and responsibilities concerning SNF/RAW disposal. "This approach [was] enabled to entrust JAVYS, as a state-owned company, by duties of [the Slovak national] waste management agency" (Mihók and Mršková 2013: 6). JAVYS also holds authorisations for decommissioning of the NPPs A1 and V1 in Bohunice.

The national nuclear fund (NNF) is a key source of finance for decommissioning NPPs, SNF and RAW storage outside of NPP premises, SNF and RAW disposal and treatment of seized (i.e captured) RAW. Its establishment, history and system of accumulation of finances are briefly summarised in section 6. From an institutional point of view, the NNF is a legal body (a special purpose state fund) governed by decisions of its sole governing body, the Board of Governors, which typically meets once a month. The Board of Governors approves or rejects requests for finance submitted by JAVYS and on occasion other companies authorised to implement NPP decommissioning and/or SNF/RAW storage and/or disposal. As mentioned in section 2.4, the NNF is also responsible for preparation of key strategic and planning documents concerning the nuclear back end in Slovakia, such as the National Policy and Program requested by the Directive 2011/70/Euratom[4].

4 Siting procedures

The Slovak authorities launched the Programme of SNF Deep Geological Repository development (the DGR Programme) in 1996, but it was suspended in 2004. The Strategy of Nuclear Energy Back-End, approved in 2008, stated that "from a short-term point of view, two actions should be made as soon as possible: 1.) to start coordination of the [DGR] Programme in order to update conceptions and planning documents, and 2.) to continue geological investigations suspended around the year 2004" (MoE 2008: 66). Although these two tasks were a part of

4 This remained valid after JAVYS was entrusted with the duties of a national agency for SNF/RAW management in 2011. These duties were confirmed in 2013, in the new legislation approved by the Slovak Parliament in order to transpose the Directive 2011/70/Euratom into Slovak law.

the document that was submitted by the MoE and approved by the government, the MoE decided not to implement these two tasks "due to financial crisis and shortage of resources" (MoE 2009). The hiatus in the DGR Program ended in 2012, after the state-owned company JAVYS was *de facto* entrusted with the role of the Slovak national SNF/RAW agency. In June 2012, JAVYS started to publicly procure a supplier to review and update the conceptions and planning documents prepared within the DGR Programme before 2004. In November 2012, a consortium of companies lead by the joint stock company ÚJP Praha from the Czech Republic was selected to carry out this task under the project "Deep repository – selection of locality, the 1st phase" (Office for Public Procurement 2012). Within this project, the following three documents were submitted to JAVYS and subsequently to the National Nuclear Fund (NNF) in late 2015: 1.) Strategy for work with the public in relation to the deep repository development and analysis of potential economic and non-economic instruments to foster implementation of deep repository, 2.) Informational and promotional materials about deep repository development in Slovakia and 3.) Update of the feasibility study for deep repository in Slovakia (MoE 2016). In late 2016, the following additional documents were submitted to the NNF by JAVYS: 4.) Criteria for selection and evaluation of localities of DGR and 5.) Proposal of legislation for stimulation of affected municipalities during implementation of investigation works and after localisation of DGR. An updated version of the document, Update of the feasibility study for deep repository in Slovakia, dated 6 December 2016, was also submitted (MoE 2017a). All of these documents are available to the public only in hard copy accessible and by means of personal visits at the NNF premises (ibid.).

At the time of this writing, JAVYS was preparing public procurement for the first part of the project, Deep repository – selection of locality, the 2nd phase (Office for Public Procurement 2017a), for which working versions of the documents were disclosed on 3 February 2017 (Office for Public Procurement 2017b). This part of the project concerns preparation for geological investigations, a framework programme for the DGR development and research, and an implementation proposal for a system of economic stimulation of affected local communities (Office for Public Procurement 2017a). Further mid-term planning information has not been made publicly available, as JAVYS has not sent its own document, Proposal of workplan for DGR development in Slovakia in years 2017 – 2023, to the MoE, and at the same time JAVYS is not liable under the Freedom of Information Act (MoE 2017a).

4.1 Procedures and criteria for site selection

Official information regarding procedures and criteria for DGR site selection in Slovakia are not available in the joint National Policy and National Programme document approved in 2015, but a short summary on the issue is publicly available in the Nuclear back-end strategy policies approved by the government in 2014 (MoE 2014), resp. in 2008 in case of the preceding strategy (MoE 2008).

Within the original DGR Programme implemented from 1996 to 2004, criteria for assessment of location suitability were chosen and applied using archived geological information and maps. In the first stage, which resulted in identification of 15 localities for further research, a total of 58 characteristics were taken into account. They included aspects of long term development of the territory, geological risks, geological structure, hydrological characteristics, geochemical aspects, engineering-geological properties, occurrence of natural sources of RAW materials and legislative protection for the area (MoE 2014: 53). In the second stage, the list of localities suggested for further geological research was reduced to five surveyed localities: the central part of the Tríbeč mountains (46 km^2), the southern part of Veporské vrchy mountains (78 km^2), the southwest part of Stolické vrchy mountains (24 km^2), the eastern part of Cerová vrchovina mountains (87 km^2), and the western part of Rimavská kotlina basin (85 km^2) (ibid.). The first three are in a granitoid environment; the remaining two are in a sedimentary rock environment. Feasibility studies were made for these two geological environments from 1996 to 2001.

During the DGR Programme hiatus from 2004 to 2012, limited academic research related to the suitability of selected environments for the DGR in Slovakia was carried out by the State Geological Institute of Dionýz Štúr and Slovak Universities (see for example: Slaninka et al. 2007; Adamcová et al. 2009). This research has not impacted the above list of five surveyed localities. Continuation of geological research in these five localities in relation to the DGR Programme is also dealt with in the document titled 'Concept for geological research and exploration of the territory of the Slovak Republic for the period 2012 to 2016 (with an outlook until 2020)', approved by the Government Resolution 73/2012.

4.2 Compensation mechanisms and socio-economic impact

The legislation and practice of added value approaches used from 1993 to 2012 was summarised in the IPPA project Deliverable no. 4.4 (Mihók 2013: 16–20). This document briefly explains the payments from the national nuclear fund implemented from 2000 to 2002, nuclear facility location fees/taxes implemented from 2002 to 2009, and relevant parts of real estate tax collection implemented since 2005. As of early 2017, the legislation and practice remained as it was reported in the IPPA Deliverable. The only new development related to an objective set "to develop and prepare implementation of a system of economic stimulation of localities affected by development and operation of repositories" in the National Programme approved by the Government in 2015. This objective was assigned to the MoE, JAVYS and the National Nuclear Fund (NNF), with a deadline of 2018 (MoE 2015: 16, task 19). Correspondingly, the MoE declared that solutions should not be focused primarily on economic stimuli; instead, a complex system of informing and working with the public should be developed for long-term use (ibid.). The first official 'Proposal of legislation for stimulation of affected municipalities during implementation of investigation works and after localisation of DGR' was submitted to the NNF in late 2016. Its content is available only in hard copies accessible by means of personal visits to the NNF premises (MoE 2017a). The update of this document is planned for the second phase of the currently ongoing Slovak DGR Development Programme (Office for Public Procurement 2017a).

Regarding the preparation and adoption of the National Programme in 2015, the MoE did not accept the proposal from the Association of towns and communities of Slovakia (ZMOS) to prepare and approve a new specific law on payments for RAW. ZMOS explicitly required that the content of a new law should be in line with the principles approved by ZMOS in 2012 and submitted to the MoE in 2012 and again in 2013. The principles requested by ZMOS were summarized in English in the IPPA Deliverable no. 4.4 (Mihók 2013: 25f), along with the related results from the IPPA project's research (ibid.: 27-33) and summary of stakeholder discussions held in 2013 (ibid.: 34-37). In 2013, the MoE stated that the principles proposed by ZMOS were 'not acceptable', justifying this decision solely with a financial reasoning (ibid.: 38). In 2015, the MoE declared that the [resubmitted] proposal of ZMOS "goes beyond the framework of the National Programme and therefore cannot be dealt with [in the process of preparing and approving the National Programme according to the Directive 2011/70/Euratom] (MoE 2015, Annex no. 6: 3).

5 Information and participation

During the communist era and to some extent also during the 1990s, [Czecho]Slovakia lacked legislation allowing public access to information and decision making. This was emphasized in 2013 during the discussion about compensation mechanisms held within the IPPA project, when the mayor from a municipality near NPP Bohunice made a reference to the fact that in the 1970s, "when the NPPs were sited ..., no one was asking the citizens about anything. Now the citizens are asking the mayors, and request information, resp. submissions to be made within EIA procedures" (Mihók 2013: 35).

Slovakia only adopted its first Freedom of Information Act (FOIA) in 2000 (211/2000 Coll.). It contained very strict clauses in favour of transparency (Mihók 2011). However, very significant amendments were made to this act in 2010, after which a majority of information, particularly in the nuclear sector, was designated confidential or classified information. Furthermore, one of these FOIA amendments "enabled the Nuclear Regulatory Authority (NRA) to inform that new investment projects in the nuclear sector have been proposed for reviews only at the notice board located at its headquarters in the capital city Bratislava. Changes made to the FOIA through amendment of the EIA Act in 2010 concerned nuclear sector information only, that is, no changes were made regarding information or documents related to any other industrial sector. The NRA officials argued that information about nuclear facilities should be kept confidential for the public's own sake, since it could be easily abused by terrorists" (ibid.: 229f). Even though the clauses banning public disclosure of the majority of nuclear sector information were formally amended in late 2011, the boundary between publicly accessible and confidential information remained unclear. In effect, this continued to favour non-transparency for nuclear sector information (Mihók and Mršková 2011: 5[5]).

Research concerning stakeholder participation in the nuclear sector EIA and SEA procedures, carried out within the IPPA project in 2011, identified five different environmental NGOs as participants in the nuclear sector in EIA and SEA procedures carried out from 2007 to 2009 (Mihók and Mršková 2011: 15). After 2010, however, only the Slovak office of Greenpeace remained active in EIA, SEA or similar procedures open to NGOs. Proposals and comments submitted within the interdepartmental observation procedure concerning nuclear back-end strategy updates and the official meeting to discuss these proposals and comments held on 18 February 2013 are the last instances of participation by Slovak environmental NGOs in the SNF- or RAW-related decisions and planning which could be identified. In 2015, not a single Slovak, foreign or international NGO asserted its right to participate in preparation of the joint National Policy and Programme document by means of submitting comments or proposals within the interdepartmental observation procedure. The decrease of environmental NGOs' participation in the nuclear sector EIA, SEA and similar procedures can, to some extent, be explained by changes in funding opportunities for environmental NGOs, which were outlined in the IPPA project Deliverable 2.4 (ibid.: 11-14).

5 FOIA amendments adopted after 2011 do not address nuclear sector information transparency.

The only actor representing the public (resp. elected representatives of the public) in the interdepartmental observation procedure for the joint National Policy and Programme document in 2015 was the Association of towns and communities of Slovakia (ZMOS), whose submissions concerning compensations and socio-economic impacts is dealt with in section 4.2. The author is unaware of any Slovak environmental or other NGO that has recently attempted to represent affected communities or the general public in matters and procedures concerning the commercial nuclear sector[6].

With regards to the joint National Policy and Programme document, neither domestic nor transboundary SEA procedures were held as a part of its preparation, consultation and adoption[7]. The decision not to carry out a SEA procedure for the National Policy and Programme was justified with the argument that these documents contained the same content as the previously approved nuclear back-end strategy. This decision by the Ministries of Economy and Environment (MoE and Ministry of Environment 2015) is also publicly available in its English translation[8]. A similar decision not to carry out the SEA procedure was made in 2013 with regards to an update of the nuclear back-end strategy. The latest SEA procedure concerning SNF final solution and RAW disposal plans held in Slovakia concerned the nuclear back-end strategy adopted in 2008 (MoE 2008), for which a joint national level and transboundary SEA procedure was held from 2 April 2007 to 15 May 2008 (MoE and Ministry of Environment 2013).

6 Costs and financing

In December 1992, very shortly before Slovakia's establishment as an independent state, Slovak authorities were asked to prepare principles of a nuclear fund-related law by the end of March 1993 (Resolution of the Slovak Government 930/1992, task 1c). Based on these principles, Act 254/1994 Coll. establishing the first Slovak nuclear fund was approved by the parliament in August 1994. The name of this fund during 1995 to 2006 was the State Fund for Decommissioning

---

6 Slovakian staff of CEE Bankwatch Network were active in matters concerning proliferation of NPPs, however only on Ukrainian territory, i.e. from the position of a potentially affected foreign country.

7 NGOs have the right to participate in the interdepartmental observation procedures concerning any document submitted to the government; the signed support of at least 500 citizens is needed to oblige authorities to consult the NGO representatives in person.

8 Available at: http://www.njf.sk/dokumenty/politika_a_program/DECISION_-_Ministry_of_the_Economy_and_the_Environment_of_the_Slovak_Republic.pdf (last accessed 25 January 2017).

of Nuclear Power Installations and Management of Spent Nuclear Fuel and Radioactive Waste (abbreviated as ŠFLJEZ in Slovak language texts and as SNIDF in English language translations, and often informally called the Decommissioning Fund). This original act was replaced in 2006 with the completely new Act 238/2006 Coll. on the National Nuclear Fund (abbreviated as NJF in the Slovak language texts and as NNF in English language translations).

The original Act on the Decommissioning Fund obligated NPP operators to contribute to the fund the equivalent of 10 percent of their revenues from sales of electricity generated by the NPPs in the preceding year. This one component mode of ŠFLJEZ/SNIDF finance generation was implemented from 1995 to 2001. The amendment of the 'Decommissioning Fund Act' in 2001 (560/2001 Coll.) introduced a new mode of finance generation composed of two components: the fixed contributions, in a value equivalent to ca. 11,618 Euro annually per every MW_e of installed electric capacity and variable contributions, stated as a percentage of revenues from sales of electricity generated by the NPPs in the preceding year, as in the original Act. The percentage of variable contribution was originally set to 6.8 percent in 2001 and applied as such from 2002 to 2006, after which it was decreased to 5.95 percent with the aforementioned Act 238/2006 Coll. and applied as such since July 2006. The fixed contribution value remained ca. 11,618 Euro per MW_e until 2012, when it was increased to 13,428 Euro per MW_e. It has been adjusted for inflation annually since (Act 550/2011 Coll. Amending the Act on NNF). From 1 July 2017 to 30 June 2017, the value of fixed contribution was ca. 14,598 Euro per MW_e of installed capacity of commercial nuclear reactors (NNF 2017).

Neither 'the Original Act on Decommissioning Fund' approved in 1994, nor 'the Act on the NNF' approved in 2006 obliged owners (resp. holders of licenses to operate or decommission non-reactor nuclear installations, such as SNF/RAW storage, processing and/or disposal facilities) to contribute to the fund (MoE 2015: 30).

In 2000, the Slovak government took cognizance of the classified document titled 'Prognosis of complete financial support for back end of nuclear power engineering in the SR' (Resolution no. 429/2000). This document showed that contributions to the Decommissioning fund "had to be increased in order to cover the full costs related to SNF treatment, including final disposal. Mandatory contributions to the Decommissioning Fund were subsequently increased with the entry into force on 1 January 2002 of Act 560/2001 Coll." (European Commission 2013: 12f). According to this author's own calculations, published in 2010[9], the increase of contributions adopted in 2001 did not have any significant mitigating impact on the fund's deficit, assessed by the author at ca. 2.4 – 3 billion Euro as

9 Also available in English upon email request to peter.mihok@umb.sk.

of the end 2009 (Mihók 2010: Annexes 1 and 2). The prognosis documents was to be updated by the authorities every two years (Government Resolution 429/2000, part B2), but this was not carried out during 2000 to 2006 period. Subsequently, this task was replaced with the task to prepare and approve nuclear back-end strategies (see section 2.4), which include financial plans for the implementation of these strategies (MoE 2017a).

In order to decrease the NNF's deficit, the Slovak Government introduced a levy on the final customers of electricity in 2010; it entered into force on January 2011 (Regulation of the Government of the Slovak Republic 426/2010). This levy was originally set at 3 EUR/MWh and was adjusted for inflation annually. Detailed information in English concerning this levy is available in the publicly accessible European Commission report addressing whether this measure constituted state aid in a manner that was compatible with the *Acquis Communautaire* (European Commission 2013). The Commission concluded "that the distortion of competition resulting from the aid granted under measure [as notified by Slovakia] is outweighed by the positive contribution of the measure to the achievement of the Euratom Treaty objectives" (ibid.: 23).

The significant amendment of the National Nuclear Fund Act, planned for 2018, should ensure that value of the contributions made by the NPP operators to the fund are no longer set by Parliament but instead by an independent body, presumably the National Nuclear Fund (MoE 2015: 15).

7 Concluding remarks

There have been several delays in Slovakia's efforts to define and implement a SNF final solution, including during the establishment of the national SNF/RAW agency as envisaged by the Atomic Act, with the amendment of the act on the nuclear fund in order to ensure sufficient financing for dealing with the nuclear back end, and in relaiton to the adoption of a law on compensation to municipalities affected by SNF/RAW facilities. All of these delays can perhaps be, directly or indirectly, related to the hiatus in Slovakia's deep geological repository development programme (2004 to 2012) officially justified by "financial crisis and shortage of resources" by the Ministry of Economy (MoE 2009).

The question arises as to whether a lack of finance can explain delays in the adoption of laws or the establishment (or entrusting) of a legal body to take responsibility concerning the final SNF solutions. This question is especially relevant when considering the 2004 to 2012 period, in which the MoE fostered completion of the Mochovce NPP Units 3 and 4. Two facts are worth noting in this regard: the MoE controls 34 percent of the shares owned by Slovakia in the project's investor, the SE company; and finalisation of Mochovce NPP will be financed without any contract for difference (as for ex. the Hinkley point NPP project in the United Kingdom) or similar financing scheme.

The Slovak DGR Programme shares a part of its history with the Czech Republic. As the links between the two countries remained strong even after their separation (also due to a very minimal language barrier and other factors.), it is reasonable to assume that Slovak authorities closely follow the development in the Czech DGR Programme. The complications faced in the Czech Republic (see: Bursík 2015; Vojtechová and Steinerová 2013; Vojtechová 2009) may also be potential factors to consider in attempts to understand the delays in Slovakia's nuclear waste governance. Nevertheless, after its restart in 2012, the Slovak DGR development programme seems to be progressing steadily.

Acknowledgments

This chapter benefited from the results of the IPPA project, which received funding from the European Atomic Energy Community's Seventh Framework Programme under Grant Agreement no. 269849.

References

Adamcová, R.; Frankovská, J. and Durmeková, T. (2009). Engineering geological clay research for a radioactive waste repository in Slovakia. In: *Acta Geologica Slovaca,* 2, 71-82.

Bennett, L. L. (1994). *Nuclear Power in the World. Its Present Status and Development Trends*. Regional seminar on "nuclear energy for better life". Sofia, Bulgaria, 16 -18 May 1994 (Revised: 18 May 1994). International Atomic Energy Agency, Vienna (Austria). http://www.iaea.org/inis/collection/NCLCollectionStore/_Public/25/062/25062539.pdf, last accessed 31 January 2017.

Bursík, M. (2015). The Trouble with Democracy. The Challenges of Nuclear Waste Governance in the Czech Republic. In: Brunnengräber, A.; Di Nucci, M. R.; Isidoro Losada, A. M.; Mez, L. and Schreurs, M.A. (Eds.). *Nuclear Waste Governance. An International Comparison,* Wiesbaden: Springer VS, 249 – 264.

European Commission (2013). *State aid SA.31860 (N 506/2010) – Slovakia. Partial financing of decommissioning of two already shut down nuclear plants (A1 and V1).* C(2013) 782 final. Brussels, 20.02.2013.

International Atomic Energy Agency (IAEA) (2017a). *Nuclear Share of Electricity Generation in 2015*. The Power Reactor Information System (PRIS). https://www.iaea.org/PRIS/WorldStatistics/NuclearShareofElectricityGeneration.aspx, last accessed 6 February 2017.

IAEA (2017b). *Slovak Republic (PRIS)*. https://www.iaea.org/PRIS/CountryStatistics/CountryDetails.aspx?current=SK, last accessed 6 February 2017.

Mihók, P. and Mršková, A. (2013). *Slovak experience with novel approaches in risk communication and public participation*. IPPA project Deliverable no. 2.5, Euratom Contract no. FP7-269849. http://www.ippaproject.eu/, last accessed 15 June 2017.

Mihók, P. (2013). *Compensations and community benefits for municipalities located near nuclear facilities in Slovakia: historical overview and stakeholders views in 2013*. IPPA project Deliverable no. 4.4, Euratom Contract no. FP7-269849. http://www.ippaproject.eu/, last accessed 15 June 2017.

Mihók, P. and Mršková, A. (2011). *Review of public participation in SEA and EIA procedures in the nuclear sector in Slovakia*. IPPA project Deliverable no. 2.4, Euratom Contract no. FP7-269849. http://www.ippaproject.eu/, last accessed 15 June 2017.

Mihók, P. (2011). Risk communication and knowledge management in a nuclear sector: a case study about the experience of Slovak non-governmental organisations. In: *Knowledge Management Research & Practice*, 9, 228 – 235.

Mihók, P. (2010). *Finančný deficit záverečnej časti jadrovej energetiky na Slovensku. Banská Bystrica (Slovakia): Ústav vedy a výskumu Univerzity Mateja Bela v Banskej Bystrici*. ISBN 978-80-557-0098-4.

Ministry of Economy (MoE) (2017a). Answer to the request for information in accordance with the Freedom of information Act (no. info 40-2017, dated 20 February 2017, resp. requested supplementary information dated 21 February 2017, sent by email to Peter Mihok).

MoE (2017b). Answer to the request for information in accordance with the Freedom of information Act (no. info201-2017, dated 6 October 2017, sent by email to Peter Mihok).

MoE (2016). Answer to the request for information in accordance with the Freedom of information Act (no. info14-2015, dated 18 January 2016, sent by email to Peter Mihok).

MoE (2015). *Návrh Vnútroštátnej politiky a Vnútroštátneho programu nakladania s vyhoretým jadrovým palivom a rádioaktívnymi odpadmi v SR ako aktualizácia strategického dokumentu Stratégia záverečnej časti mierového využívania jadrovej energie v Slovenskej republike*. Document no. 05837/2015-1000-29004, approved by the Government of the Slovak Republic on 8 July 2015 in the Resolution no. 387/2015.

MoE and Ministry of Environment (2015). *Decision No. MH SR: 09498/2015-4100-28524* (resp. Decision No. MŽP SR: 2909/2015-3.4/hp). Bratislava, 15 June 2015.

MoE (2014). *Návrh Stratégie záverečnej časti mierového využívania jadrovej energie v SR*. Document no. 1464/2013-1000, approved by the Government of the Slovak Republic on 15 January 2014 in the Resolution no. 26/2014.

MoE and Ministry of Environment (2013). *Decision No. MH SR: 2727/2013-4100* (resp. Decision No. MŽP SR: 2909/2013-3.4./hp). Bratislava, 30 April 2013.

MoE (2009). Answer to the request for information in accordance with the Freedom of information Act (no. info625/2009-1010, dated 19 November 2009, sent by email to Peter Mihok).

MoE (2008). *Návrh Stratégie záverečnej časti jadrovej energetiky (Nové znenie)*. Document no. 998/2008-1000, approved by the Government of the Slovak Republic on 21 May 2008 in the Resolution no. 328/2008.

Ministry of Environment (2017). Answer to the request for information in accordance with the Freedom of the information Act (no. ŽOI 39/2017, dated 22 February 2017, sent by email to Peter Mihok).

Ministry of Environment (2016). *Záverečné stanovisko (1604/2016 – 3.4/hp) vydané Ministerstvom životného prostredia Slovenskej republiky podľa zákona č. 24/2006 Z. z. o posudzovaní vplyvov na životné prostredie a o zmene a doplnení niektorých zákonov v znení neskorších predpisov*. http://www.enviroportal.sk/eia/dokument/239035, last accessed 8 February 2017.

National Nuclear Fund (NNF) (2017). Answer to the request for information in accordance with the Freedom of information Act (dated 19 January 2006, sent by email to Peter Mihok, not numbered).

NNF (2015a). *The national programme for handling of spent nuclear fuel and radioactive wastes in the Slovak Republic'*. http://www.njf.sk/dokumenty/politika_a_program/The_National_Programme_for_handling_of_SNF_and_RW_in_SR.pdf, last accessed 10 May 2017. Note: this is the English language mutation of a part of the document referred as MoE (2015).

NNF (2015b). *The national policy for handling of spent nuclear fuel and radioactive wastes in the Slovak Republic'*. http://www.njf.sk/dokumenty/politika_a_program/The_National_Policy_for_handling_of_SNF_and_RW_in_SR.pdf, last accessed 10 May 2017. Note: this is the English language mutation of a part of the document referred as MoE (2015).

Nuclear Regulatory Authority (NRA) (2015). *Report of the Slovak republic compiled in terms of article 14 par.1 Council Directive 2011/70/Euratom*. Report submitted to the European Commission (Ref. Ares(2015)3540608 - 27/08/2015).

Office for Public Procurement (2017a). *Oznámenie 1070 - POS (Úrad pre verejné obstarávanie – ÚVO). Vestník č. 18/2017 - 25.01.2017*. https://www.uvo.gov.sk/vestnik/oznamenie/detail/350492, last accessed 24 February 2017.

Office for Public Procurement (2017b). *Súťažné podklady-pracovná verzia (Zákazka: IPR 100TUND20008 - Hlbinné úložisko - výber lokality, 2. etapa - 1. 4asť)*. https://www.uvo.gov.sk/vyhladavanie-dokumentov/detail/771292, last accessed 24 February 2017.

Office for Public Procurement (2012). Announcement No. 14161 – VUS. (Úrad pre verejné obstarávanie – ÚVO, Oznámenie [o výsledku verejného obstarávania 4.) 14161 – VUS. Vestník č. 225/2012 - 23.11.2012. https://www.uvo.gov.sk/vestnik/oznamenie/detail/187367, last accessed 20 January 2017.

Organisation for Economic Co-operation and Development (OECD) (2013). *Nuclear Legislation in OECD and NEA Countries. Regulatory and Institutional Framework for Nuclear Activities: Slovak Republic.* https://www.oecd-nea.org/law/legislation/slovak.pdf, last accessed 15 February 2017.

Regulation of the Government of the Slovak Republic No. 426/2010.

Resolutions of Government of the Slovak Republic No. 930/1992, 190/1994, 429/2000, 275/2008, 328/2008, 948/2008 and 73/2012.

Slaninka, I.; Hók, J. and Franzen, J. (2007). Vývoj a stav hlbinného úložiska RAO v Slovenskej Republike z hľadiska geologického poznania. In: *Acta Geologica Slovaca,* Special issue no. 1, 17 – 23.

Vojtechová, H. and Steinerová, L. (2013). *Critical Evaluation of Knowledge and Experience from RISCOM Implementation and Proposal 'Changes in the Communication Strategy in the Czech Republic'.* IPPA project Deliverable no. 2.3, Euratom Contract no. FP7-269849. http://www.ippaproject.eu/, last accessed 15 June 2017.

Vojtechová, H. (2009). *Evaluation, testing and application of participatory approaches in the Czech Republic.* Application of the RISCOM model in the Czech Republic. ARGONA project Deliverable no. D14, Euratom Contract no. FP6-036413. http://www.argonaproject.eu/, last accessed 15 June 2017.

VÚJE (2013). *Interim storage of spent nuclear fuel in Mochovce. Intention pursuant to Act No. 24/2006 Coll. on Environmental Impact Assessment and amending other laws.* http://www.enviroportal.sk/eia/dokument/197029, last accessed 8 February 2017.

An Arranged Marriage

Nuclear Waste Governance and Nuclear Energy in Slovenia and Croatia

Leo Šešerko[1]

Abstract

As the plans for the first Yugoslavian Nuclear power plant (NPP) Krško were drafted in the 1970s, there was no mention of nuclear waste. Similarly, the purchase contract with Westinghouse made no mention of it. After the disintegration of Yugoslavia in 1990, the previous contract between the two states, Slovenia and Croatia, extended the primary 50:50 provision of ownership to nuclear waste. This agreement is the origin of both cooperation and entanglements about nuclear waste. Slovenia and Croatia have decided to postpone the construction of a final depository for used nuclear fuel until the NPP Krško I, and the pending NPP Krško II has been definitively put out of operation. The first strategy for spent fuel management was adopted in 1996. It was based on the agreement between the governments of Slovenia and Croatia. However, cooperation has proven difficult as each side wants to make as much profit from the NPP as possible and to take over as little cost regarding used fuel as it can. Two separate final repositories for nuclear waste are probable.

The actual distribution of responsibilities through the end of the operation of the nuclear power facilities in Slovenia between the Agency for Radioactive Waste of Slovenia (ARAO) for low and medium active waste and the NPP Krško for used fuel is unsatisfactory and creates risks. The ARAO is a state public institution, and is responsible for handling all waste, preparing strategies, and developing technical capacity as well as operating facilities for storage or disposal. The NPP is primarily a profit-driven company that operates the plant and whose purpose is generation of electricity. These two functions challenge one another.

---

1 Environmental Protection College, Trg mladosti 7, 3320 Velenje, Slovenia, info@vsvo.si; E-Mail: leo.seserko@gmail.com

A. Brunnengräber et al. (eds.), *Challenges of Nuclear Waste Governance*,
Energiepolitik und Klimaschutz. Energy Policy and Climate Protection,
https://doi.org/10.1007/978-3-658-21441-8_9

1 Introduction

The NPP Krško started operating in the autumn of 1981 and has been operating commercially since 1983. It is a Westinghouse two-loop Pressurized Water Reactor with a nominal output power of 727/696 MWe (gross electrical power/net electrical power). The plant is owned by state-owned Slovenian and Croatian electrical power companies, GEN energija d.o.o. and Hrvatska Elektroprivreda - HEP d.d., respectively. Ownership is divided 50:50 between Slovenia and Croatia. The plant is operated by the public/private enterprise Krško NPP d.o.o. The NPP Krško is the major producer of radioactive waste in Slovenia and Croatia. All operational radioactive waste and spent nuclear fuel produced through the NPP are stored on site. Solid radioactive waste is treated and then packed into steel drums, which are then stored in the solid radioactive waste storage facility.

Between 1993 and 2002, tensions surrounding the opposing interests of Croatia and Slovenia in relation to the NPP came to a head. For the HEP and Croatia, the main disputes were supposedly reduced employment rights for Croatian workers in the NPP, participation of Croatian managers at the senior level, payment for the steam generators' replacement project imposed by the Slovenian PM Drnovšek, the application of different accounting standards for Croatian and Slovenian stakeholders, and Slovenian insistence on HEP's contribution into the Slovenian decommissioning fund. For Slovenia, the main disputes were HEP's failure to pay for the electricity it received from the NPP Krško, HEP's refusal to contribute to the Slovenian decommissioning fund, HEP's refusal to approve all projects from the modernization program, which were seen as necessary to further secure operation of the NPP by the Slovenians, the amount of pooled depreciation, and the Croatian decision to cancel the construction of a second NPP on Croatian soil, as established in the original agreement. Disputes culminated on 30 July 1998, as the government of Slovenia, the Slovenian transmission company ELES and the NPP Krško disconnected transmission lines to Croatia and terminated all deliveries of electricity from the NPP Krško to HEP (Tomšić 2010). The government of Slovenia issued a decree on the transformation of the NPP Krško in line with Slovenia's new company laws. The decree was to remain in force until the implementation of a new bilateral agreement between Slovenia and Croatia was reached. This new agreement was finally signed on 19 December 2001.

The NPP was designed to operate until the end of 2023. Since the spring of 2015, however, the Slovenian government has sought to legally extend the operation lifetime for 20 years (until 2043), despite its unfulfilled promise to hold a national referendum on the nuclear projects. In a joint commission on NPP issues that same year, representatives of the Croatian government agreed to the lifetime extension of 20 years. The construction of a second reactor at the same site is being proposed by the Slovenian government. According to the ESPOO convention, any operation extension requires transboundary verification, that is approval or rejection from neighboring countries; it is rather uncertain, however, whether Austria, as a neighbouring country, would support it.

In 2003, a new agreement between Croatia and Slovenia on the status of Croatian investments in the NPP and other legal relationships related to the NPP Krško, its use and decommissioning was adopted and ratified. It represents the first legal regulation on nuclear waste between the two states after 22 years of operation. The contract's more concrete language should help to prevent some of the vagueness which resulted in nonfulfillment of earlier agreements from the time of the NPP's establishment under the Yugoslavian government.

Following the agreement, in 2004, the first program for NPP decommissioning was established. However, there were numerous challenges to renewing the program every five years as required. These issues included differing opinions on general costs, payments to local communities around the NPP site, and each state's level of engagement in NPP reconstruction.

There are two entities in Slovenia in charge of nuclear waste. Low and middle radioactive waste is managed by the Slovenian national agency for radioactive waste management, ARAO; spent fuel and highly radioactive waste are to be managed by the Slovenian electrical power company GEN energija d.o.o. as long as the NPP is in operation. This raises questions about possible conflicts of interests and responsibilities, as a provider of nuclear power is also a producer of spent fuel and responsible for spent fuel management.

Because of the two (state) owners of the NPP Krško, two separate funds for decommissioning and management of nuclear waste were established; the ARAO was established as a state-owned public institution for radioactive waste management in 1991 and the fund for decommissioning the NPP Krško was established in 1995. There is no joint understanding of the use of the collected funds. For instance, Slovenia's government supports financing of the surrounding local communities to assure compliance by the neighbouring population, while the Croatian government does not see this support as necessary. Moreover, there are disparities in collected funds, which are generally seen as insufficient. These disparities resulted from the lack of a united 'master plan' for managing all kinds of nuclear waste together. Additionally, the levy of 0.30 Euro cents per kWhe, which is paid by the Slovenian electrical nuclear power company (and to the Croatian electric company HEP) is not sufficient to finance the construction of a LILW depository. Beyond these funding issues, the two governments could not agree on construction of a joint LILW depository. As a result, each intends to construct a depository for its "half" of the waste. However, both sides were unanimous in postponing construction of a deep geological repository for spent fuel until the NPP's end of life and leaving spent fuel in temporary storage at the NPP site. They also agreed to extend the operation of the NPP for another 20 years, until 2043.

Another open question is the "revolving door" between the nuclear industry and the government energy administrations. For instance, the Director of the Energy Directorate of the Ministry of Infrastructure and Space /MzI/ in Slovenia is a former collaborator of Mr. Danijel Levičar of the GEN Energija d.o.o. which is the owner of the NPP Krško. Through such relationships, the nuclear energy sector has a decisive influence not only in the domain of nuclear energy, but also in relation to competing energy industries, particularly solar, wind, geothermal and bio energy sources. The energy sector also has influence over energy pricing and subsidies. Thus, it is no surprise that the nuclear industry presents itself as the cheapest and most favored energy, while subsidies for solar panels and wind turbines are blocked. In this way, nuclear industry collaborators are in a position to influence key political decisions in state energy politics.

At the same time, there have been no further steps taken towards an agreement on the decommissioning program and nuclear waste management. The agreement adopted in 2004 is out of date, as it was planned to be updated every five years. Among several unanswered questions is the involvement of the local population and its right to be informed and integrated in decision making, as defined in the Aarhus convention. At this time, the local population had neither been consulted about the twenty-year extension of the operation life time of Krško 1 nor about the plans to construct an additional dry used nuclear fuel depository and a LILW depository at the NPP site, although these plans were distributed to the public by the press. Many speculate that the Slovenian government of Miro Cerar is testing public reactions and may eventually change course if there are significant negative reactions. But there have not been any such reactions and the press has shown a complete lack of dinterest in the issue. Following the Chernobyl and Fukushima accidents, public opinion has been against nuclear. However, given the intensive state supression of nuclear opposition in the media and engaged antinuclear activists being forced out of their jobs, there has been only a weak public rejection of the nuclear industry's plans. As the Croatian government also supports a lifetime extension, it is increasingly unlikely that the prevailing public rejection of the nuclear industry in Slovenia after the Fukushima accident will be taken into consideration. This is the case both at the local level, in the vicinity of the NPP, and at the national level. A referendum for public decision-making was promised in the coalition declaration of the newly elected Slovenian government, but never occurred.

This leaves policymakers and scholars with a range of open questions: what amount of waste is expected to be stored at the NPP site, as the Croatian side rejects the Slovenian proposal to construct a common LILW depository? What amount of waste will need to be managed, and consequently what scale and design for a depository is needed? Which actor will possess and manage the waste? What will happen with Croatian nuclear waste that is already deposited onsite at the NPP?

Similarly, numerous questions arise regarding the permitting and oversight of any planned waste disposal facilities. According to Andrej Stritar, the head of the Slovenian Nuclear Safety Administration (SNSA), construction of an additional new dry depository for spent fuel at the NPP site, which is needed for the life time extention of the NPP Krško from 40 to 60 years until 2043, can be permitted without an environmental impact assessment /EIA/or approval from the local community Krško.

Based on an earlier agreement between the Slovenian and Croatian governments but without any consultation with affected residents or consultations with interested NGOs or the general public, on 2 February 2017, the Slovenian Agency for Environment (ARSO) issued a decision that "for prolongation of the operation period of the NPP Krško ... there is no need to implement an environmental impact assessment (EIA) and acquire an environmental approval." (ARSO 2017) As ARSO is part of the Ministry of the Environment and Spatial Planning (MOP), Minister of the Environment Irena Majcen is at least partially responsible for the rejection of an EIA in relation to the prolongation of the operation time of the NPP through 2043. Minister Majcen has been blamed by environmentalists for years for the ineffective reaction of the government to industrial toxic contamination in the location of Cinkarna Celje in central Slovenia which had even attracted the scrutiny of the EU Commission. Political allies of industrial pollution and industrial pollution pressure groups in former eastern European countries and in numerous other countries are trying hard to place their adherents in key positions in environmental ministries and agencies. This has been the case in Slovenia even though after the declaration of independence in 1990 it was one of the more prosperous former socialist countries with a well established political culture and state of law. But after becoming a member of the EU in 2004 Slovenia like other former eastern European countries has seen only weak participation of citizens on fundamental political questions and continued strong development of nuclear production of electricity.

On 6 March 2017 the Legal Information Centre for NGOs in Slovenia filed a complaint in court produced by four NGOs against the decision of the ARSO and the MOP to not implement a EIA procedure for the life time extention of the NPP Krško. The ARSO had concluded that the prolongation of the operation time will not have any harmful impact on environment. Such an assessment however, is contingent upon whether an NPP is operating or not as an operating NPP poses more potential harm than one which is not in operation even though all NPPs bare risks. The NPP in question is of the earliest generation so that after 40 years of operation the potential of greater risks with continued operation should be considered. With regard to the ESPOO and Aarhus conventions ARSO did not correctly evaluate the potential risk of an environmental accident for the population and for future generations and other living creatures. ARSO argued that other governmental bodies, like the Ministry of Health, found no reason to be concerned about possible harm from an extension of operation. This however, shows negligence on the part of the Minister of Health, Milojka Kolar Celarc, a former economist, expert in taxes and alternatively for 10 years state secretary in the ministry of finances and official in an insurance company.

Slovenia expressed in its Declaration on Activities Republic of the Republic of Slovenia in the Institutions of the European Union from January 2016 to June 2017 its view that "On the potential further development of nuclear energy there must be a democratic consensus." (DeUDIEU1617 (2016) (Chapter Environment, Nuclear Safety) It is international standard and a willingly accepted obligation to take into consideration public opinion as called for in Directive 2014/87/ EURATOM. Article 8 requests member states in accordance with national legislation and international law to enable residents to effectively participate in the process of decision-making in relation to the issuing of a permit for a nuclear facility. In Slovenia this is to be assured by the procedure of issuing an environmental protection consent within an environmental impact assessment. Slovenia has around 200 hazardous industrial facilities that needed to go through such a procedure. ARSO decided in these cases that an EIA was needed, but in the case of the NPP Krško, despite the potential for a Fukushima or Chernobyl size accident and the far greater harm and more lasting consequences such an accident would have than could occur in one of the 200 industrial facilities, ARSO surprisingly followed the unprofessional opinion of the SNSA and Ministry of Health and did not ask for an EIA.

There have been additional signs of the aging of the NPP. Seven broken fuel rods were found during the last inspection laying at the bottom of the reactor; they were then replaced by simple metal rods. On 16 February 2017, the NPP was automatically shut down because of a defective "control valve for main supply water in the classic part of the power plant". It was just on 3 day later that ARSO rejected the EIA. Two months later, on 21 April 2017 at 3 a.m., a defective steam-relief valve woke the residents in the vicinity of the NPP Krško out of their sleep. The NPP had to be powered down again. Residents who lived just a few hundred metres distance from the fence of the NPP, strayed through the streets to the closest village, in an effort to find out what is going on and whether they needed to evacuate immediately or not. They could get no explanation on their telephones about what was going on. The guard at the entrance to the NPP could only tell them, he had no more information than they did.

The GEN, the formal owner of the Slovene half of the NPP (100 percent state owned) intends to ignore the decline of nuclear energy in Europe and as in several other eastern European EU member states started preparations and investments for construction of a second reactor at the site in Krško. During the search for a LILW repository across Slovenia more than 10 years ago it became obvious that because of the firm opposition of the local population it would be impossible to site any new major nuclear object anywhere else in the country except in Krško. The GEN fostered preparations for construction of NPP Krško II and engaged a consortium of firms to do the necessary research. In 2006 the Institut de radioprotection et de sûreté nucléaré (IRSN) contracted several subcontractors to conduct a seismic hazard and fault line assessment.

The findings and recommendations of the IRSN were not what the government expected: "IRSN's opinion on this matter is that this new and serious finding [about Libna trench, r. L.Š] does not allow concluding in a favorable manner as regards the suitability of the Krško II sites for the implantation of a new nuclear power plant. Acknowledging the fact that the feasibility of designing a new reactor against fault surface displacement is questionable, and consistent with IAEA and NRC recommendations, IRSN believes that GEN should consider revising its strategy for the Krško II project and further examine the possibility to search for an alternative site." (Oona Scotti 2013)

After these recommendations were proposed, the GEN suspended the contract with the IRSN and started looking after a more cooperative institute. History repeats itself for a second time. When the research for Krško I was done almost 50 years ago, the investors skipped the geological institute of Ljubljana considering it unreliable; they found their needed expertise in Skopje, Macedonia.

The findings and recommendations of the IRSN are important, as there is just one nuclear site in Slovenia where nuclear facilities are located in overly close proximity to one another. In these crowded conditions, if a major accident happens to one, they will all be affected. If the site is deemed unsuitable for the NPP reactor Krško II, the same seizmic conditions would impact reactor Krško I's extension, the additional dry fuel repository and the LILW repository. This makes it easier to explain why the ARSO rejected having an EIA procedure. Under the law they might then all have to be rejected.

The NPP Krško acquired an operation permit in 1984 without time limitation. ARSO followed the wrong opinion of the Ministry of Health, when it determined that prolongation of the operation time does not represent a change in the existing situation and thus poses no further threat to the security of the population and environment. The SNSA changed the operation allowance in 2013 so that the NPP must now perform a periodical security revision and may operate further if the results are positive. But for all other nuclear facilities the operation admission is limited to 10 years by the Law on Protection Against Ionizing

Radiation. Only the biggest and the most radiation hazardous facility in the country has no such operation time limitation. This is disrespectful of article 7 of the Convention on Nuclear Safety which obliges member contractors to set a legislative and regulatory framework to govern the safety of NPPs.

The ignoring of Slovenian and neigbouring states' citizens and civil society (NGOs) in the EIA process harks to the autocratic past. Exclusive corporative networks dominate over the state of law and the right of free speech. The nuclear energy industry's predominance and arrogance and the hypocritical character of today's key political institutions suggest democracy remains an illusion despite the social changes since the 1990s or more broadly over the last 200 years .

The regulation on financial compensation to communities near a NPP is defined in the law proposed by the Slovenian Ministry of Finance and by the SNSA as a compensation for the impact, expected and actual, for the loss of property value or for stress or health burdens as a result of the presence of the NPP. This compensation is smaller than the compensation for the proposed LILW repository, which was introduced in 2010. Behind this disproportionality there is the intention to promote and advance the nuclear industry, as well as the interests of professional groups which receive benefits through nuclear projects, high tech development in a region where industrial enterprises, banks and airports are sold under the EU commission pressure to foreign investors and their research centers moved out of the country to the seat of the enterprise. Nuclear reactors seem to be attractive as a lasting remnant of scientific industrial development.

The disproportionality in funding includes immediate forcing out of a job any person opposing nuclear power and their close relatives, not only from nuclear companies but also from companies cooperating with them and from all public administrative jobs. Further, only local NGOs that cooperate with the nuclear industry are supported by public financing, while those opposing it are cut from any public backing. Local public media faces the same censoring – only those that express full support for the nuclear industry and are not critical are able to survive.

2 Nuclear waste disposal in Croatia and Slovenia

2.1 Historical background

After the Second World War, the Yugoslavian leader Tito ordered the establishment of a nuclear weapons program. He saw the development of nuclear power as key to overall economic development and as a hard currency for use in international politics. Yugoslavian nuclear research was initially conducted in close collaboration with Norway, as Tito was not trusted by Stalin after his clash with the Communist Information Bureau (Cominform) in 1948 after which Yugoslavia was expelled from the Soviet Bloc.

Tito terminated the nuclear weapons program after a critical accident at Vinca's heavy water RB reactor killed one person on 15 October 1958 and provoked radiation poisoning of a further five persons. He opted instead for a 'peaceful' nuclear power program and supported the establishment of research facilities for the study of the entire nuclear fuel cycle. There was a pilot-scale uranium mine at Zletovska Reka in Macedonia and a larger, fully operating deep underground mine at Žirovski vrh, a densely populated region of Slovenia. Mining took place between 1968 and 1990 without any health protection for the miners, who were directly breathing radioactive dust in, and those living nearby the mine. Mining operations were ultimately shut down, though not because of significant health concerns, which were dismissed by the mining management. Rather, Croatia did not want to continue paying for half of the production, given that it could purchase uranium for less. Slovenia was forced to either pay the difference of 10 million USD per month and cover all production costs for its own uranium cake or shut the mine down. This finally happened during the democratic political changes of 1990-91. In the public media, the closure was portrayed as the result of political pressure from the new Green party of the time which was not dissuaded by the job losses threatened by mining proponents.

2.2 *The early opposition to nuclear in Slovenia*

Free public discussion about the construction of a NPP was not tolerated in Yugoslavia. Similarly, the question of what should be done with the remaining nuclear waste was not addressed in public or media discussions, as the regime did not tolerate any disturbing questions that could hinder their project. Both the public and nuclear experts were excluded from the negotiations about the purchase of the Westinghouse NPP. The negotiations were led by formal investors in the Sava River Power Plants Ljubljana and Elektroprivreda (Electrical management) Zagreb without expert knowledge in nuclear power. Obviously, neither the scientific competences of the negotiators nor the question of what to do with used fuel in the long term played an important role in the process. It was President Tito himself, who was the deciding authority behind the negotiations. The IJS senior management and its former director Milan Osredkar felt hurt for years afterwards because their opinions were disregarded during the negotiations for the purchase of the NPP.

The problematic character of the supposedly 'peaceful' nuclear power was unmasked by the Chernobyl disaster in what is now the Ukraine in 1986. The accident had a major international impact in central Europe[2]. A little more than a year after the Chernobyl accident, in November 1987, Italy held a referendum which led to a decision to abandon nuclear power plans. In Slovenia, there were huge antinuclear demonstrations in Ljubljana and Krško which led to the rise of the Green party. In the first democratic elections in April 1990, the Green party won almost 9 percent of the general vote within the proportional election system and joined the coalition of newly founded political parties. As a condition to joining the opposition parties' coalition, the Greens advocated shutting down nuclear power.

Despite the success of opposition movements in neighboring countries and the growth of nuclear opposition in Slovenia, plans for nuclear development proceeded. After 1992, Drnovšek's government wanted to move ahead with both construction of a LILW depository and with reconstruction of the NPP Krško with the purchase of new steam generators after only 10 years of operation of the old ones.

His next step to foster nuclear power in Slovenia was a public invitation for local communities to candidate as a possible location for a LILW depository, with the promise of adequate financing. Soon, a number of communities (Velika Polana, Zavrč, Šmartno pri Litiji, Sevnica, Brežice, Krško, Poljčane, Lenart, Kostanjevica na Krki), all within economically depressed central and eastern Slovenia and largely concentrated in the vicinity of the NPP Krško, announced their willingness to compete to host a depository. Of the 36 potential locations, five were chosen. Three were in one of the most depressed regions of the country, the western Haloze (Nadole, Kočice, Sedlašek), and the others two were close to the location of the NPP (Presladol, Poklek and Blanco).

Local support for the repository was not, however, widespread. After a brief period, the process of choosing a LILW depository location was brought to a halt. Around the country, mayors speculated about what amount of money their community could gain from hosting a depository, but were later confronted with such strong popular opposition that they had to give up their plans with the exception of the three communities in the vicinity of the NPP site, where there were already large populations of NPP workers. These communities were promised sizable compensation which contributed to their pro-nuclear stance; they tended toward intense political repression of individuals and public opinion opposed to nuclear power.

In 1995, a third of the members of the Slovenian parliament signed a bill calling for a referendum to shut down the NPP Krško. At the last moment, this bill was obstructed by the then-liberal prime minister Janez Drnovšek.

2 Several years prior, in 1978, Austria had held a referendum on the Zwentendorf nuclear power plant. Voters narrowly rejected it, with 50.5 percent voting against it.

2.3 The Slovenian national inventory

The Slovenian nuclear program includes the NPP Krško and the research reactor TRIGA MARK II, situated in the suburbs of the capital Ljubljana. These two reactors have produced approximately 450 tons of heavy metal (t_{HM}) in spent fuel. "The latest estimate of the total number of spent fuel assemblies generated in their 40-year operation is now 1553 (620 metric tons of metallic uranium). For the extended lifetime until 2043, a total number of 2281 fuel assemblies (or 912 metric tons) has been calculated." (Železnik and Kegel 2011: 704.3) The spent fuel is stored at the NPP Krško, while the used fuel from the TRIGA research reactor was returned to the US in 1999. As the operator of a nuclear facility is responsible for handling used fuel waste, the heat-generating nuclear waste is stored in the spent fuel pool at the NPP. The ARAO will take over spent fuel after the NPP and the TRIGA research reactor stop operation definitively. It will be also responsible for potential recycling, for planning and construction of an interim repository, and packaging and depositing of spent fuel. After construction of the repository, the ARAO will also be responsible for its operation, management and long term monitoring and maintenance. At the TRIGA, there is also a central depository for nuclear waste from small producers in operation, also managed by the ARAO.

Additionally, there are the Jazbec mine waste pile and the Boršt mill tailings site of the closed uranium mine at Žirovski vrh, where uranium ore was exploited underground. The Jazbec mine waste pile covers an area of 51,000 m^2 and amounts to 1,828,000 tons of mine waste with an average concentration of 53 g U3O8/t and 48,000 tons of red mud from raffinate neutralisation with the specific activity of 65 kBq of Th-230/kg.

The closed uranium mine is situated approximately forty kilometres northwest of Ljubljana in a densely populated countryside. Both waste piles are assumed to be statically unstable. Hard rain in November 1990 caused an earth slide at the Boršt mill tailings disposal site, which is situated on a hillside. Some 7.4 million tons of material became unstable and started sliding at a rate of about 0.5 – 1 mm/day.[3] With the addition of drainage, the sliding was stopped. Ultimately, the site was deemed unsafe and in 2002, the NSA rejected permits for the construction of the Boršt mill tailings disposal site.

---

3 "Geophysical investigations and research boreholes determined the depth of the landslide of 20 up to 60 m. Deformations stopped after the draining tunnel had been excavated in 1994. Present measurements have confirmed the landslide stability. But the question of earthquake stability of the landslide still remains open." Karmen Fifer Bizjak: Stabilnost plazu ob potresu v območju jalovišča Boršt / Earthquake Stability of the Landslade in the Region of the Uranium Mine Tailing Deposit, Zbornik referatov 1. Šukljetovih dnevov 1, 2000, Ed. Gaberc, Ana Marija Majes, Bojan, p. 113-120.

Both repositories are to be handed over to ARAO. "The Žirovski Vrh Uranium Mine was in operation from 1984 to 1990. Its lifetime production was 610,000 tons of ore corresponding to 452.5 tons (U3O8 equivalent) of yellow cake." (Vrankar 2005: 127.1).[4]

2.4 The (interim) storage site

Used fuel from the NPP Krško is first deposited into a designated water pool, which is housed in the same building as active systems for cooling the pool. Following the Fukushima accident in 2011, the NPP prepared a study to assess different storage options for spent fuel with the aim of reducing the risk of a nuclear incident in the NPP and improving the safety of the storage pool. The NPP then made a proposal to construct an additional dry storage building for spent fuel by 2018, with an operating life of 60 years in order to ensure continuous operation and sufficient storage capacity in case of the extension of operation time for the NPP Krško 1 and the construction of the NPP Krško 2. The construction and operation of the repository will be financed by the NPP. After a period of dry storage is provided for further processing, used fuel will be packaged and disposed of in a deep geological repository after 2065. However, these plans are in question as it is not yet clear whether Croatia will join these plans or manage its own repository. What is certain is that Croatian stakeholders support the lifetime extension of the NPP Krško 1 for 20 years, until 2043.

2.5 The waste management strategy

2.5.1 The plan for the LILW repository

With the adoption of the Ordinance of National Spatial Plan for the LILW Repository at the Vrbina site in the Krško municipality the process of selecting the location and type of depository was completed. It was December 2009. In January 2011, two experts from IAEA evaluated the project with regard to the geotechnical aspects of construction. They identified critical elements of the suggested design, defined missing data from the field studies, and assessed its long term qualities (safety). In separate assessments they rated the proposal for a LILW depository at Vrbina very negatively. "The geological conditions of the selected site were found to be generally unfavourable ...The worst finding,

4 During its operation, miners were inhaling toxic dust in the deep mine areas without the use of respirators or other lung protection devices.

however, was that the ground water level is a mere 3m below ground level meaning that the construction and operation of the repository will take place below ground water level which clearly does not comply with IAEA requirements for the safe design of a waste repository." (IAEA 2011b) Such conditions are unsursprising, given that the NPP was sited along the Sava river bed which provides cooling for the reactor. While proximity to the NPP proved to be important for limiting public opposition to the depository, geotechnically it represents a problem.

The IAEA experts, Robert Chaplow and Jaroslav Pacovsky, exposed several additional unsolved problems in relation to the proposed site and repository[5] (IAEA 2011a). Due to the unclear ownership of the land where the repository is to be built, the exact site was inaccessable for sample taking. Samples were thus taken in the vicinity of the proposed site rather than at the site itself; futher impacts of the "agressive underground water" during construction were not specified. The experts further found that basic information needed for monitoring in advance of and during construction was missing.

They also argued that after roughly 300 years, the proposed silos and waste packages would lose their ability to act as physical barriers. The ARAO assumes that degraded concrete will retain the abiliy to act as a chemical barrier, absorbing certain radionuclides and retarding their migration away from the silos. However, the IAEA experts argued that this assumption is unrealistic and found that the waste is not specified well and the procedures for filling the silos were not sufficiently defined. They made a similar finding in regard to the materials used, which should be certified for 300 years; the certification of materials is not mentioned in the project documentation. Pacovsky further reported that communication between the developers of individual elements of the preliminary repository project did not correspond to the up-to-date standards of the project.

This report could not be found on the ARAO's official web site for 2011, in contrast to the report of the Integrated Regulatory Review Service (IRRS) of the IAEA in the same year. Moreover, there was no public discussion by experts about what the two IAEA reports meant for the LILW repository project in Vrbina and the associated risks.

5 "In addition, it became apparent that the calculations carried out by ARAO were based on information obtained from areas remote from the actual proposed repository site." See: Chaplow (2011: 4).

2.5.2 The plan for the used fuel repository

The main obstacle for a clear and well-defined radioactive waste disposal plan and reason for the two variants (see Fig. 1) is the uncertainty around whether Croatia will construct its own repositories or join Slovenian activities and thus construction of the LILW and spent fuel repositories. The start and duration of the main activities related to spent fuel storage for both variants are depicted in Figure 1.

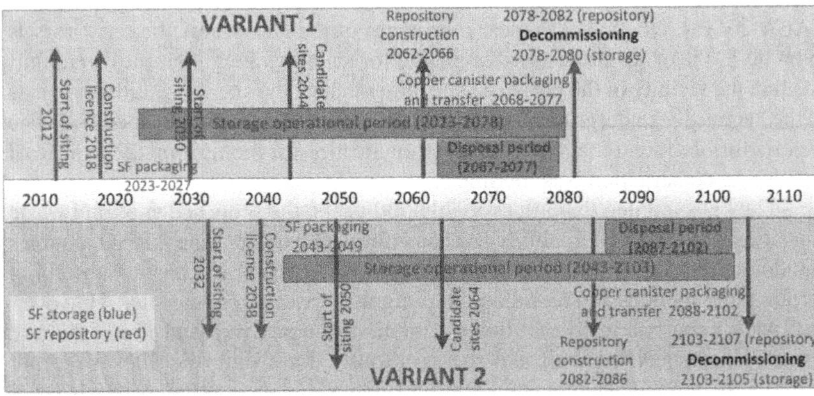

Figure 1: Timeline of activities related to spent fuel storage and repository for 2 variants. Source: Železnik, N. and Kegel, L. 2011: 704.4) http://www.djs.si/proc/nene2011/pdf/704.pdf.

The ARAO has prepared the "Revised Reference Scenario for Geological Disposal Facility in Hard Rock" a semi-generic description of the geological repository considered suitable for the NPP spent fuel, high-level waste, and some long-lived low- and interim-level radioactive waste (LL-LILW). It discusses all phases of repository development, including research activities, siting, construction, operation, and closure. The analyzed scenarios are prepared in two variants. The plan for spent fuel disposal was based on assumptions that there will be no reprocessing and that the repository will be constructed in a hard rock environment at a depth of 500 meter. The entire disposal system plan is based on the Swedish KBS-3V concept, developed by the Swedish Spent Fuel Management Agency (SKB). The repository development also includes the construction and operation of an underground testing facility at the site of the future repository. A sufficient cooling period is required prior to disposal to allow optimal utilization of canister capacity (4 SF elements per copper canister).

Disposal of spent fuel starts after a 45-year storage period, according to both variants. No retrievability options are foreseen for either the LL-LILW repository or the spent fuel repository.

3 The legal and institutional framework

3.1 The legal framework

Article 72 of Slovenia's constitution states that everyone has the legal right to a healthy living environment. The state is thus responsible for promoting a healthy living environment. To this end, the conditions and manner in which economic and other activities are pursued have to be established by law. In the case of nuclear power, however, there is no way to assign full responsibility for any possible environmental damages, as they are restricted 1) to conditions defined by law and 2) responsibility for harming people or nature regards only individuals, while corporations are largely exempted.

The Vienna Convention on Civil Liability for Nuclear Damage was ratified by Croatia in 1993. To regulate nuclear waste management of the joint NPP Krško, in 2003, both countries signed a treaty on the regulation of the status and other legal relations connected to investment in the nuclear power plant, its utilization, and its decommissioning (Official Gazette of RS, no. 23/2003). Croatia adopted the Law on Radiological and Nuclear Safety in accordance with the EUROATOM Treaty in 2013. Additionally, Croatia adopted the Strategy for the Disposal of Radioactive Waste, Spent Sources and Spent Nuclear Fuel in 2014, to regulate the obligation to take over used fuel from the NPP Krško.

The Law on Ionising Radiation Protection and Nuclear Safety (ZVISJV-NPB6) and the secondary legislation is in accordance with the standards of the IAEA. Slovenia was also fully supportive during the preparations and adoption of the directive 2009/71/EUROATOM. In 2011, Slovenia adopted the EU Radioactive Waste and Spent Fuel Management Directive and, until January 2013, harmonized its national legislation with it.

3.2 The institutional framework

The Ministry of Environment and Spatial Planning (MOP) is the regulatory authority responsible for the development and implementation of the waste management policy. It has a supervisory function with respect to the Administration of the Republic of Slovenia for Nuclear Safety (URSJV) in the Ministry of Health, the Office for Civil Protection and Disaster Relief at the Ministry of Defense and the Ministry of Interior.

The SNSA is responsible for the control of nuclear safety, nuclear facilities and radioactive sources in the country, with the exception of resources in human and veterinary medicine, which is in responsibility of the URSVS.

The SNSA performs administrative and development tasks in the fields of radiation and nuclear safety, radiation practices and the use of radiation sources, environmental protection against ionizing radiation, physical protection of nuclear materials and facilities, nuclear non-proliferation and security of nuclear materials, radioactivity monitoring in the environment, and liability for nuclear damage. It also carries out inspection duties in the above areas and is involved with the State Headquarters of Civil Protection in the determination of protective measures for the population and public information regarding emergency radiological or nuclear events.

The SNSA further performs professional, administrative, supervisory, and developmental tasks in the fields of radiation use in medicine and veterinary medicine, protection of human health against the harmful effects of ionizing radiation, systematic inspection of working and living environments where humans have been exposed to natural radiation sources, monitoring of radioactive contamination of food and drinking water, reduction and prevention of harmful effects of non-ionizing radiation, and auditing and authorization of radiation protection experts.

The Energy Directorate, as a body within the Ministry of Infrastructure (MZI), is responsible for the development of the broader energy policies, laws, and use of nuclear energy. The directorate is preparing a comprehensive energy policy regarding energy supply and the granting of mining rights for the exploration and exploitation of all kinds of mineral resources. It monitors the state-owned energy enterprises, including the GEN energija, the Slovenian owner of the NPP Krško. In this way, the Directorate also has an impact on the nuclear safety of the facility, as its long-term provision depends largely on the stability of the NPP's business and financial position. The Energy Directorate also monitors the operation of the fund for the decommissioning and disposal of radioactive waste from the NPP, and in this way makes nuclear electricity presumably one of the cheapest and cleanest energy sources on the market. These low costs and clean sources are realized via legally and administratively consolidated low tax rates for nuclear waste operators and low costs for future nuclear waste management, omission of or state-paid insurance costs and the initiation of administrative obstacles for the integration of solar, wind and geothermal energy in the electricity grid, as well as active support for construction of public infrastructure for centralized production of electricity entities like NPPs, enforced by the government and prepared by the energy directorate.

The Ministry for Infrastructure (MzI) has a supervisory function and is being integrated into the work of the Croatian-Slovenian intergovernmental commission established under the BHRNEK agreement.

3.3 Procedures and criteria for site selection

The licensing process for a second NPP in Slovenia will follow in ten phases. These comprise: 1) Preparation of the National Strategic Spatial Plan by the MOP on the basis of national programs; 2) Issuing of an energy licence for the NPP by the Energy Agency and of an operating permit for the NPP by the Ministry of Economy before the start of the National Strategic Spacial Plan process; 3) Issuing of the National Spatial Plan (NSP) which covers the general siting process for all facilities of national importance. The NSP is led by the MOP but involves other stakeholders, including the space arrangement authorities (various ministries, also SNSA, municipal bodies and agencies, non-governmental organizations), the public, involved municipalities, neighbouring countries (signers of ESPOO convention), etc. Part of the NSP for a new NPP is also the Special Safety Analysis, which is the key document regarding nuclear and radiation safety in the siting process; 4) Conducting of an Environmental Impact Assessment (EIA) and obtaining environmental protection consent: the investor prepares the EIA report including information required by competent ministries (including SNSA). The EIA and the project of intended activities, after being publicly displayed, are the basis of application for the environmental protection consent, which is issued by the Environmental Agency; 5) Consent for construction is issued by the SNSA based on the overview and approval of the Preliminary Safety Analysis Report, design for construction, decommissioning program, waste management program, etc. The consent for construction represents the basis of application for a construction license; 6) The investor applies for a license after receiving the consent for construction from the SNSA. After the license is issued, the construction of the NPP may start; 7) The expectations of the European Commission (EC) are examined. The investor gathers information regarding requirements pertaining to radioactive waste disposal as required by the Treaty establishing the European Atomic Energy Community (EURATOM). The SNSA provides this information to the EC, which is obliged to give its opinion within six months time; 8) Before the license to use the facility will be issued, a technical check and a trial operation must be performed. The trial operation requires the consent of the SNSA. Once the SNSA has issued its consent, the MOP must issue a decision on the start of the trial operation which is to be supervized by the SNSA; 9) The MOP issues the license for using the facility after it verifies that parameters regarding environmental impacts from the trial operation meet the prescribed limits. License for use of the facility is the final part of the general process for licensing objects; 10) The investor applies for an operating license from the SNSA after first receiving a license for the use of the facility. The investor must present the Final Safety Analysis Report, report on the trial operation, decommissioning program, quality management documentation, etc. The SNSA reviews and decides about the approval of the submitted documentation and on the issuing of the operating license. This overarching position of the SNSA is in a certain way overshadowed by the fact that the longtime director of the office, Andrej Stritar is a strong supporter of nuclear power[6].

The historical, social and political changes that took place between the 1970s and the 1990s distinctly impacted the procedure and criteria for nuclear site selection. The site selection of the NPP Krško in the 1970s took place without any participation of the concerned population, but instead with persecution by the police and blacklisting of individuals who opposed the nuclear plans. These steps were orchestrated by President Tito. However, with the democratic political awakening of the population in the following decade, such top down methods of site selection became ineffective. With the rejection of nuclear power in neighboring Italy and Austria, it is not surprising that in the 1990s no site for a LILW repository could be found anywhere outside of Vrbina (near the operating NPP site). Political proponents of nuclear power in Slovenia and Croatia came to realize that it would not be possible to find new sites for nuclear facilities outside of existing sites. As a result, the proposed NPP Krško 2 is to be built at the same site as Krško 1. Finding an optimal site on solid ground, unthreatened by flooding, seismic instability, fires, or mudslides, that would also win the approval of concerned populations proved impossible. The chosen site at Vrbina does not have optimal physical conditions, but it was deemed acceptable because the dominant local political forces favor nuclear power and have been backed by successive national governments.

The siting of nuclear objects should be executed according to the National Strategic Spatial Plan and the Ionising Radiation Protection and Nuclear Safety Act (Article 64). There is however confusion about whether the used fuel interim dry depot, which was invented by the NPP after the Fukushima accident and should be constructed by 2018 also should be based on the requirements of the National Strategic Spatial Act. It is also unclear whether postponing the operation of the NPP until 2043 requires a new procedure, or whether it is enough for the government to simply declare its intention to extentd the operation time of the existing NPP facility. In the latter case, there would be no need for either a

6 Andrej Stritar, declared during a crowded public forum on nuclear power in the 1980s at Cankarjev dom that nuclear waste is so safe that he was ready to put it under his bed.

strategic spatial plan or for an environmental protection assessment. Even the construction of a second reactor could be intepreted in this way, which would favor the interests of investors.

The nuclear industry's intention is to resolve one of its 'main problems' that Slovenia does not have any regulations or prescribed procedures, criteria, or set of conditions for determining the parameters which nuclear technology and facilities must meet. Establishing prescribed project parameters would do much to facilitate the procedure of consent acquisition for investors and shorten the procedure itself, as the spatial planning stakeholders would then merely need to check whether the design meets the requirements and conditions. These conditions are, for the time being, determined on an ad hoc basis during the preparation of the National Spatial Plan.

3.4 Compensation mechanisms and socio-economic impact

As soon as the National Spatial Plan for the LILW repository was adopted in 2011, the municipalities of Krško and Brežice and the ARAO suspended their cooperation in the siting, construction and operation of facilities with the local nuclear partnership, which had been in effect since 2007. The partnership's purpose was to establish mutual understanding and cooperation between investors, governmental bodies, and local authorities in Krško and Brežice on the one side and concerned local populations on the other regarding operating the existing nuclear facilities and siting new ones, along with managing compensation. Once the plan was adopted, investors assumed they could drop their pretense of collaboration with the local population. They secured an exclusive majority on the community council, so they could support any proposal put forward by the mayor of Krsko to comply with the ARAO and the nuclear facilities' interests, and ultimately to elect a mayor loyal to them.

The residents' wish for continuation of the local partnership was ignored. Since communication between the nuclear industry and its political representatives in state and local administration was no longer deemed necessary, the residents were not accepted as active NGOs and financial support for their organisations was withdrawn; they thus became inactive. In 2015, the Local Nuclear Partnership was reestablished to give voice to the needs and interests of concerned residents. This included residents of all municipalities of the Posavje region, which receive compensation for exposure to the NPP (Krško, Brežice, Sevnica, Kostanjevica na Krki and Kozje) and residents from other regions. However, the attitude of the local communities and the owners of the nuclear facilities to this newly established partnership remain negative.

These communities receive financial compensation according to the Decree on the Criteria for Determining the Amount of Compensation for Limited Use of Space in the Area of a Nuclear Object (Uredba o merilih za določitev višine nadomestila zaradi omejene rabe prostora na območju jedrskega objekta) which can be used by mayors to take political and administrative control over the communities and to silence opposed residents. They are paid exclusively from the Slovenian decomissioning fund, as the Croatian side is of the position that its decomissioning fund is sufficient only to pay for the Croatian nuclear waste respository on Croatian soil. As it is negotiated between the two states, the Croatian share of the nuclear waste will be removed first once the NPP Krško I stops operation in 2043 or, if the NPP Krško II is built, around 2070. The Croatian decomissioning fund is intended only for nuclear waste maintenance once this nuclear waste is delivered to Croatia. It may seem illogical that the Croatian investors in NPP Krško do not participate in financial compensation to the local communities around Krško; however this is a strategic move on the part of Slovenian investors to ensure that Croatian investors support the lifetime extension of NPP Krško I and possible construction of NPP Krško II.

The ARAO pays the communities 5,647,211 Euro per year for the LILW repository, although there is not yet a repository in sight. The NPP pays taxes into this compensation fund, however more than 50 percent of the money is transfered back to the state budget (see Table 1 and Table 2).

At the local level, residents face the risk of exposure to possible health hazards and economic disadvantages but also receive better local infrastructure. For employees and representatives of the nuclear facilities in public administration, there are also economic advantages. Local political and democratic culture is under constant pressure from the interests supportive of the nuclear facilities. Public disagreement with these interests catapults an individual or a group into the margins of local society and even into a precarious economic or social status.

Table 1: Communities receiving annual compensation or tax payments

Community	Compensation		Tax on use of building Site		Together
Article of the Decree	Article 7 (1)	Article 7 (2)	Article 8 (1)	Article 8 (2)	
Krško	5.698.440 €	57.560 €	153.570 €	2.357.167 €	8.266.737 €
Brežice	-	-	153.570 €	1.715.368 €	1.868.938 €
Sevnica	-	-	153.570 €	210.347 €	363.917 €
Kostanjevica	-	-	153.570 €	28.642 €	182.212 €
Kozje	-	-	153.570 €	39.626 €	193.196 €

Source: Decree on the criteria for setting compensation level payable for limited use of space within the area of a nuclear facility.
www.mop.gov.si/.../visina_nadomestila_raba_prostora_obmocje_jedrske_objekte.doc

Table 2: Annual compensation and tax on use of building site

	Compensation		Tax on use of building Site		Together
	Article 7 (1)	Article 7 (2)	Article 8 (1)	Article 8 (2)	
NPP, compensation	57.560 €	51.229 €	-	-	5.227.789 €
NPP, tax	-	-	767.850 €	4.351.150 €	
The ARAO, compensation for the LILW repository (Vrbina)	5.640.880 €	6.331 €	-	-	5.647.211 €

Source: www.mop.gov.si/.../visina_nadomestila_raba_prostora_obmocje_jedrske_objekte.doc

The majority of political parties represented in the parliament are not willing to openly admit to their support of nuclear power but promote it nevertheless as much as possible. Increasingly, the majority of members of successive governments prefer not to speak publicly on nuclear issues and instead hold up the opinions of 'nuclear experts' who are chosen, according to their endorsement and promotion of nuclear energy. In the official statements and documents on energy, nuclear energy is presented as the most sustainable and cheapest energy option for the long run.

Although there has been significant growth of sustainable sources of energy (SSE) and espcially solar power since 2009, when Slovenia adopted the Regulation on Support for Electricity Produced from Renewable Energy Sources (Ur.

l. RS, no. 37/2009) (Brecl and Topič 2014), there are still very effective administrative obstacles, orchestrated by nuclear power representatives in the government, to prevent increased use and promotion of wind, solar, and geothermal energy. Costs and financing

In the ReNPROG (2015), the MOP states that the majority of funds for radioactive waste and spent fuel management should come solely from the sale of electricity from the NPP Krško: "The state budget and taxpayers with these costs have not been and will not be burdened". However, this statement has never been underpinned by serious calculations. The program for decommissioning and disposal of low and intermediate level radioactive waste and spent nuclear fuel from the NPP Krško estimates the costs of decommissioning between 1.15 and 1.7 billion Euros. The decommissioning program further defines the contribution for the fund as 3 Euros per megawatt hour, or 0.3 cents per kWh of electricity produced by the NPP and sold in Slovenia. The Slovenian Fund for used fuel and disposal of radioactive waste from the NPP Krško was founded in December 1994, 13 years after the NPP's operation began in 1981. The HEP only began payments into the Croatian fund for decommissioning in 2006, but pays an annual rate of 14.25 million Euro. At the moment, the budget of the Croatian fund is higher than the budget of its Slovenian counterpart. The two funds are not comparable, either in regard to their financial sources nor in regard to their management and use of finances. They are alike only in that they are funded in order to enable the decommissioning of the NPP Krško and refer to the 2002 Treaty between the Government of the Republic of Slovenia and the Government of the Republic of Croatia on the status and other legal relations connected to investments in the nuclear power plant, its exploitation and decommissioning.

Table 3: Nominal costs of all DP activities in various Slovenian scenarios (million Euro 2009)

Activity	NPP 2023	NPP 2043
D&D	518,93	554,22
SF storage	319,23	417,79
SF disposal	1136,45	1330,15
LILW joint disposal 2018	532,78	742,43
LILW joint disposal 2038	-	606,34
LILW separate Slovenian	491,00	697,79
LILW separate Croatian	271,70	271,70

Source: http://www.rs-rs.si/rsrs/rsrs.nsf/I/K73D90123DFE26972C12578F50019CAD8/$file/SkladNEK_SP06-09.pdf

However, the treaty does not clarify whether the states will construct a single joint repository for used fuel or separate repositories. Slovenia prefers a joint deep repository project, as it would be financially favourable for both countries. Croatia accepts joint interim repositories until the final decommissioning, along with the lifetime extension of the NPP Krško, but criticizes the Slovenian side for not signing the second draft revision of the decommissioning of the NPP in 2010. Croatia criticized that Slovenia unilaterally decided to construct a LILW repository. Given the likely lifetime extension of the NPP Krško until 2043, and that the 2011 Directive 2011/70/ Euratom committed all member states to work out their respective policy for radioactive waste management and spent nuclear fuel by August 2013 and implement that policy by August 2015, Croatia argued that it is only responsible for 50 percent of the radioactive waste and spent nuclear fuel from the NPP Krško for which it will continue to research locations for a LILW repository on Croatian territory. A possible location has been indicated at Trgovska gora, close to the Bosnian border. Therefore, a joint LILW and deep repository for used fuel from the NPP Krško appears rather unlikely.

On 20 July 2015, a meeting of the intergovernmental commission of Croatia and Slovenia confirmed plans to extend the operation time of NPP Krško from 40 to 60 years (until 2043), even though there are clear signs of reactor aging and downtown Zagreb, the capital, is just 24.5 km away. However, the meeting came about five years too late. During the previous meeting in 2010, the Slovenian side made a strategic mistake by refusing to accept and confirm Revision 2 of the Program of Decommissioning and Radioactive Waste and Spent Nuclear Fuel for the NPP. This refusal was due to the opposition from the Slovenian council of experts - i.e. 'nuclear experts' who all belong to the nuclear lobby. They were opposed to the decommissioning program because the dues to the decommissioning fund would have risen from 3 Euro per MWh to at least 6 Euro per MWh. As a result, from 2010 onwards the payments into Slovenia's decommissioning fund are too low to cover the expected final costs. It is not surprising that Croatia supported the extension of the NPP operation until 2043, because under these financial conditions the NPP is likely to be very profitable and nuclear power can be said to be the cheapest of all energy sources. However, even with this profitability there will not be enough money collected for the management of the radioactive waste and used fuel (see Figures 2 and 3).

According to international standards, the ultimate financial responsibility lies with the country where a NPP is situated - that is, if the money runs out, the Slovenian taxpayers will have to pay the difference. At this moment, the costs for decommissioning have apparently slightly decreased (by approximately 400 million Euro nominal value).

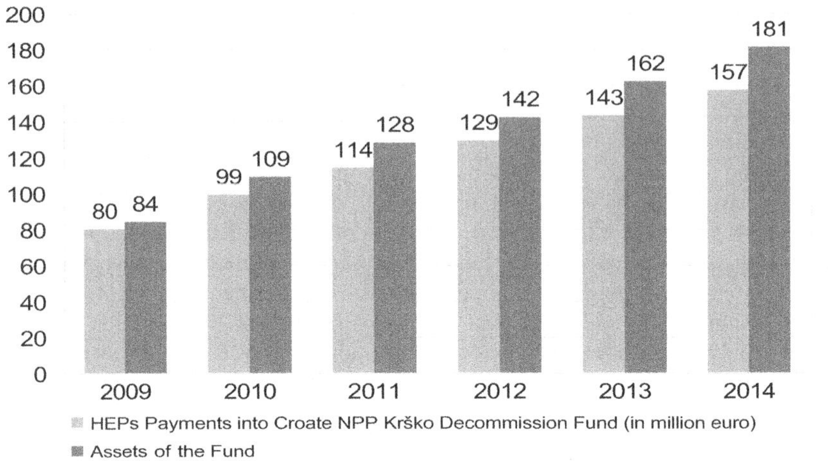

Figure 2: Assets of the Croatian NPP Krško Decommission Fund (in million Euro).
Source: http://www.fond-nek.hr/images/Prilog20-skracena%20verzija.pdf, p. 13.

These funds will be reallocated to the construction of the dry storage for used fuel at the NPP site as operating expenses (previously they were included in the program for decommissioning and disposal of LILW and used fuel). Thus, there will be even fewer assets for the actual decommissioning of the NPP.

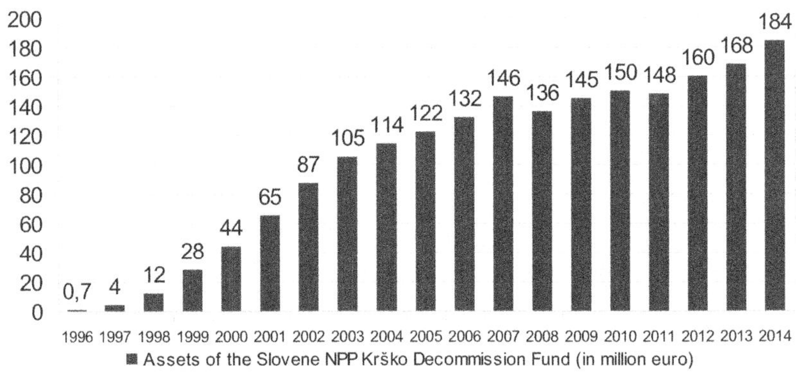

Figure 3: Assets of the Slovenian Decommission Fund, Sklad za financiranje razgradnje NEK in odlaganje radioaktivnih odpadkov iz NEK, (Slovenian Fund for financing of decommissioning of NPP Krško and depositioning of radioactive waste from the NPP).

4 Conclusions

The cooperation of two states as joint owners of NPP Krško has proven to be difficult. The two states have had opposing interests from the beginning, such as with the uranium mine Žirovski vrh and the subsequent mill tailings site in Slovenia. The owners of the plant also had different opinions regarding the new steam generators bought just 13 years after the start of operation. Today, it seems likely that there will be two LILW depositories built and two separate final deep ground depositories for the used fuel. Both owners of the NPP have however agreed to extend the NPP's operation through 2043. This extension was agreed upon despite key unanswered questions about radioactive waste management and the fact that there are clear indications of reactor aging, such as seven broken fuel rods found in the reactor during the last inspection. Although mandatory nuclear waste management provisions were initially adopted, 20 years after the start of operations, the payments into two separate decommissioning funds are not coordinated and not sufficient to fully cover the expected costs of radioactive waste management and decommissioning. There have also been conflicting opinions among the owners regarding compensation payments to the surrounding population. There was a local partnership to represent the interests of the affected residents for a short period of time, but it was disbanded after the industry's favored plan to construct a LILW depository next to the NPP was accepted. The local population was not consulted when the NPP was planned or during its operation. As of 2015, a new local nuclear partnership is trying to organize itself to represent the interests of the affected population.

References

ARAO (2011a). *Program of NPP Krško Decommissioning and SF & LILW Disposal*, revision 2 - Module 1, Module 1, Introduction, APO Ltd Doc. No.: 25-09-3435/06; ARAO Doc. No.: 10-1/01-01-027-00.

ARAO (2011b). *Program of NPP Krško Decommissioning and SF & LILW Disposal*, revision 2 - Module 7, Scenario Development and Costs Estimate, APO Ltd Doc. No.: 25-09-3435/06; ARAO Doc. No.: 10-1/01-01-027-00.

ARSO (2017). *MOP. Sklep za poseg: Podaljšanje obratovalne dobe Nuklearne elektrarne Krško s 40 na 60 let do leta 2043.* (Decision for intervention. Prolongation of the operation time of the NPP Krško from 40 to 60 years until 2043), Nr. 35405-286/2016-16, 17.2.2017.

Brecl, K. and Topič, M. (2014). Energy and economic yield of photovoltaic systems: reactive-power impact. In: *Elektrotehniški vestnik*, ISSN 0013-5852. [Slovenska tiskana izd.], 81(1/2), 9-14.

Čarman, M.; Jemec Auflič, M. and Komac, M. (2014). *Landslides at a uranium mill tailing site Boršt (Slovenia) detected by radar interferometry*, Berlin, Heidelberg: Springer, 527-536. http://lps16.esa.int/posterfiles/paper0491/Landslides%20at%20a%20uranium%20mill%20tailing.pdf, last accessed 11 June 2017.

DeUDIEU1617 (2016). *Deklaracija o usmeritvah za delovanje Republike Slovenije v institucijah Evropske unije v obdobju januar 2016-junij 2017.* (Declaration on orientations of the Republic of Slovenia in the institutions of the European Union in the period January 2016–June 2017) (Official journal RS, no. 19/16). http://www.pisrs.si/Pis.web/pregledPredpisa?id=DEKL42, last accessed 11 June 2017.

Fifer Bizjak, K. (2000). *Stabilnost plazu ob potresu v območju jalovišča Boršt.* (Earthquake Stability of the Landslade in the Region of the Uranium Mine Tailing Deposit), Zbornik referatov 1. Šukljetovih dnevov 1. In: Gaberc, Ana Marija and Majes, Bojan (Eds.), 113-120.

Fond za financiranje razgradnje i zbrinjavanja radioaktivnog otpada i istrošenog nuklearnog goriva Nuklearne elektrane Krško (2015). *Croatian decommission fund for financing of decommissioning and removing of radioactive waste and used fuel of NPP Krško 2014.* http://www.fond-nek.hr/images/Prilog20-skracena%20verzija.pdf, last accessed 19 January 2017.

IAEA (2011a). *End of mission report on »LILW Repository Mission 2 – Presentation of Results 3«, 2011-01-18 to 2011-01-20, (IAEA-TCR-05872)*, Developing a New Iteration of Decommissioning, Spent Fuel and Low and Intermediate Level Waste Management Programmes for the Krško Nuclear Power Plant, p. 4.

IAEA (2011b). *End of Mission Report on »LILW Repository Mission 2 – Presentation of Results«*, 2011-01-18 to 2011-01-20, IAEA-TCR-05859.

ReNPROJG (2006). *Resolucija o Nacionalnem programu ravnanja z radioaktivnimi odpadki in izrabljenim jedrskim gorivom za obdobje 2006-2015.* (Resolution on the 2006-2015 National Programme for Managing Radioactive Waste and Spent Nuclear Fuel), Official Gazette of the Republic of Slovenia, No. 15/06. http://www.pisrs.si/Pis.web/pregledPredpisa?id=RESO42, last accessed 19 January 2017.

ReNPROG (2015). *Resolucija o nacionalnem programu ravnanja z radioaktivnimi odpadki in izrabljenim gorivom za obdobje 2016-2025.* (Draft on Resolution on the 2006-2015 National Programme for Managing Radioactive Waste and Spent Nuclear Fuel).

Scotti, O. (2013). *Presentation of IRSN report on possible NPP Krsko II site, What are implications of this report?.* Insitut de radioprotection et de sûreté nucléare, Event on nuclear in Ljubljana, Slovenia, 2.12.2013, at the Faculty of Social Sciences, 35 pages.

Siniša C.; Polona T.; Osojnik, I.; Kališnik, M. and Stritar, A.(URSJV) (SNSA) (2009). *Licensing Process for the New Nuclear Power Plant in Slovenia and the Role of*

SNSA. International Conference Nuclear for New Europe 2009, Bled, Slovenia, 14-17 September. http://www.djs.si/proc/bled2009/htm/pdf/1302.pdf, last accessed 11 June 2017.

Sklad za financiranje razgradnje NEK in odlaganje radioaktivnih odpadkov iz NEK (Slovenian Fund for financing of decommissioning of the NPP Krško and depositioning of radioactive waste from the NPP) (2014). http://www.sklad-nek.si/, last accessed 11 June 2017.

Stritar, A.; Cimeša, S.; Podjavoršek, M.; Mühleisen, A.; Vojnovič, D.; Rašula, J.; Konda, P. and Uršič. M. (IJS) (SNSA) (2008). *Use of PSA for Risk Informed Regulatory Review and Assessment*. International Conference Nuclear Energy for New Europe 2008, Portorož, Slovenia, 8-11 September. http://www.nss.si/proc/port2008/pdf/402.pdf, last accessed 11 June 2017.

Tomšić, Ž. (2010). *Experience from sharing of the Krško NPP between Croatia and Slovenia*. INPRO Dialogue Forum on Nuclear Energy Innovations, Vienna, 4-7 October. https://www.iaea.org/INPRO/2nd_Dialogue_Forum/Tomsic_INPRO-Vienna-10-2010-final2.pdf, last accessed 11 June 2017.

Viršek, S.; Veselič, M. and Rupret, M. (ARAO) (2007). *Umeščanje odlagališča nizko in srednje radioaktivnih odpadkov v Sloveniji* (Siting of a LILW repository in Slovenia), Posvetovanje rudarskih in geotehnoloških strokovnjakov ob 40. Skoku čez kožo, 19-29.

Vrankar, L. (2005). *Regulatory View of the Close-Out of the Uranium Ore Mine Žirovski Vrh*. International Conference Nuclear Energy for New Europe, Bled, Slovenia. http://www.iaea.org/inis/collection/NCLCollectionStore/_Public/37/104/37104789.pdf?r=1.

Železnik, N.; Kralj, M.; Mele, I.; Stropnik, P.; Lokner, V.; Levanat, I. and Rapić, A. (2010). *Revision 2 of the Program of NPP Krško Decommissioning and SF&LILW Disposal*. International Conference Nuclear Energy for New Europe,Portorož, Slovenia, 6-9 September.

Železnik, N. and Kegel, L. (2011). *Spent Fuel Management in Slovenia: Current Status and Future Plans*. 20th International Conference for New Europe, Bovec, Slovenia, 12-15 September.

Too much to Handle

Radioactive Waste Management in a Post Nuclear Accident State: Ukraine

Olexi Pasyuk[1]

Abstract

In 1986, Ukraine experienced a major nuclear accident at the Chornobyl[2] nuclear power plant (NPP); over three decades later, this event continues to define Ukraine's waste management situation. Today, radioactive waste at the Chornobyl NPP site and surrounding exclusion zone (CEZ) constitutes over 98 percent of total solid radioactive waste. Spent nuclear fuel is excluded from this figure as it has special legal status and is not considered to be radioactive waste. Following Ukraine's independence from the Soviet Union, its institutional system to manage nuclear waste problems has continually changed and has not reached the state of clear responsibilities and distribution of roles between various institutions. However, the need for this clarity is recognized by experts and proposals have been made to centralise the management system. EU and IAEA funding enables research on the waste management system most suitable for Ukraine, including deep geological disposal (DGD), regulatory system improvements and physical infrastructure. Adaptation of the Ukrainian standards and practices to the European standards will be accelerated in view of the EU-Ukraine Association Agreement. Because of the ongoing military conflict with Russia, Ukraine lost control over its research reactor in Sebastopol and nuclear waste collection center in Donetsk.

---

1 Centre of Environmental Initiatives 'Ecoaction', www.ecoaction.org.ua; E-Mail: opasyuk@bankwatch.org

2 We use Chornobyl as correct Ukrainian town name rather than commonly known Chernobyl. The later one is the Russian translation of the town and power plant name and means 'wormwood'. Russian language was dominant in USSR and used as a basis for transliteration to English. Ukrainian official names of the plant and the town are transliterated as Chornobyl and we follow this rule.

A. Brunnengräber et al. (eds.), *Challenges of Nuclear Waste Governance*, Energiepolitik und Klimaschutz. Energy Policy and Climate Protection, https://doi.org/10.1007/978-3-658-21441-8_10

1 Introduction

The 1986 Chornobyl nuclear accident led to the biggest release of radioactive materials from a destroyed nuclear reactor in history. Ever since, dealing with radioactive waste from the reactor's destruction has been the primary concern of the Ukrainian state and has garnered attention and financial support from the international community. Solving this problem required the nuclear industry to learn from its mistakes in attempting to put the waste under some level of control.

The CEZ defines Ukraine's waste strategy. The absence of a local population and the reality that most of the country's nuclear waste is located in this area makes it suitable as a final disposal site (storage on the ground and a future deep geological disposal (DGD). The final disposal of the nuclear waste is looked at taking a very long perspective. Ukrainian nuclear experts argue that there are richer and more advanced countries performing research on strategies for final nuclear waste disposal, and thus Ukraine simply needs to wait for the time when best practices have been defined by others.

Ukraine has traditionally sent its spent nuclear fuel to Russia for reprocessing. However, ever-growing costs led to the decision to build its own capacities that will allow for nuclear waste storage in the country for a period of 50 to 100 years. The Association agreement with the European Union means adaptations of the Ukrainian legislation to the relevant EU and Euratom Directives. This might require Ukraine to accelerate its decision making process on numerous aspects of nuclear waste and spent nuclear fuel management to meet the requirements of EU Directives (see legal framework section).

2 Nuclear waste disposal in Ukraine

2.1 Historical background

The Ukrainian nuclear industry originates from uranium mining development in the late 1940s as the USSR was developing its military nuclear program. The VostGOK (Eastern ore processing) plant was established in 1951 and produced the first uranium concentrate in 1959. Various facilities built as elements of the USSR nuclear industry appeared in Ukrainian cities during Soviet times and continue to function today. The Turboatom company produced the first turbines for test nuclear reactors in 1956 and became the main facility providing turbines for the nuclear power plants built by the USSR.

The construction of the first reactor in Soviet Ukraine, the Chrnobyl NPP, was started in 1970 and completed in 1977. By the time the Soviet Union collapsed in 1991, Ukraine had 15 operating nuclear reactors and 3 more in advanced stages of construction. There are 15 reactors in operation in Ukraine today.

The Chornobyl accident on 26 April 1986 which destroyed reactor number 4 has had a great impact not only on the global nuclear industry, but also on the destiny of USSR. In the moment of proclaimed transformation of Soviet society towards more openness, the massive popular antinuclear movement joined with the Ukrainian movement in calling for independence. Mikhail Gorbachev, then the President of the USSR, suggested that the accident "was perhaps the real cause of the collapse of the Soviet Union five years later" (Gorbachev 2006).

When still a part of the USSR, the Ukrainian Parliament introduced a moratorium on the construction of new nuclear reactors. This moratorium was subsequently abolished in 1993 as nuclear power was seen as a basis for energy independence. Ukraine had an ambition to develop a full nuclear cycle that would allow it to produce nuclear fuel domestically. Nuclear was seen as a solution for energy sector problems, as Ukraine received nuclear fuel for free in exchange for passing over its nuclear weapons to Russia (following the Budapest Memorandum on Security Assurances). (Trofymovych 2016: 280).

Today, Ukraine's nuclear sector continues to depend on Russia, using Russian TVEL (nuclear fuel producer) as its primary source of fresh nuclear fuel and a destination for the reprocessing of much of its spent nuclear fuel. The Ukrainian state, with support from Western governments and in cooperation with Westinghouse, has taken steps to limit this dependence by bringing in alternative fuel and building spent fuel storage facilities within Ukraine. However, the country is far from achieving independence in the nuclear sector.

Major accidents like Chornobyl as well as the overall safety and image of the Ukrainian nuclear industry were of major concern to governments and the nuclear industry in Europe and the USA. Various programs backed with the financial support of the EU and USA were implemented following Ukrainian independence. EU technical assistance programs like Tacis (which received 170 million Euro in funding since 2007 from the Instrument for Nuclear Safety Cooperation) were given unusual rights to invest in industrial equipment, in contrast to their historic role as advisors. The European Commission has supported numerous projects to develop Ukrainian nuclear safety and waste management systems.

Management of the infrastructure and waste in the Chornobyl exclusion zone and development of the new shelter above the destroyed reactor became a truly international effort. Various projects financed by international donors and managed by the European Bank for Reconstruction and Development (EBRD) allowed implementation of massive projects as New Safe Confinement, which Ukraine would have a had difficult time implementing by itself. The various failures in this process (for example, Areva's failure to build proper Spent Nuclear

Fuel Storage. See section 2.4.2), however, have demonstrated how unprepared the nuclear industry is to deal with the challenges brought by accidents of Chornobyl scale.

Ukraine has leveraged the Chornobyl accident for its broader nuclear ambitions, communicating that it would not close operating Chornobyl reactors until it received funding from the G7 and EC governments for two new reactors: Khmelnitsky 2 and Rivne 4. After the long saga of the Khmelnitsky and Rivne reactors' (known as K2R4) project appraisal, the last operating unit (no. 3) at Chornobyl was shut down in December 2000. New Soviet design rectors were completed in 2004 with financial support from the EBRD and Euroatom. Conditions of the loans included that Ukraine develop an effective nuclear safety system.

External technical and financial support as well as Ukraine's active participation in relevant international treaties continue to improve the nuclear waste management system and regulatory framework. However, since the majority of nuclear waste originated from the Chornobyl accident, international practice is not fully applied. Critically, the state rather than the nuclear operator company is financially responsible.

Russian military intervention in eastern Ukraine and Crimea has further complicated the situation. Ukrainian authorities have lost control over the storage of the nuclear waste in Donetsk and its research nuclear reactor in Sebastopol, although they remain responsible for nuclear safety at these sites. Most concerningly, the warfare involves heavy artillery and missiles just 200km away from the biggest European nuclear power plant at Zaporizhzhya and its dry spent nuclear fuel storage.

2.2 The national inventory

2.2.1 Categorization

Ukrainian legislation defines radioactive waste as radioactive materials created by human activity that cannot be further used (The Parliament of Ukraine 1995). Spent nuclear fuel does not formally fall into this category. The nuclear industry and the government believe that SNF contains valuable elements as uranium and plutonium that might be used in the future. However, different methods to categorize waste are used for different purposes. Besides, the introduction of new categorizations of waste were done without cancelation of previous ones.

The Ministry of Health defines five different approaches to classifying radioactive waste in its Main Sanitary Rules for Radiation Safety in Ukraine (Ministry of Health 2005). These approaches are: 1.) State of matter. Here waste is categorized by aggregate state into solid and liquid waste' 2.) Type. Based on period of radioactivity. Short-lived RAW should have a potential radiation dose rate below 1mSv/year after 300 years from the moment of disposal. Consequently, monitoring of such a disposal site can be stopped or simplified. Such waste can be disposed of at the near-surface facilities. Long-lived RAW will have a potential radiation dose above 50 mSv/year and should be stored in deep geological formations; 3.) Groups. RAW is also classified by the exemption level shown in Table 1; 4.) Categories. There are three categories of RAW based on specific activity range in kBq/kg: low level waste, intermediate level waste and high level waste. See Table 2; 5.) Kinds. Classifies waste based on the half-life of the radionuclides in the waste (short-lived includes radionuclides with a half-life of no more than 10 years; middle-lived includes radionuclides with a half-life between 10 and 100 years; andlong-lived radionuclides with a half-life of over 100 years and short-lived RAW are further split into sub-categories referring to radionuclides with half-lives measured in days, weeks or years.

Table 1: RAW classified by the 'exemption level'

RAW group	Solid RAW	Exemption level kBq/kg
1	Transuranic alpha-emitting radionuclides	0.1
2	Alpha-emitting radionuclides	1.0
3	Beta- and Gamma- emitting radionuclides (Except those in the group 4)	10
4	3H, 14C, 36Cl, 45Ca,53Mn, 55Fe, 59Ni, 63Ni,93 mNb, 99Tc, 109Cd,135Cs, 147Pm, 151Sm,171T m, 204Tl	100

Source: Ministry of Health 2005

Table 2: Solid and liquid RAW categories based on specific activity

<table>
<tr><td colspan="2" rowspan="3">RAW Category</td><td colspan="4">Solid RAW specific activity range, kBq/kg</td><td rowspan="3">Liquid RAW specific activity as a multiplier PC_B^{ingest}</td></tr>
<tr><td colspan="2">Alpha-emitting radionuclides</td><td colspan="2">Beta- and Gamma- emitting radionuclides</td></tr>
<tr><td>Group 1</td><td>Group 2</td><td>Group 3</td><td>Group 4</td></tr>
<tr><td>1</td><td>Low level waste</td><td>$>10^{-1} <10^{1}$</td><td>$> 10^{0} < 10^{2}$</td><td>$> 10^{1} < 10^{3}$</td><td>$> 10^{2} < 10^{4}$</td><td>$> 1 < 10^{2}$</td></tr>
<tr><td>2</td><td>Intermediate level was</td><td>$\geq 10^{1} <10^{5}$</td><td>$\geq 10^{2} < 10^{6}$</td><td>$\geq 10^{3} < 10^{7}$</td><td>$\geq 10^{4} < 10^{8}$</td><td>$\geq 10^{2} < 10^{6}$</td></tr>
<tr><td>3</td><td>High level waste</td><td>$\geq 10^{5}$</td><td>$\geq 10^{6}$</td><td>$\geq 10^{7}$</td><td>$\geq 10^{8}$</td><td>$\geq 10^{6}$</td></tr>
</table>

Source: Ministry of Health 2005

Additionally, the Ministry of Health regulation enables classification by the technologies that lead to the creation of RAWs as well as accidental sources. Beyond these very different approaches to RAW categorization from the Ministry of Health, there are categories defined by other legal provisions. This circumstance creates confusion and makes it difficult for various actors to come to shared conclusions on appropriate waste management strategies.

Because of this situation, the EU Instrument for Nuclear Safety Cooperation project (U4.01/08-C) has been working in Ukraine to help update radioactive waste classification for disposal purposes. The suggested classification divides radioactive waste into classes to meet the requirements for disposal in four types of repositories: surface repository (landfill-type facilities with limited regulatory control); near-surface repository with a system of engineered barriers; underground repository located at intermediate depths; and DGD. The authors of the proposed system argue that moving to this classification system will make waste management cheaper, enabling simpler storage for low-level waste.

System developers suggest to divide radioactive waste into the following classes: non-radioactive waste; natural radioactive materials (NORM); very low-level waste (VLLW); low-level waste (LLW); intermediate-level waste (ILW); high-level waste (HLW); and spent radiation sources (SRS). The updated classification is expected to result in considerable efficiency as radioactive waste is classified according to optimum disposal method and types of repositories. Draft legislation has been developed for adoption by Ukraine's parliament (Proskura 2014); however, the law remains absent from the parliament database.

2.2.2 RAW accounting

The fifth State Inventory of RAW was carried out in 2013. There are state registers of radiation sources and of radioactive waste, which exchange information in the process of this inventory. For example, 719 spent radioactive sources (SRS) were transferred to the Radon facilities in 2014 as RAW and 241 user stored 3715 SRS in 2014. There were 11,784 radioactive sealed sources in the register at the end of 2014 (State Nuclear Regulatory Inspectorate of Ukraine 2015a).

The most comprehensive overview of the radioactive waste stored and managed in Ukraine is available in English in the National reports on Compliance with the Obligations under the Joint Convention on the Safety of Spent Fuel Management and on the Safety of Radioactive Waste Management. The last of these reports was published by the State Nuclear Regulatory Inspectorate of Ukraine in 2008. Results of the fifth State Inventory of RAW cannot be found online.

The 2009 Radioactive Waste Management Strategy of Ukraine stated that 27 m^3 of solid and 35.1 m^3 of liquid radioactive waste is produced for every 1 bn kWh of nuclear electricity, which equates to 370 m^3 of solid and 480 m^3 of liquid waste per year. Collectively, Ukrainian solid waste storage sites were at 30-70 percent capacity and liquid waste storages at 21-76 percent capacity. The nearly full storage units of the high-concentration salt solution at NPP sites was pointed out as a particularly important problem (Cabinet of Ministers of Ukraine 2009).

Table 3: Types of radioactive waste and their management in Ukraine

Type of Liability	**Long Term Management Policy**	**Funding of Liability**	**Current Practice/ Facilities**	**Planned Facilities**
Spent fuel	Decision deferred	Funds set aside during operation	Reprocessing abroad and interim long-term storage	Interim Storage Facility -2 at ChNPP; centralized Interim Storage Facility; deep geologic repository
Nuclear fuel cycle waste	Treatment complexes at each NPP; final disposal at a centralized repository (*Vektor site)*	National RAW Fund (fees collected from electricity tariff)	On site storage, treatment at the NPP RAW complexes; Liquid Radioactive Waste Treatment Plant for ChNPP	Treatment complexes for all operating NPPs completion, Industrial Complex for Solid Radioactive Waste Management for ChNPP completion;
Application waste	Final disposal at centralized repository, (*Vektor site)*	National RAW Fund (fees paid by waste producers)	Storage and limited treatment at Radon facilities	Centralized processing and disposal facility, (*Vektor site)*
Decommissioning liabilities	ChNPP – plans in place, operating NPPs – strategy in place	Decommissioning Fund (fees collected from electricity tariff) ChNPP – state budget	ChNPP – final shutdown and conservation stage	
Disused sealed sources	Historic and orphan sources – centralized repository; new - return to producers/ centralized repository	National RAW Fund (fees paid by waste producers); state budget for legacy	Storage at Radon facilities	Centralized processing and long-term storage (*Vektor site)*; deep geologic repository

Source: State Nuclear Regulatory Inspectorate of Ukraine 2015a

Estimates for the total amount of solid RAW and liquid RAW in Ukraine are 2,960,000 m^3 and 42,340 m^3, respectively. The distribution of nuclear waste at various sites is as follows:

Table 4: Share of Solid Radioactive Waste in Ukraine by location

Location	Share of all Solid Waste (%)
CEZ (Temporary waste Localization Sites)	72.4
Shelter over Destroyed Chornobyl 4 unit	20.2
Radioactive Materials Storages	5.8
Chornobyl NPP site	0,1
Nuclear Power Plants	1,3
"Radon" Waste Storages	0.2

Source: Gramotkin 2016[3]

Table 5: Share of Liquid Radioactive Waste in Ukraine by location

	Share of all Liquid Waste (%)
Chornobyl Nuclear Power Plant	47.2
Other Nuclear Power Plants	43.9
"Shelter"	5.9
"Radon" Waste Storages	1.9
Research Nuclear Reactors	1.1

Source: Gramotkin 2016[1]

The estimated amount of the radioactive waste stored at Chornobyl NPP storages is 21,000 m^3 of liquid radioactive waste and 2,500 m^3 of solid radioactive waste. Another 500 m^3 of liquid RAW and 225,000 m^3 of solid RAW will be produced at the process of Chornobyl NPP decommissioning (Gramotkin 2016)[1].

2.2.3 RAW at nuclear power plants

The National Nuclear Energy Generating Company, Energoatom, implemented the "Comprehensive Program for Radioactive Waste Management" in 2012-2016. As a result of this program, there is ongoing construction of RAW Treat-

3 The author of the document is not confirmed. The document is not published and was obtained by email. Figures correspond well to other sources.

ment Plants NPP sites in order to minimize RAW and prepare for disposal or long-term storage in the centralized near-surface disposal facilities at the Vektor site, once it is built (State Nuclear Regulatory Inspectorate of Ukraine 2015a: 20).

Energoatom NPPs produce on average 27 m^3 SRAW and 35m^3 LRAW per 1 bn kWt·h of electricity. The lack of space at the temporary storage sites is becoming a problem in view of plans to extend the lifetime of the reactors. The situation is made even more difficult by the presence of liquid radioactive waste: according to current legislation, it is not suitable for disposal and there are no facilities to solidify this waste (Kondratiev 2016: 41).

Chornobyl NPP has three reactors that continued to operate after the accident at unit 4 in 1986. The last one was shut down in 2000. Activities to decommission units 1, 2 and 3 of the plant are ongoing. High-level RAW are collected in special containers (KTZV-0.2) and are held in temporary storage organized at the building originally used to store fresh nuclear fuel. Low and medium level waste is sent directly to Buriakivka storage (Kondratiev 2016). In the process of decommissioning the Chornobyl site, 0.03 m^3 (0,012 t) of solid radioactive waste was accumulated in 2014. Overall 3,783 m^3 of high-level and long-live RAW is stored there with a cumulative activity of 8.59 TBq (State Nuclear Regulatory Inspectorate of Ukraine 2015b).

At the Chornobyl site, there is an Industrial Complex for Solid Radioactive Waste Management (ICSRM) constructed by RWE NUKEM GmbH with the European Commission support, as well as a Liquid Radioactive Waste Treatment Plant (LRWTP) built with EBRD support. The European Commission has also financed construction of the Long-Length Waste Cutting Facility at Chornobyl NPP (LICF Project). All of the facilities were completed in recent years.

2.2.4 RAW at destroyed Chornobyl reactor

2006 estimates suggested that there are 400,000 to 1,740,000 m^3 of RAW located in the Object Shelter (OS) (also known as the 'Sarcophagus') at the site of the destroyed Chornobyl NPP unit 4. High level of radiation and destruction caused by the explosion makes it difficult to have precise estimates. At the beginning of 2005, their total activity was known to be about $4.1 \cdot 10^{17}$ Bq.

Over 10 percent of the total amount of the OS RAW is high level waste (HLW), a significant amount of which is concrete, metal structures and equipment and other materials from the reactor. Over 2,800 t of HLW are fuel content materials (FCM), including lava-like FCM, fragments of the reactor active zone, reactor graphite and fuel dust.

At the OS, there is constant accumulation of atmospheric water, condensate and liquids of technological origin. Liquid RAW (LRW) has arisen from the interaction of water with radioactive materials.

Annually, up to 900 m^3 of LRW are pumped from the accessible OS rooms and transported to the onsite treatment and storage system for LRW. In the OS operation, including transformation of the OS into an environmentally safe system (OS stabilization stage), considerable amounts of solid RAW have arisen and have been disposed of at the Buryakivka Radioactive Waste Disposal Site (RWDS) (Ministry of Ukraine of Emergencies and Affairs of population protection from the consequences of Chornobyl Catastrophe 2006).

Radionuclide and chemical composition of LRW depends on its location. Water inside the OS contains Cs134, Cs137, Sr90, Pu239-240 and Am241 as well as organic and membrane-forming compounds (State Nuclear Regulatory Committee of Ukraine 2008b). According to the state register of radioactive waste in 2007, the following radioactive waste is located inside the OS and at its site:

Table 6: Radioactive waste is located inside the OS and at its site

№	Type of RAW; (location)	Physical state	Category of activity	Volume, m^3	General activity, TBq	Nuclide composition%
1.	Solid radioactive waste[1], located inside Shelter and at Shelter industrial site, occurred as a result of accident and works on elimination of accident consequences	Fresh and spent fuel assemblies, lava-like FCM, dust, metal equipment, construction -and-assembly elements, etc.	Intermediate and high-level	530,400 – 1,737,400	740,000 (20 MCi)	Mix of radionuclides (uranium, caesium, strontium, cobalt, transuranium elements – plutonium, americium and others)
2.	After-accident waste[2] located inside Shelter	Liquid RAW	Intermediate and low-level	2,500-3,000 [2]	12,4 (335 Ci)	Mix of radionuclides: uranium, caesium, strontium, plutonium and others.

Notes by SNRCU: 1. Data in table is approximate and based on the results of research. 2. Amount of liquid waste changes every year depending on atmospheric precipitation that enters the shelter (State Nuclear Regulatory Committee of Ukraine 2008b: 93).

A major international effort to make the Chornobyl accident site as safe as possible, the Shelter Implementation Plan (SIP), was implemented in 1997 and managed by the EBRD with financial support from the USA, European Commission and various national governments.

The SIP effort was first focused on developing proper monitoring systems inside of the Object Shelter, developing surrounding infrastructure and then stabilising the sarcophagus to manage the risk of an accidental collapse. One of the walls was stabilised with steel structures to take some of the weight off the roof.

In November 2016, a milestone in the completion of the extraordinary New Safe Confinement (NSC) was achieved when an arch-shaped structure weighing 36,000 tonnes and standing 108 meters tall was moved from the construction site over the SO. The structure includes massive cranes and is designed in a way that allows for disassembly of the old sarcophagus and further retrieval of the nuclear waste inside the destroyed reactor.

NSC construction started in 2010. The cost of the NSC is now at 1,504 million Euro (1,424 million Euro plus 80 million Euro of unexpected costs), according to the Ukrainian Accounting Chamber. It is designed to be in operation for 100 years. Overall costs of the Shelter Implementation Plan are expected to amount to 2.1 billion Euro; the plan is to be completed by 2017. It is funded by contributions from more than 40 countries and organisations.

2.2.5 RAW in the Chornobyl exclusion zone

The Chornobyl accident created a great amount of RAW that requires very special attention. Dealing with the nuclear waste storage systems created in the rush after the accident is complex, as these systems do not meet current nuclear safety requirements. Hundreds of thousands of cubic meters of radioactive waste are stored at more than 600 interim radioactive waste locations and under the Chornobyl Shelter. A significant share of this waste is considered long-lived radioactive waste (State Nuclear Regulatory Inspectorate of Ukraine 2015a). Buryakivka, Pidlisny and Chornobyl NPP Stage III are major sites of radioactive waste disposal in the CEZ.

Buryakivka RWDS has been in operation since 1987. Buryakivka RWDS is composed of 30 near-surface storage modules (trenches) for RAW disposal. The main engineering barrier which provides for radionuclide storage is a special clay protective layer, 1 meter thick. Since the Buryakivka RWDS began operation, approximately 1,330.5 thousand tons (665.25 thousand m^3) of Chornobyl origin RAW were located in the trenches with a total capacity 2.53 E15 Bq (as of late 2012).

Pidlisny RDWS and Chornobyl NPP Stage III RDWS were constructed during the years following the Chornobyl accident. These facilities contain the most dangerous high-activity and long-lived emergency RAW. According to the State Nuclear Regulatory Inspectorate, in the future, all RAW will be removed and re-disposed of in geological storage facilities. Before beginning construction of these geological storage facilities, the safety of existing facilities is to be maintained and improved. Accordingly, activities to protect waste disposal sites from degradation and support necessary localizing functions of the engineering barriers of these storage facilities, as well as to create additional barriers and to improve monitoring systems, were carried out in 2012 (State Nuclear Regulatory Inspectorate of Ukraine 2013).

As most of Ukraine's nuclear waste is located in the CEZ, where there is also an availability of nuclear infrastructure and a small local population, this zone was chosen to host all of the country's RAW. The Vektor complex at the edge of the 30-km zone is supposed to accumulate all of the waste from various facilities. The design of the complex envisions storage of the 533,644 m^3 of RAW. The first stage of the complex development envisions two facilities and supporting infrastructure with a total volume of 19,200 m^3. Construction started in 2000 and was stopped in 2010 due to lack of funding. Thus, some of the elements constructed at the beginning of this process have started to deteriorate (State Nuclear Regulatory Inspectorate of Ukraine 2016: 66).

2.2.6 RAW from non-NPP sources

Non-NPP Radioactive waste is systemically managed via the Radon facilities which are situated at six locations across the country. There were 539,728 ionizing radiation sources with a total 2,86E+16 Bq and 5864 m^3 of RAW, totalling 7,28E+15 BQ at the Radon sites in 2014 (State Nuclear Regulatory Inspectorate of Ukraine 2015b: 50). In light of the outdated standards at these storage sites there is a plan eventually to move the waste to the centralised storage at the Vektor complex. First, however, approval is required for the operational and technical approaches to extract the waste from existing wells at Radon sites, as these sites were not designed for waste retrieval. Funding and possible personnel exposure to radiation during extraction operations are key limiting factors.

In 2014, there were 14 cases of unexpected radioactive materials identification – in most cases radioactive sources found in the scrap-metal brought to metallurgy plants (State Nuclear Regulatory Inspectorate of Ukraine 2015b: 47). Ukraine has over a dozen uranium tailing sites created by the mining and enrichment industry. Some, like Balka Scherbakivska, are operational, while others are closed. There was no proper management of the sites following the collapse of the Soviet Union, and thus the nearby population continues to be at risk. Various programs to manage the risk of radioactive pollution were designed by state bodies, but often lack funding for implementation. According to the Radiation Safety Standards of Ukraine, waste from the uranium mining industry is not considered RAW.

There are four RAW disposal sites remaining from the former activities of the USSR Army. These are maintained by the Ministry of Defence and State Border Guard Service of Ukraine (State Nuclear Regulatory Inspectorate of Ukraine 2014: 31).

2.3 Spent fuel

2.3.1 Chornobyl reactors

There is spent nuclear fuel stored in 21,284 spent fuel assemblies (SFA) at the site of Chornobyl NPP. There is no fresh nuclear fuel at the Chornobyl site. The majority of SFAs (21,231.5)[4] is stored in the cooling pool of the Wet Spent Fuel Storage Facility (ISF-1), which was commissioned in 1986. The remaining 52.5 SFAs (which are damaged) are stored in the cooling pools of unit 1 and 2 of Chornobyl NPP (ChNPP) (State Nuclear Regulatory Inspectorate of Ukraine 2015b). The State Specialized Enterprise Chornobyl NPP is in charge of Chornobyl SNF.

The ISF-1 is licensed through 2025, following an earlier decision by the nuclear regulator. A new dry type storage facility (ISF-2) is under construction at the ChNPP site in order to provide for safe long-term storage of all spent nuclear fuel. This construction is sponsored by the international community as a part of safe Chornobyl plant decommissioning efforts. It was originally planned to be completed in 2004. The construction was started by French company Areva, but the contract was cancelled as the storage technology was shown to be inadequate for the Chornobyl SNF. Construction was subsequently taken over by the US company Holtec and is currently at a late stage of completion.

2.3.2 Research reactors

Spent fuel from the WWR-M research reactor of NASU INR (Kyiv) is stored in a SF storage facility at the research reactor. There has been no decision made on

4 Some spent fuel assemblies are broken, thus 0,5 of SFA

its future. The Ministry of Education and Science is responsible for the SNF's management.

The research reactor IR-100 at the Sebastopol Nuclear Energy and Industry Institute (Sebastopol) has no SF – based on the definition of spent fuel in the Joint Convention (State Nuclear Regulatory Committee of Ukraine 2008b: 13).

Following Russia's annexation of Crimea, the Ukrainian nuclear regulator has lost contact with the management of the Sebastopol Nuclear Energy and Industry Institute and invalidated its licence for reactor operation (State Nuclear Regulatory Inspectorate of Ukraine 2015b: 10).

2.3.3 Operating commercial reactors

Ukraine has arrangements with the Russian Federation for dealing with spent fuel from Khmelnitsky, Rivne and South Ukraine NPPs. The spent fuel is transported to the Mayak facility at Chelyabinsk oblast of the Russian Federation. The WWER-440 fuel (from the units 1 and 2 of Rivne NPPs) is reprocessed there, while WWER-1000 fuel is stored. Mayak is expected to finalise the WWER-1000 reprocessing process development and will start WWER-1000 fuel reprocessing in 2017. The contract for the processing of Ukrainian WWER-440 fuel suggests that Ukraine should start receiving radioactive waste accumulated from reprocessing in 2018.

The high radioactive waste created from reprocessing in Russia should come back to Ukraine. However, the storage for the disposal of this waste is not yet constructed. The government plan is to build an appropriate facility at the Vektor complex.

To reduce the cost associated with SNF management, a dry storage facility was built at Zaporizhzhya NPP. The feasibility study done by Energoatom and approved by the Cabinet of Ministers of Ukraine in 2009 showed the economic viability of storing SNF in Ukraine, rather then sending it to the Russian Federation. Building a single centralised storage site was chosen as the most efficient approach (State Nuclear Regulatory Inspectorate of Ukraine 2015b: 39). The decision has been criticised by environmental NGOs which suggest that storage systems should be built at NPP sites and transferred to the site of final disposal once it is ready.

2.3.4 Alternative Westinghouse fuel

Almost all of the fuel used by the Ukrainian NPPs is produced by the Russian TVEL manufacturer. Since the late 1990s, Ukraine has cooperated with the

Westinghouse company to become an alternative producer of nuclear fuel for WWER reactors. So far, the fuel produced at the Westinghouse factory in Sweden has been tested at the South Ukraine NPP and the tests have recently been extended to include the Zaporizhzhya NPP.

We can expect that the handling of spent Westinghouse fuel should not be different from that produced by TVEL. However, this fuel will not be sent to Russia for reprocessing and instead needs to be stored in Ukraine. There is the dry storage at Zaporizhzhya NPP. But for the South Ukraine NPP, fuel should stay on site until the Centralised Dry Storage is commissioned in the CEZ and the spent fuel of all Ukrainian NPPs (but Zaporizhzhya) is transported there.

2.4 The (interim) storage sites

2.4.1 Radioactive waste at NPPs

Nuclear power plants in Ukraine manage and store their radioactive waste at their sites. The key deficiency of the current practice is the absence of facilities to prepare radioactive waste for long-term storage or disposal. It is expected that this waste will be transported to the centralised Vektor storage once it is fully operational.

Most of the solid radioactive waste is generated at NPPs during maintenance, repair and modernisation work. About 80 percent of solid waste is rags, insulation materials, metal and concrete pieces of equipment and buildings. Liquid waste is made up of the trapped water coming from the leakages of the primary circuit, cooling pool, discharges from the deactivation rooms and test laboratories, etc. Still bottoms that are formed in the process of evaporation of radioactive waters, as well as sorbents and molten salt formed by further evaporation processes are also considered to be liquid waste.

Solid waste is sorted primarily based on the dose rate and put in temporary storage on site, both in bulk and in 200 liter steel barrels. There were no plans to use installations of the deep evaporation of liquid waste at the time of the NPP's design, so there is no dedicated storage for salt fusion cake. This is instead housed in solid waste storage systems.

The lack of space at the temporary RAW storage systems is a problem in view of the plans to extend reactor lifetimes. The situation is further compounded by the challenges of the liquid radioactive waste: it is not suitable for disposal (according to current legislation) and there are no facilities to solidify it.

Energoatom is implementing a program to build facilities for solid radioactive waste management at the power plants. Such complexes are in late-stage construction at Zaporizhzhya and Rivne power plants. Construction of solid waste treatment facilities at Khmelnitsky plant is about to start. There is no final decision to build such a facility at the South Ukraine NPP (Kondratiev 2016).

All NPPs have systems for liquid radioactive waste storage. LRW is stored in stainless steel containers with an alarm system to detect leakage. The containers are stored in concrete spaces covered with stainless steel sheets. NPPs have evaporation facilities with different levels of efficiency.

Solid wastes are sorted according to gamma radiation intensity and are transported to the temporary storage units on site. Some NPPs have facilities to reduce the volume of waste by compacting with pressure (ZNPP and SUNPP) or incineration (ZNPP). The storage units are concrete buildings divided into different sections by waste type. There are fire alarm and automatic firefighting systems as well as filtered ventilation systems (State Enterprise 'National Nuclear Energy Generating Company 'Energoatom'' 2014: 12).

2.4.2 Spent nuclear fuel

Currently Ukraine has two facilities for the temporary storage of SNF: Intermediate Spent Fuel Storage (ISF) 1 (wet) at the Chornobyl NPP and ISF (dry) at the Zaporizhzhya NPP. There are two more storage facilities under construction: ISF-2 (dry) at the Chornobyl NPP and Centralised ISF for the fuel of the WWER reactors at multiple Ukrainian NPPs. While the first is about to start operations, the second is still a greenfield project.

ISF (dry) at Zaporizhzhya NPP (only for SNF generated at Zaporizhzhya) – in operation

Zaporizhzhya NPP was the first to reach the end of available space at the spent fuel pools on site. The resulting Dry Type Spent Fuel Storage Facility project was started in 1996 and the first stage of the storage was completed in 2001, with a capacity for 100 ventilated storage casks. The second stage was completed in 2011. The storage facility is designed to fit 380 casks with 9,000 fuel assemblies. There were 124 casks on site as of 1 January 2015.

The storage facility was designed by the US company Duke Engineering & Services. The cask includes 24 fuel assemblies that spent five years at the spent fuel pool and have low energy production (below 1 kWh) (State Nuclear Regulatory Inspectorate of Ukraine 2015b: 38).

ISF-1 (wet) at Chornobyl NPP – in operation

The Chornobyl SNF of the RBMK type was supposed to stay in the cooling pools next to reactors for no less than 1.5 years and then stored in the wet SNF storage. There are 21,284 fuel assemblies at Chornobyl NPP site as of 1 January 2015. There are 52.5 damaged SFAs in the pools of the reactors 1 (32) and 2 (20.5) and 21,231.5 undamaged SFAs in the ISF-1.

Chornobyl NPP State Enterprise is responsible for implementing the action plan to improve safety of ISF-1. ISF-1 was supposed to be free of fuel and closed in 2016, but will need to operate longer as the replacement (ISF-2) was not commissioned in time. The life-time of ISF-1 will end in 2025, based on the safety reassessment conducted in 2011 (State Nuclear Regulatory Inspectorate of Ukraine 2015b: 42).

ISF-2 (dry) at Chornobyl NPP - under construction

The contract to build new spent fuel storage for the Chornobyl NPP fuel was signed by Ukrainian government with Framatome (now Areva) in 1999. The spent nuclear fuel from Chornobyl's no. 1, 2 and 3 reactors was to be stored there for at least 100 years. It was expected that facility will be ready by the year 2005. The project was a part of the special fund managed by the European Bank for Reconstruction and Development (EBRD). However, in 2004, it became clear that Areva's technical solution was not suitable for the Chornobyl reactors' fuel and construction was stopped. Areva had to pay a fine.

Areva's contract was taken over by Holtec International in September 2007. The new facility would retain the concrete structures built by Vinci and Bouygues, as well as some equipment. Work at the site only resumed in October 2014, 14 years after its start and 11 years after the shutdown of the construction managed by Areva. The total cost of the storage facility has more or less quadrupled. Today, total costs are estimated at more than 300 million Euro.

The design of ISF-2 suggests that the fuel assemblies will be divided into two parts. Each part will be placed in special cartridges. Then, each of the 186 cartridges will be put into hermetic steel containers with helium. These containers will be stored in concrete storage modules, where they can stay for 100 years. The design allows for extraction of the containers, to see if they are still hermetically sealed or if repacking is necessary (State Nuclear Regulatory Inspectorate of Ukraine 2015b: 43).

CSFSF (dry) for the SF from Khmelnytska, Rivne and South Ukraine NPPs - site is selected, designing is ongoing

The Centralized Spent Fuel Storage Facility (CSFSF) is designed to have a capacity for 12,500 SFA of WWER-1000 and 4,000 SFA of WWER-440 type and to operate for 100 years. The decision to locate central storage in the CEZ is spelled out in the law 4384-VI approved on 9 February 2012 by the Parliament of Ukraine.

The Cabinet of Ministers approved the process for allocating 45.2 hectares of land between the relocated villages of Stara Krasnytsia, Buriakivka, Chystogalivka and Stechanka of Kyiv oblast to Energoatom to build the storage and connecting railroad (State Nuclear Regulatory Inspectorate of Ukraine 2015b: 40). In 2014, Energoatom was assigned to operate CSFSF. In that same year, the State Agency of Ukraine for Exclusion Zone Management obtained a special permit for preparatory works on CSFSF construction.

The current contract with Holtec International requires the company to supply the specified process equipment to store 2,511 SFA of WWER-1000 and 1,105 SFA of WWER-440, supply spent fuel handling and storage technology and implement other activities to support the establishment of the CSFSF. In 2015, the operator (NNEGC Energoatom) with the support from IAEA experts, developed the "Plan for Equipment Infrastructure Requirements for the Spent Fuel Transfer Process in Ukraine: NPPs to CSFSF" (State Nuclear Regulatory Inspectorate of Ukraine 2015a: 42).

2.5 Waste management strategy (with current waste disposal plan/concept)

Implementation of state strategy in the area of radioactive waste management is performed in accordance with the Strategy on Radioactive Waste Management in Ukraine (up to 2060), the National Environmental Program on Radioactive Waste Management and the National Program on Chornobyl NPP decommissioning and Shelter Transformation into an Environmentally Safe System.

The strategy includes organizational and technical measures directed at the management of so-called "post Chornobyl" waste, localized in the CEZ at the Chornobyl NPP site. In the Exclusion Zone at the Chornobyl NPP site and at the site of Vektor Complex, a number of new facilities for RAW management are being constructed: for removal of RAW from existing temporary storage facilities; sorting, processing and conditioning of RAW; new facilities for RAW storage; and near surface storage facilities for conditioned RAW (State Nuclear Regulatory Inspectorate of Ukraine 2013).

The strategy is designed for 50 years. This is the time needed to develop an entire infrastructure for radioactive waste storage and mitigate the long-term nature of radioactive waste threats.

The strategy will be implemented in three stages. The first stage (10 years) includes developing: a proper legal base, a national institution to manage waste disposal, financing mechanisms, and the containers and vehicle fleets to transport the waste. In addition, it includes designing, construction and operation of the storage systems for short-term low- and medium- level waste, as well as storage systems for high level and long-term low- and medium level waste; identification of a site to dispose high level and long-lived low- and medium level waste in deep geological formations, and the creation of new and modernization of the existing facilities for processing, conditioning and packaging of radioactive waste in accordance to the requirements for the waste to be stored and disposed.

The second stage (30 years) includes: completing work to clean up dangerous storage of radioactive waste of Chornobyl origin as well as from the state corporation UkrDO 'Radon' and national defence programs; exclusion, conditioning and transfer for the ultimate disposal of the operational radioactive waste from NPP storage units as well as the waste generated from NPP (including Chornobyl NPP); disposal of all short-lived low and medium level waste; storage of the high level and long-lived low- and middle- level waste; design, construction and commissioning of the storage system for high level and long-lived low- and middle- level waste; and, design of the technologies, equipment and work to extract radioactive waste from the Chornobyl unit 4 OS.

The third stage (10 years) includes: operation of the equipment to condition and pack radioactive waste of all types and categories; transportation to the storage units for disposal; operation of the storage units to dispose radioactive waste of all types and categories; finalising the disposal of radioactive waste created in the process of the Chornobyl NPP decommissioning and turning the OS into an environmentally safe site; and, conducting activities to rehabilitate radioactively polluted areas. (Cabinet of Ministers of Ukraine 2009)

The main aspects of the RW management system development are: on-site treatment of the NPP RW to a condition where it can be disposed of or stored for a long time; collection, conditioning, temporary storage, transportation of RW created by medicine, science, industry at the specialized regional enterprise of the State Company UkrDO Radon; centralized disposal of the low- middle level short-lived RW and long term storage of the long-lived and high-level RW of all origins at the storage units of the Vektor complex; disposal of the long-lived and high-level RW in geological storage systems; creation of the national RW Management Organization; providing sustainable and sufficient financing of RW management activities; and, development of the legal base and international cooperation. (Datsenko 2015)

2.5.1 Vektor facility

The strategy for nuclear waste management is based on the idea of having one centralized storage for different types of waste at the Vektor complex. Different onsite facilities will handle radioactive waste from Chornobyl NPP and the CEZ as well as from sealed radioactive sources. There will be centralized near-surface disposal facilities for solid RAW. An engineered near-surface disposal facility for solid radioactive waste at the Vektor Site is in operation. Two near-surface disposal facilities (SRW-1 and SRW-2) are under construction and a management structure has been established. In addition, a Centralized Long-Term Storage Facility for Spent Sources (CLTSF) is under construction. In addition, a special facility is being designed for long-term storage of vitrified high-level RAW to be returned after reprocessing of SF from NPPs and two facilities are being designed, one for long-term storage of long-lived RAW, and the other for long-term storage of high-level RAW from Object Shelter. (State Nuclear Regulatory Inspectorate of Ukraine 2015a: 12)

The construction of the facilities at Vektor face constant delays because of a lack of funding. This has delayed the infrastructure for glassification of the waste to be returned from Russia which was supposed to have been completed in 2010 (Energoatom letter to Chornobyl trade unions) and the completion of the five storage units which were supposed to have been completed by 2012.

2.5.2 Spent nuclear fuel

Ukraine has not made a decision on the final management of SNF beyond long-term storage; it is following a strategy of delaying on making a decision. Additionally, there is no clear unified plan for SNF management by the state. The Spent Nuclear Fuel Management and Centralised Storage Design, Location and Construction Law only deals with fuel from the WWER reactors and states that it will be managed by the Chornobyl NPP SSE along with the separate storage of spent fuel storage from the Chornobyl reactors. However, the construction of the Centralised Dry Storage is managed by Energoatom.

While there is a special fund for radioactive waste management, there is no such fund for spent nuclear fuel. Rather, the expectation is that the nuclear operator is in control of the situation and will solve problems as part of operational costs.

2.5.3 Final disposal plan

The process of identifying a site for deep geological disposal (DGD) in Ukraine started in 1993. From 1996 to 2003, a screening of possible sites across Ukraine was conducted. In 2000-2006, complex research on two of the preferred sites was implemented along with research into the conceptual design of the geological disposal and RAW isolation technologies.

Designers of the DGD assume that it will be a place for SNF storage as well as classified RAW. There is also an assumption that Ukraine will be building new nuclear power plants until 2030 as described by the Energy Strategy of Ukraine (now outdated, as new versions of the strategy cancelled proposals to construct 20 new reactors by 2030) (Shybetskyi 2011).

Preliminary investigations have shortlisted sites for a DGD for high- and intermediate-level wastes, including those arising from Chornobyl decommissioning and clean-up. It is assumed that approximately 59,000 m3 of long-lived waste must be disposed of in the geological repository. With 95 percent of the total volume of long-lived waste stored at the CEZ, research has focused on the use this area for final disposal. Most of this research has been conducted within the framework of international technical assistance projects with the aim of providing a scientific basis for future decision-making. Based on geological and geophysical investigations conducted in 2001-2003, two areas within the Ukrainian Shield were selected for deep boreholes: Veresnia and Tovsty Lis (see Figure 1).

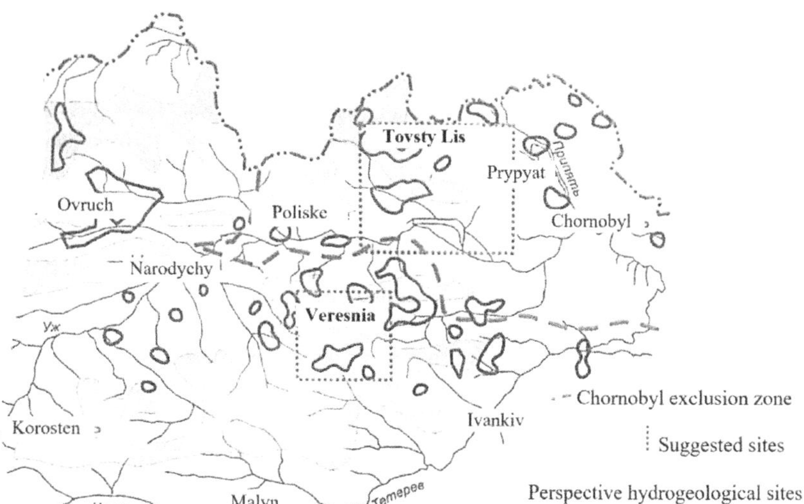

Figure 1: Suggested geological disposal sites based on hydrogeological and geodynamic characteristics. (Source: Shestopalov 2006: 231)

3 The legal and institutional framework

3.1 The legal framework

Ukrainian legislation in the field of radioactive waste management attempts to mirror international rules and best practices. A number of laws were developed as a part of the technical assistance provided by the European Union. The need to implement these provisions is a challenge for the regulator. In one case, there was even a typographical mistake in the figures that define the category of the waste written into a law.

In 2015, the regulator presented a legislative concept for how to match the strict Soviet regulations with the Western approach, in which some provisions are simply recommendations.[5]

Different aspects of waste management are covered by a number of Ukrainian laws: On the Use of Nuclear Energy and Radiation Safety; On Radioactive Waste Management; On Protection of Human Against Impact of Ionizing Radiation; On Decision Making Procedure for Sitting, Design, Construction of Nuclear Facilities and Objects for Radioactive Waste Management of State Importance; On Physical Protection of Nuclear Facilities, Nuclear Materials, Radioactive Waste and Other Sources of Ionizing Radiation; and, On Permissive Activity in the Area of Nuclear Energy Use.

RAW management is further defined in norms, regulations and standards for nuclear and radiation safety. There are requirements regarding radioactive waste management until their disposal as well as in regard to their near surface storage; general radiation safety requirements; general provisions for the safe disposal of radioactive waste in geological repositories; and, general provisions for NPP safety. (State Nuclear Regulatory Committee of Ukraine 2008b)

The Law on the Use of Nuclear Energy and Radiation Safety (1995) is the key legislation of the sector. It defines the decision-making process regarding the location, construction and decommissioning of RAW management facilities. Guiding principles for the decision-making process include: 1) the probability of an ionising radiation impact, 2) the number of people that would be effected and 3) assuring that individual doses of radiation should be the lowest practically achievable.

The Law on Radioactive Waste Management (1995) further defines the principles of state policy. These include: prioritizing the health and safety of people and the environment from radioactive impacts (in accordance with state norms of the radiation safety); minimizing the level of waste production (to the

5 Selected Ukrainian regulations on nuclear waste are available in English at: http://www.snrc.gov.ua/nuclear/en/doccatalog/list?currDir=119835

extent practically achievable); avoiding uncontrolled accumulation of radioactive waste; involving local communities and authorities in decision making; and clearly defining the responsibilities of the involved parties for safe RAW management. The law states that the state's radioactive waste management fund should cover the cost of radioactive waste management.

In 2009, the On the National Program for the Chornobyl NPP Decommissioning and Transformation of the Shelter Object into an Ecologically Safe System Law replaced the previous Comprehensive Program.

Other programs and strategies which specify steps to be completed by the state in the field of RAW management include the National Ecological Program of Radioactive Waste Management[6]; the Radioactive Waste Management Strategy in Ukraine[7]; and the State Program for Safe Storage of Disused High-Level Sources[8].

The Radioactive Waste Management Strategy of Ukraine addresses DGD as well as the terms of reference for the development and approval for geological storage (GS); research on the selected GS sites, the development of technology to store radioactive waste and to build the storage unit; the design and safety analysis of the storage facility; and the building, licensing and commissioning of geological storage (including construction of the underground research lab as a pilot section of the storage). (Cabinet of Ministers of Ukraine 2009)

In 2012, the parliament passed a law dedicated to the creation of a centralised storage facility. The law states that centralised storage will be part of the Special Enterprise Chornobyl NPP. It specifies that the storage facility should have the capacity to hold 16,259 nuclear fuel assemblies. These will contain fuel from the Ukrainian NPPs and specifically the WWER-440 and WWER-1000 reactors. It requires the company to allocate 10 percent of the storage costs to the development of the social infrastructure in the nearby city of Slavutych as well as Ivankiv and Polissya rayons of Kyiv oblast.

International agreements

Ukraine is party to the Joint Convention on the Safety of Radioactive Waste Management and the Safety of Spent Fuel Management and the Convention on Nuclear Safety, both of which fall under the IAEA.

6 Approved by the Law of Ukraine No. 516-VI, 17 Sept 2008.

7 Approved by the Order of Government No. 990p, 19 August 2009.

8 Approved by the Order of Government No. 1092, 3 August 2006.

EU Association Agreement

Following the signing of the EU-Ukraine Association agreement, the government developed an action plan. Varios Directives are relevant for the implementation of the plan[9]: Council Directive 2013/59/EURATOM of 5 December 2013, defining basic safety standards for protection against the dangers arising from exposure to ionizing radiation[10]; Council Directive 2006/117/Euratom of 20 November 2006 on the supervision and control of shipments of radioactive waste and spent fuel and Council Directive 2014/87/Euratom of 7 August 2014, amending Directive 2009/71/Euratom and establishing a Community framework for the nuclear safety of nuclear installations.

The State Nuclear Regulator Inspectorate created an interagency working group to implement the provisions found in the Directives.

3.2 The institutional framework

In accordance with Ukrainian law, the State Nuclear Regulatory Inspectorate of Ukraine implements functions of the national regulatory body on nuclear and radiation safety. It sets safety criteria, requirements and conditions for nuclear energy use (normative documents, standards), grants permits and licenses to carry out activities in this area (licensing), and supervises implementation of legislation, norms, rules and standards of nuclear and radiation safety (supervision).

Responsibility for the management of WWER reactors' SNF is assigned to the nuclear operator Energatom and the Ministry of Energy and Coal. Responsibility for the management of Chornobyl (RBMK) reactors' SNF is assigned to the Special State Enterprise Chornobyl NPP and the Ministry of Environment and Natural Resource. SNF from the research reactors is stored on sites and is the responsibility of the Ministry of Education and Science.

In 2010, the Ministry of Emergencies (no longer in existence) created a single national utility responsible for long term storage and disposal of RAW called the Specialised State Enterprise Central utility to manage radioactive waste (SSE CEMRW) a main utility of the Ukrainian State Corporation Radon (USC Radon). USC Radon specializes in scientific and technical research, design engineering, technical and project documentation and production of special equipment and devices. It also specializes in the design, construction, exploitation and technical and radiation monitoring of existing storage points for radioactive waste

9 For details, see: http://goo.gl/Lexqfa

10 and repealing Directives 89/618/Euratom, 90/641/ Euratom, 96/29/ Euratom, 97/43/ Euratom, and 2003/122/ Euratom

at all stages, including those located in the Chornobyl Exclusion Zone. USC Radon includes 6 regional centres to collect radioactive waste from medical, scientific and other facilities, as well as operates the utility that handles radioactive waste at uranium mines.

The State Exclusion Zone Agency (SEZA) is currently responsible for the long-term storage and disposal of radioactive waste and thus SSE CEMRW; this agency reports to the Ministry of Environment and Natural Resources. However, the Ministry of Energy and Coal is responsible for preliminary treatment of the waste produced by NPPs with the exception of Chornobyl NPP.

More specifically, the State Exclusion Zone Agency is responsible for: management of RAW long-term storage and disposal, including disposal of the radioactive waste at existing storage units and the development of a DGD; the State Fund to Manage Radioactive Waste, development of a method to use the funds and sponsorship of the statewide environmental program on the radioactive waste management; ensuring state registry of radioactive waste and its storage systems, state inventory, also on radioactive materials managed by other state agencies; organizing activities to identify, create and support stable functioning of the system for physical protection of RAW, as well as the safety of collection, transportation, processing, storage and disposal stages; archiving documentation that defines the utilities which manage RAW, control of warning signs and fences around the utilities with RAW etc.; making decisions on the closure or continued use of RAW storage units following approval of the State Nuclear Regulatory Inspectorate; ensuring development of the study plans and expert study programs on RAW management; and organizing training, retraining and advanced skills development for personnel in the field of radioactive waste. (Cabinet of Ministers of Ukraine 2014)

4 Siting procedures

The Ukrainian Law On Decision Making Procedures for Siting, Design, and Construction of Nuclear Facilities and Radioactive Waste Management Objects of National Importance (Law No. 2861-15) was introduced in 2005. It is the key document that defines the decision-making process in siting nuclear waste storage and final disposal sites.

4.1 Procedures and criteria for site selection

According to this law, the government of Ukraine is responsible for proposing a specific law on siting any individual nuclear facility of national importance; this law is subject to approval by the parliament.

The state nuclear regulator adopted requirements and procedures for the selection of the nuclear waste sites in 2008 (State Nuclear Regulatory Comittee 2008a). It is up to the nuclear operator to plan and implement activities for the waste facility siting. An environmental impact assessment (EIA) is required for the preselected sites.

Investigations of alternative sites are also required. The procedures set a rather lengthy list of requirements for the sites, including geological, hydrological, geochemical, anthropogenic, sociological and other conditions. Site preference depends on the ability to provide maximum isolation and safety and consideration of socio-economic factors. The procedures emphasize a conservative approach to this site criteria assessment.

Also relevant in site selection is the nuclear regulator's General Requirements on Radioactive Safety for Deep Geological Disposal Sites (29.05.2007 No. 81). It is based on the requirements for nuclear waste sites, but goes into more specific details. It also includes provisions for the elimination of the transboundary impacts and reducing the financial burden on future generations.

As stated previously, the CEZ is the top choice for the location of the new waste storage facilities. There are thus governmental decisions in place and work in progress to build the Vektor facility for radioactive waste and spent nuclear fuel storage. Most of the detailed research on options for final disposal in geological formations is also happening in the CEZ.

4.2 Compensation mechanisms and socio-economic impact

The Law on the Use of Nuclear Energy and Radiological Safety suggests that the volume of financing allocated to socio-economic development is determined for every site separately and is to be defined in the law that allows construction of the site. These costs are allocated in the construction budget and are disbursed as a reflection of the actual expenditures on a monthly basis to the local budgets. In the case of centralised spent nuclear fuel storage, the allocation to social programs accounts for 10 percent of the project cost.

The local population has additional rights for being within the observation zone of the RAW management utility, including specific social infrastructure like shelters and personal protection items (iodine tablets) in case of emergencies. The funds are to be allocated from the State Radioactive Waste Management Fund. The Cabinet of Ministers defines which local administrations are eligible for socio-economic compensation for the impacts of every particular utility.

5 Information and participation

The parliament has to approve legislation for the construction of any new facility that manages nuclear materials. It can only do so if the siting is approved by the local authorities. The local authorities "accept the decision on agreement after conducting of local advisory questioning of citizens of Ukraine (advisory referendum) on this issue", according to law No. 2861-15 on decising making procedures. The procedures for nuclear waste site selection require the nuclear regulator to provide the information necessary to conduct public hearings to local authorities (State Nuclear Regulatory Committee of Ukraine 2008c: 51).

A public consultation process organized by Energoatom in 2008 regarding the siting of the Centralised Spent Fuel Storage is the only instance where a consultation process was started. One public hearing was conducted at Slavutich, the satellite town of Chornobyl NPP on the other side of the Dnipro river. In the Ivankiv region, which borders the CEZ and the proposed site of the spent nuclear fuel storage the company limited its public engagement to an information round table, as it did in Kyiv.

Despite the requirement to have a public referendum on the siting of nuclear facilities, there is no evidence that it was ever held for the Centralised Spent Fuel Storage. Instead, the legislation was adjusted in 2009, removing the need for public consultation for the facilities located in the CEZ as they related to efforts to construct a Centralised Spent Fuel Storage facility. It is up to the state body responsible for the management of the exclusion zone to agree on construction there. The law to build a centralized SNF storage was approved by Parliament in 2012.

The existence of the CEZ gives the nuclear industry a significant opportunity to limit public consultation on waste management facilities. The absence of the local population makes it attractive, in addition to the presence of existing infrastructure for waste management and transport as well as proximity to the majority of Ukrainian nuclear waste. Locating the DGD in the CEZ will, most likely, not require public consultation either. Unfortunately, this also means further transformation of the CEZ into a nuclear waste dumpsite.

6 Costs and financing

The costs of nuclear waste management on the NPP sites are covered by Energoatom operational costs. Future costs of radioactive waste management are covered by the environmental tax accumulated in the RAW Management Fund. The state guarantees that it will manage the waste of the companies that have paid the tax.

In 2009, the State RAW Management Fund was created with by Law 17.09.2008 515-VI. Money is collected for this state fund through pollution fees assigned to RAW creation and temporary storage by its producers. The fund is managed by the State Exclusion Zone Agency. Since 2011, the amount and the method of payment to the fund is determined by the tax code (article 249 Section VIII).

The fund receives around UAH 600 million annually (20 million Euro at the exchange rate of March 2016). The nuclear operator Energoatom has transferred around UAH 3.1 billion (over 166 million Euro) into the fund between 2009 and 2014. The exchange rate has changed from 10 UAH: 1 Euro in 2009 to 25 UAH: 1 Euro in 2014 (Energy and Coal Ministry of Ukraine 2015).

The use of the fund is defined by the Cabinet of Ministers Order (20.05.2009 No. 473). The order suggests that the fund be used for the implementation of the State Environmental Program on Radioactive Waste Management.

The 2010 budgetary law widens the scope of the fund's use and allows for spending on other tasks. As a result, there is not enough money for the activities that the fund should finance – namely, radioactive waste management. In its letter to the Chornobyl United Trade Union Organization, Energoatom has stated that the program on waste management was financed at the level of 8-10 percent.

International donors have paid for a number of RAW management utilities (the Liquid Radioactive Waste Treatment Plant and the Industrial Complex for Solid Radioactive Waste Management) (Chornobyl NPP 2017). Industry specialists raise concerns over the absence of a single state policy, leading to duplicated facilities built at the sites of NPPs and at the centralised waste management facility.

Involvement of international donors is partly a result of the 1995 Memorandum of Understanding among the G7 nations, the EU and Ukraine, which established international support to ensure timely and safe closure of the Chornobyl NPP (G-7 and Ukraine 1995).

EBRD was asked to manage the Chornobyl-related funds as it had experience managing a Nuclear Safety Account (NSA) that operated to improve safety of the nuclear reactors in eastern Europe. Its responsibility was extended to Ukraine in 1995. Since then, close to 2.5 billion Euro has been received for EBRD-managed Chornobyl projects from 45 donors. The EBRD contributes 715 million Euro of its own resources in support of the work to transform Chornobyl into an environmentally safe and secure site.

While EBRD positions its involvement as a tool to improve nuclear safety, some of its loans are seen by environmentalists as questionable, as it supports the Ukrainian nuclear industry's staying afloat. The most recent example of this is the EBRD 300 million Euro loan (backed by another 300 million Euro from Euratom) for the Nuclear Safety Upgrade Program. This loan provides crucial funding to implement ongoing efforts to extend the lifetime of the Ukraine's ageing nuclear reactors (Holovko 2012).

Additionally, the EU has been providing various grants to improve radioactive waste management in Ukraine. As a part of the INSC U4.01/08-B project, EU experts have calculated that with the existing radioactive waste classification, the total cost of storing all waste in two types of storage systems would be UAH 750 billion (25 billion Euro). This estimate can be compared to the UAH 684.5 billion allocated as overall state budget expenses in 2016. It is clear that Ukraine will not be able to allocate sufficient funds to store all waste. The new system of radioactive waste classification proposed by the EU project on waste classification, is expected to cut the cost by a factor of 10 and be in accordance with international standards (State Exclusion Zone Agency 2016).

Ukraine has systematically attempted to cut costs on spent nuclear fuel reprocessing by Russia. It initially built a SNF storage system at Zaporizhzhya NPP and is slowly moving forward with plans to build a centralized spent fuel storage for other NPPs. Although the costs for construction and maintenance of the storage facility have not been published, according to Energoatom the operation of the storage allows Ukraine to save USD 40 million per year by not sending SNF to Russia for reprocessing (State Enterprise 'National Nuclear Energy Generating Company "Energoatom" 2006).

The Cabinet of Ministers ordered to build a Centralised Spent Fuel Storage estimated UAH 1.59 billion (160 million Euro) for infrastructure construction and an annual maintenance cost of 3.67 billion UAH (370 million Euro) (Cabinet of Ministers of Ukraine 2009). According to the president of Energoatom, the cost of production and 100 years of operation for one cask amount to around USD 2.2 million, compared to USD 15 million cost of sending the same volume of SF to Russia for reprocessing (Ukrainian News Agency 2015).

7 Conclusions

The development of radioactive waste management in Ukraine has been impacted by the transition of the state management system in the post-Soviet era, the Chornobyl nuclear accident and a constant lack of funding. In the 1990s, the nuclear industry was too attractive for the bankrupt Ukrainian state to consider the future costs of nuclear waste. The nuclear industry was not obligated to repay capital investments (this was done by the collapsed USSR), it did not bear financial responsibility for the Chornobyl catastrophe and it received free nuclear fuel from Russia.

The scale and the cost of the Chornobyl nuclear waste problem overshadows the problems of the waste accumulated at Ukraine's additional 15 operating commercial reactors. It also complicates the governance of radioactive waste management.

At the same time, the CEZ has become extremely attractive for the development of the centralised nuclear waste and spent nuclear fuel management systems, as well as a final disposal site. With the absence of a sizeable local population to consult, existing infrastructure and trained personnel, and its proximity to the majority of the country's nuclear waste, the CEZ is seen as an ideal place to focus operations. Thus, plans to construct a centralised spent nuclear fuel storage and ongoing research for a final deep geological disposal are ongoing.

The change of legislation in 2009 enabled construction of the new facilities in CEZ that are not related to the mitigation of the Chornobyl nuclear accident, which were previously forbidden. Further changes removed requirements to consult with local populations for proposed facilities located in the CEZ, which are required for corresponding decisions elsewhere in the country.

Ukraine's waste management system suffers from the government's focus on solving day-to-day tasks rather than long-term objectives. This absence of strategy is particularly problematic for radioactive waste management. For example, it took just a few years since its creation to undermine functioning of the Radioactive Waste Fund by widening the scope of activities it can finance and directing the funds accumulated to other purposes. Due to the lack of funding, activities envisioned by the State Waste Management Strategy are systematically delayed.

Nevertheless, Ukraine is progressing with the development of the legal, institutional and scientific structures needed to fulfil recommendations of the IAEA and adopt European best practices. Most of these activities are carried out with EU and IAEA financial support. Implementation of the provisions of the EU-Ukraine Association agreement is expected to drive the process further, as Ukraine will adjust its legal framework to relevant EU Directives.

References

Cabinet of Ministers of Ukraine (4 February 2009). *Approval of the Feasibility Study to Invest in the Construction of a Centralized Repository for Spent Nuclear Fuel of WWER Domestic Nuclear Power Plants.* http://www.kmu.gov.ua/control/uk/cardnpd?docid=192437006, last accessed 9 June 2017.

Cabinet of Ministers of Ukraine (19 August 2009). *Radioactive Waste Management Strategy of Ukraine.* http://zakon2.rada.gov.ua/laws/show/990-2009-%D1%80, last accessed 8 June 2017.

Cabinet of Ministers of Ukraine (22 October 2014). *State Exclusion Zone Management Agency Provisions #564*. http://zakon2.rada.gov.ua/laws/show/564-2014-%D0%BF, last accessed 8 June 2017.

Chornobyl NPP (2017). *Current Projects*. http://chnpp.gov.ua/en/decommissioning-projects, last accessed 8 June 2017.

Datsenko, V. (2015). *Radoactive Waste Management. Presentation at the Public Council of the State Nuclear Regulatory Inspection of Ukraine*, 22.09.2015. Kyiv.

Energy and Coal Ministry of Ukraine (2015). *Letter to the National Ecological Centre of Ukraine*, 2 February 2015.

G-7 and Ukraine (20 December 1995). *Memorandum of Understanding between the Governments of the G-7 Countries and the Commission of the European Communities and the Government of Ukraine on the Closure of the Chernobyl Nuclear Power Plant*. Ottawa. http://www-bcf.usc.edu/~meshkati/G7.html, last accessed 8 June 2017.

Gorbachev, M. S. (17 April 2006). Turning point at Chernobyl. In: *Daily Times*. http://www.gorby.ru/en/presscenter/publication/show_25057/, last accessed 8 June 2017.

Gramotkin, I. (2016). *Memorandum on radioactive materials management*.

Holovko, I. (2012). *Ukraine – Nuclear Power Plant Safety Upgrade Programme*, Kyiv: CEE Bankwatch Network. http://bankwatch.org/sites/default/files/briefing-Ukraine-NPPs-13Nov2012.pdf, last accessed 8 June 2017.

Kondratiev S. (2016). Radioactive Waste Management at Ukrainian Nuclear Power Plants. State and Challenges. In: *Nuclear and Radiation Safety*, 2(70), 41-46. http://sstc.kiev.ua/documents/journal/2016/2/9_2_2016_text.pdf, last accessed 8 June 2017.

Ministry of Health (2005). *Main Sanitary rules of the Radiation Safety in Ukraine*. http://zakon3.rada.gov.ua/laws/show/z0552-05, last accessed 8 June 2017.

Ministry of Ukraine of Emergencies (2011). *Twenty-five Years after Chornobyl Accident: Safety for the Future*, Kyiv: KIM. http://chernobyl.info/Portals/0/Docs/ua-25-chornobyl-angl-c.pdf, last accessed 8 June 2017.

Ministry of Ukraine of Emergencies and Affairs of population protection from the consequences of Chornobyl Catastrophe (2006). *20 years after Chornobyl Catastrophe*. Future outlook.

Proskura, N. S. (2014). Radioactive Waste Classification for Long-term Disposal Safety. In: *Nuclear and Radiation Safety*, 2(62), 37-43. http://sstc.kiev.ua/documents/journal/2014/2/9_2_2014_text.pdf, last accessed 8 June 2017.

Shestopalov V.M., B. W. (2006). *Geological disposal of radioactive waste in Ukraine (problems and solutions)*. Kyiv.

Shybetskyi, I. (2011). *Methodology of the Site Investigation Program for Safety Case Development at Geological Disposal of Radioactive Waste.* http://znp.igns.gov.ua/materials/19n/Shybetskyi.pdf , last accessed 8 June 2017.

State Enterprise 'National Nuclear Energy Generating Company "Energoatom" (2006). *Dry Cask Spent Nuclear Fuel Storage at ZNPP.* http://www.energoatom.kiev.ua/ua/actvts/stroitelstvo/cssnf/press/668-suhe_shovische_vdpratcovanogo__yadernogo_paliva_zaes/, last accessed 8 June 2017.

State Enterprise 'National Nuclear Energy Generating Company 'Energoatom" (2014). *Radioactive Waste Management at SE "NNEGC "Energoatom""NPPs.* http://www.energoatom.kiev.ua/files/file/radioactive_waste.pdf, last accessed 8 June 2017.

State Exclusion Zone Agency (2016). *Explanatory Note to the law Suggested Changes to the legislation in the field of the radioactive waste management.* http://dazv.gov.ua/images/pdf/zakony/POJaSNJuVALNA_ZAPYSKA.pdf, last accessed 8 June 2017.

State Nuclear Regulatory Comittee (2008a). *On approval of requirements to site selection to accommodate the radioactive waste disposal facility.* http://zakon3.rada.gov.ua/laws/show/z1166-08, last accessed 8 June 2017.

State Nuclear Regulatory Committee of Ukraine (2008b). *On Compliance with the Obligations under the Joint Convention on the Safety of Spent Fuel Management and on the Safety of Radioactive Waste Management.* http://www.snrc.gov.ua/nuclear/doccatalog/document?id=119923, last accessed 8 June 2017.

State Nuclear Regulatory Committee of Ukraine (2008c). *Radioactive Waste Storage and Disposal Siting Requirements.* http://zakon3.rada.gov.ua/laws/show/z1166-08/page, last accessed 8 June 2017.

State Nuclear Regulatory Inspectorate of Ukraine (2013). *Report on Nuclear and Radiation Safety of Ukraine in 2012.*

State Nuclear Regulatory Inspectorate of Ukraine (2014). *National report of Ukraine on compliance with the obligations of the Joint Convention on the safety of spent fuel management and on the safety of radioactive waste management.*

State Nuclear Regulatory Inspectorate of Ukraine (2015a). *Presentation of the National report of Ukraine on compliance with the obligations of the Joint Convention on the safety of spent fuel management and on the safety of radioactive waste management.* Vienna. http://www.snrc.gov.ua/nuclear/doccatalog/document?id=284149, last accessed 8 June 2017.

State Nuclear Regulatory Inspectorate of Ukraine (2015b). *Report on Nuclear and Radiation Safety in Ukraine in 2014.*

State Nuclear Regulatory Inspectorate of Ukraine (2016). *Report on Nuclear and Radiation Safety of Ukraine in 2015.* http://www.snrc.gov.ua/nuclear/doccatalog/document?id=324059, last accessed 8 June 2017.

The Parliament of Ukraine (1995). *The law of Ukraine on nuclear energy and radiological safety.* http://zakon2.rada.gov.ua/laws/show/39/95-%D0%B2%D1%80, last accessed 8 June 2017.

Trofymovych, L. (2016). Budapest Memorandum of 1994 and international "gurantees" of Ukraine's Security. In: *Military Resarch Bulletin*, (16), 280. http://www.asv.gov.ua/content/nauka/editions/26/2016-26/276.pdf, last accessed 8 June 2017.

Ukrainian News Agency (2015, 13 November). *Energoatom will borrow $250 million to build spent nuclear fuel storage.* https://ukranews.com/ua/news/390053-energoatom-vizme-kredyt-v--250-mln-dlya-budivnyctva-skhovyshha-vidpracovanogo-yadernogo-palyva, last accessed 8 June 2017.

IV. Nuclear Waste Governance in Canada and the Global South

A Twinned Approach

The Challenges of Nuclear Waste Governance in Canada

Cindy Vestergaard[1]

Abstract

Over the past decade, Canada has been implementing an 'Adaptive Phased Management' approach to handling its radioactive waste, involving interim storage at reactor sites and eventual final disposal in a centralised, deep geological repository. The process is described as a flexible step-by-step, phased approach supported by public engagement and scientific research at each interval along the way. Its timeline for implementation is long. From identifying a willing host community through to the construction, monitoring and eventual closure of the repository, the process will span generations, taking potentially upwards of 140 years or more to fully implement.

---

1 Stimson Center, 1211 Connecticut Ave NW, 8th Floor. Washington, DC, 20036 USA; E-Mail: cvestergaard@stimson.org

A. Brunnengräber et al. (eds.), *Challenges of Nuclear Waste Governance*, Energiepolitik und Klimaschutz. Energy Policy and Climate Protection, https://doi.org/10.1007/978-3-658-21441-8_11

1 Introduction

In 2007, after decades of public inquiries, commissions and panels on how radioactive spent fuel and nuclear waste should be managed, Canada adopted what has been termed the 'Adaptive Phased Management' approach. The APM envisions centralised containment and isolation of used fuel in deep geological disposal, allowing for monitoring and retrieval after emplacement. It also includes an option for shallow storage at the central site prior to placement in the repository if it should be so decided.

The main tenets of APM are technical and social, meaning a deep repository will only be constructed at a site based on technical and scientific demonstrations of health and environmental safety, security of nuclear materials (including meeting international safeguards obligations) and economic feasibility alongside socially-demonstrated community willingness, well-being and involvement. These twinned concepts set the framework for a long, multi-generational series of steps and phases that will require dedicated efforts and finances to maintain momentum along the way.

This chapter begins with an overview of Canada's nuclear power, followed by a historical summary of how the APM process emerged, the inventory Canada currently has (and is anticipated to have in the future), what the APM legally and institutionally entails along with how it is being implemented today and what is expected in the future in terms of finances, community engagement and siting procedures.

2 Nuclear power in Canada

Canadian industry and Atomic Energy of Canada Limited (AECL) began a Canadian line of nuclear power plants (NPPs) in the late 1950s. The CANDU (Canada Deuterium Uranium) reactor uses heavy water (deuterium oxide) as a moderator and coolant and is fuelled by natural uranium (as distinct from enriched uranium). In 1962, the first prototype (22 MWe) went operational at Rolphton, Ontario, 30 kilometres from the Chalk River facilities, with the second, larger prototype (200 MWe) beginning to generate power at Douglas Point, Ontario, in 1967. Rolphton was permanently shutdown in 1987 and Douglas Point in 1984.

Table 1: Nuclear Power Plants in Canada

Sites	Owner/ Operator	No. of Units	Installed Capacity (MWe)	Start of Commercial Operation	Planned Closure	Current Status
Bruce A and B Nuclear Generating Stations, Kincardine, Ontario	Bruce Power	8	6,232 (Bruce A & B combined)	Bruce A 1977-79 Bruce B 1984 - 1987	Refurbishment and multi-year life extension program of 6 units began in 2016 and will continue through 2053, allowing units to operate till 2064.	All units operating
Darlington Nuclear Generating Station, Clarington, Ontario	Ontario Power Generation (OPG)	4	3,512	1990-1993	Until 2055 (beginning of shutdown state 2048 and final shutdown in 2085)	All four reactors operating
Pickering Nuclear Generating Station, Pickering Ontario	OPG	8	3,100	1971 -1973 for units 1 to 4 1983 - 1986 for units 5 to 8	Initially 2020, operations extended until 2024.	Six operating; two in safe storage
Point Lepreau Generating Station, New Brunswick	New Brunswick Power Nuclear (NBPM)	1	705	1 February 1983	Until 2036	Operating
Total		**21**	**13,549 (13,553 operating)**			**19 operational**

Source: IAEA Power Reactor Information System (PRIS)

Commercial reactors began operating in 1971 in Pickering, Ontario. In total, 25 reactors were built. Six have since been permanently shutdown and today 19 are operational, located at five plants in three provinces. The Gentilly-2[2] Nuclear

2 The Gentilly-1 was permanent shutdown on 1 June 1977. See the IAEA's Power Reactor Information System (PRIS):

https://www.iaea.org/PRIS/CountryStatistics/ReactorDetails.aspx?current=47, last accessed 17 April 2017.

Generating Station, located near Trois-Rivières, Québec, was permanently shut down in December 2012, leaving Ontario and New Brunswick as the only provinces with operating plants. In Ontario, nuclear power generates more than 50 percent of the province's electricity. In 2016, Canada's total nuclear electricity production was 97,445 GWh or 15.63 percent of the country's electricity (IAEA 2017). The nineteen power reactors currently operating are listed in Figure 1.

3 Nuclear waste disposal in Canada

3.1 Historical background

Canada's approach to nuclear waste management has its roots in the late 1960s. In 1969, Canada's then nuclear regulator, the Atomic Energy Control Board (AECB) requested Chalk River Nuclear Laboratories (owned by the crown corporation Atomic Energy Canada Ltd, AECL) to conduct research on storing and disposing of nuclear waste. AECL joined with Ontario Hydro and Hydro Quebec to form a committee of waste owners to study the issue and in 1972 presented their conclusions, including the concept of a retrievable waste facility. Following on the committee's recommendation for a design study, the AECL led a 'Committee Assessing Fuel Storage' to study long-term retrievable storage of spent fuel. Its report of December 1974 recommended above-ground canister-type storage as an interim measure and deep geological disposal as a permanent one. As noted by a 1980 report on the history of the AECB, these twin concepts (interim storage and final geological disposal) would provide the direction for developing Canada's nuclear waste management (Sims1980: 146).

In 1974, the AECB published a Guide for Licensing of Radioactive Waste Management Facilities. In it, three categories of waste storage or management facilities were outlined which the Board would be prepared to license: Category A facilities for the collection, packaging and temporary storage of radioactive wastes for up to two years; Category B for storage up to 50 years; and Category C for permanent waste storage (AECB 1974: 6). All three categories were intended to be retrievable pending demonstration of the long-term integrity of any disposal facilities (ibid.: 10). Until then, the Board would only license waste storage facilities designed to allow for retrievability (Didyk et al. 1975: 10). As noted by the AECB in 1976, the expectation of eventually moving the material meant that "interim storage necessarily implies retrievability." Conversely, "ultimate storage or disposal does not, in the logical sense, necessarily imply retrievability" (Didyk 1976: 7).

In April 1977, the Department of Energy, Mines, and Resources (EMR, now NRCan) commissioned a panel to examine nuclear waste disposal. Completed in four months, the Hare report underscored nuclear waste should "not be allowed to continue to accumulate indefinitely in interim storage" (Aiken et al. 1977: 1). A year later, the Ontario Royal Commission on Electric Power Planning, known as the Porter Commission, reviewed Ontario's long-range plan for generating electricity and similarly highlighted the urgent need to develop permanent disposal facilities (Porter 1978).

Both reports paved the way for a joint Canada-Ontario Nuclear Fuel Waste Management Program in 1978 to develop a generic concept for deep geological disposal. Under the program, Ontario Hydro (now Ontario Power Generation, OPG) was responsible for developing technologies for interim storage and transportation of irradiated fuel while AECL focused on developing a concept for a spent fuel deep underground repository within the plutonic rock of the Canadian Shield. Technologies for the immobilisation of both irradiated fuel and fuel-recycle wastes were explored, leaving options for the disposal of either form (AECL 1983: 5).

In 1988, AECL submitted its concept for geological disposal to the government. The design followed the Hare report's recommendation to bury waste 500-1000 metres deep in the Canadian Shield and included the use of both geological and engineered barriers and contained no provisions for monitoring or retrieval (Ramana 2013: 198). In turn the federal Minister of Energy, Mines and Resources referred the concept for public review and in October 1989, a public inquiry named after its chair Blair Seaborn was established. The Seaborn panel was tasked to examine the safety and criteria for accepting long-term waste management and disposal and to comment on the feasibility of the AECL design. The Panel held public meetings in provinces that hosted nuclear reactors or mined, refined or fabricated uranium (Manitoba, New Brunswick, Ontario, Quebec and Saskatchewan). The Panel also developed the guidelines for AECL to carry out an environmental impact assessment.

In 1994, the AECL submitted an Environmental Impact Statement (EIS) on the effects of its conceptual repository, including interim storage, transportation and closure. As a round of public hearings discussed the EIS, the Federal Government issued a Policy Framework for Radioactive Waste in 1996 outlining key principles and respective roles of government and waste owners. It specified that Ottawa was responsible for developing policy, regulations and oversight of waste producers and owners, and for ensuring radioactive waste disposal was carried out in a safe, environmentally sound, comprehensive, cost-effective and integrated manner. Waste producers in turn were responsible for the funding, organisation, management and operation of disposal and other waste facilities in line with the 'polluter pays' principle (CNL 2013: 10).

Nine years after its establishment, the Seaborn Panel released its report, concluding the feasibility and safety of the AECL design had been adequately demonstrated at a conceptual level; but broad public support for the idea had not been demonstrated and therefore did not meet the required level of acceptability to be adopted as Canada's approach (NWMO 2005: 16). A number of the Panel's recommendations (including how the Government could reach a decision on acceptability), coupled with the 1996 Policy Framework for Radioactive Waste, provided the basis for the Canadian Parliament passing the Nuclear Fuel Waste Act (NFWA) in 2002.

The NFWA in turn requires nuclear waste owners to establish a waste management organisation to "provide recommendations to the Government of Canada on the long-term management of used nuclear fuel." In turn, the Governor in Council selects one of the options proposed and provides oversight during implementation. In the fall of 2002, utility companies (OPG, Hydro Quebec and New Brunswick Power) established the Nuclear Waste Management Organization (NWMO) and, as mandated by the NFWA, investigated three approaches based on both storage (on-site at nuclear reactor sites or centralised either above or below ground) and permanent disposal. For the latter, the Act specified the study be based on the concept described by the AECL in its 1994 Environmental Impact Statement and the recommendations of the Seaborn report (NFWA, Section 15). In its report submitted in November 2005, the NWMO recommended a fourth option: the Adaptive Phased Management (APM) approach, offering a blending of each of the three options.

APM is based on phased and adaptive decision making, incorporating flexibility to implementation to allow for a step-by-step approach, supported by research and public engagement at each interval along the way. It provides for ongoing monitoring systems and retrieval for irradiated fuel bundles. On 14 June 2007, Ottawa accepted the approach as Canada's plan, calling for centralised containment and isolation of Canada's used fuel in a deep geological repository with an informed and willing host. As noted by the Minister of NRCan:

> This approach includes the isolation and containment of the used fuel in a deep geologic repository, with an option for temporary shallow underground storage. This safe, long-term approach will ensure that used nuclear fuel is monitored and retrievable. It is designed to take advantage of emerging energy technologies, including the future possibility of recycling used nuclear fuel (NRCan 2007).

3.2 The national inventory

In March 2008, the CSA Group (formerly the Canadian Standards Association), a not-for-profit organisation comprised of government, industry and consumer groups, developed along with the CNSC a standard for a radioactive waste classification system. Four categories were recognised: high-level, intermediate-level and low-level radioactive wastes along with uranium mine and mill waste (CNSC 2017a). Every five years, the Low-Level Radioactive Waste Management Office (LLRWMO) issues the Inventory of Radioactive Waste in Canada report, on behalf of NRCan, which provides statistics based on Canada's four waste categories. Data on inventory is also reported to the International Atomic Energy Agency (IAEA) and included in the Radioactive Waste Management Database which tracks low- and intermediate-level waste (L&ILW) worldwide.

High-level waste

As of 30 June 2015, Canada had approximately 2.6 million used CANDU fuel bundles (approximately 52,000 tonnes of heavy metal (t-HM)) in storage at reactor sites.

Table 2: Inventory of spent fuel in Canada as of June 30, 2015.

Site	Current net power capacity (GWe)	Number of fuel bundles in wet storage	Number of fuel bundles in dry storage
Bruce A and B Nuclear Generating Stations	4.693	687,768	425,078
Darlington Nuclear Generating Station	3.512	335,279	153,917
Douglas Point Waste Management Facility		0	22,256
Gentilly-1 Waste Management Facility		0	3,213
Gentilly-2 Nuclear Generating Station	0.635	32,801	97,140
Pickering A and B Nuclear Generating Station	3.094	400,440	300,977
Point Lepreau Nuclear Generating Station	0.635	39,730	92,700
Chalk River Laboratories		0	4,921
Whiteshell Laboratories		0	2,268
Total:	12.569	1,496,018	1,102,470

Source: Progress Through Collaboration Annual Report, 2015b: 11 (Nuclear Waste Management Organization)

Projections for the total number of used fuel bundles produced to the end of life for the existing reactor fleet range from 3.4 to 5.2 million used CANDU fuel bundles (approximately 69,000 t-HM to 103,000 t-HM). The lower end is based on an average of 25 effective full power years (EFPY) of operation for each reactor with no additional refurbishment beyond what has already been completed. The higher estimate assumes most reactors are refurbished and life extended for an additional 25 EFPY of operation (Garamszeghy 2015: iii).

Low- and intermediate-level waste

Low- and intermediate-level waste (L&ILW) in Canada refers to all radioactive waste except spent fuel and waste from uranium mining and milling (CNSC 2015: 20). Low-level radioactive waste is minimally radioactive and contaminated equipment from the operation of a nuclear facility such as protective clothing, mops, tools and dust collectors. Intermediate/level waste consists mostly of resins, air filters and used reactor components. Most of the radioactivity subsides within 300 years. According to the CNSC, Ontario Power Generation, which owns 20 of Canada's 22 CANDU reactors, generates approximately 77 percent of Canada's operational L&ILW waste annually. CNL produces approximately 17 percent as well as accepting L&ILW for long-term management from a number of small producers and users of radioactive materials across Canada such as hospitals, universities and small industries (which make up approximately three percent of annual volume). The remaining three percent is generated by the other two CANDU reactors (in New Brunswick and Quebec) and Cameco Corporation's uranium processing and conversion facilities in Ontario (CNSC 2017b).

Uranium mine and mill waste

Over 200 million tonnes of uranium mill tailings have been generated in Canada since the mid-1950s (CNSC 2015: 22). Given the large volumes and low level of radioactivity levels of mine and mill waste, such waste is decommissioned in place (at mine and mill sites), in mined out open pits converted to tailings management facilities (TMFs). There are 20 TMFs located at closed or decommissioned uranium mines: 14 are located in Ontario (in the Elliot Lake and Bancroft areas), four in northern Saskatchewan and two in Northwest Territories (CNSC 2017b). Of these only three TMFs in Saskatchewan are operational, the others no longer receive waste material. The management of shutdown or decommissioned mines and mines, along with their adjacent waste facilities, is either under the responsibility of the former operators or within the purview of the provincial and federal governments (CNSC 2017a).

This category also forms historical low-level radioactive waste (LLRW) which was waste resulting from handling and processing from the 1930s to 1970s.

Historic waste

Canada has approximately 1.7 million cubic metres of wastes and contaminated soils in the Northwest Territories, British Columbia, Alberta and Ontario related to radium and uranium production and processing dating back to the 1930s. Known as historic low-level waste (LLW), the Government of Canada is responsible for its long-term management which is undertaken by the LLRWMO of Canadian Nuclear Laboratories Ltd (CNL). The bulk of historic LLW is located in Port Hope and Clarington, Ontario. The Government of Canada and the local municipalities agreed in March 2001 to community-developed proposals as potential solutions for the clean-up and long-term management of historic LLW in the Port Hope area, establishing the Port Hope Area Initiative (PHAI).

Canada also manages legacy liabilities resulting from nuclear research and development conducted by the National Research Council of Canada (1944 to 1952), AECL (1952 to 2014), and now CNL, on behalf of the Government of Canada. The liabilities consist of outdated and unused research facilities and buildings, a variety of buried and stored radioactive waste and affected lands. Facilities include Whiteshell Laboratories, Chalk River Laboratories (CRL) and the three partially decommissioned prototype reactors (Gentilly-1, the Nuclear Power Demonstration reactor and Douglas Point). Also included are small quantities of high-level waste, such as prototype and research reactor used fuel, by-products of the production of medical isotopes and waste from early fuel reprocessing experiments conducted between the 1940s and 1960s at CRL (CNSC 2017a).

3.3 The (interim) storage sites

After their removal from CANDU reactors, fuel bundles are stored in water pools for a minimum of 6-10 years before being placed in convection cooled concrete silos or bunkers for interim dry storage. The number of disposal sites is limited to government-approved and monitored facilities on land set aside in perpetuity for disposal purposes. As such, solid wastes and high and medium level liquid wastes are stored at locations adjacent to reactor facilities. Given such facilities produced the volumes of waste that need to be stored, they maintain a skilled, on-site staff trained in the safe handling of radioactive materials (Dewar 1969: 3; Ramana 2013). All the inventory of Canada's spent fuel is currently in interim storage, co-located at the facilities mapped below.

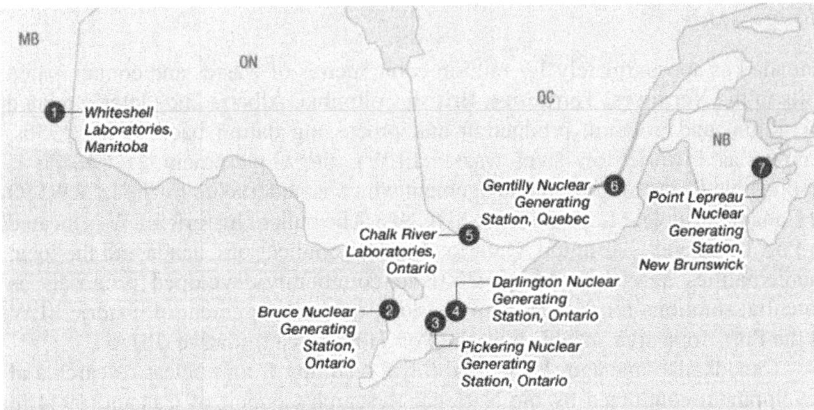

Figure 1: Map: Interim storage facilities in Canada (Source: Nuclear Waste Management Organization 2017a)

In addition, OPG's low and intermediate level waste from its CANDU reactors is stored at the Western Waste Management Facility (WWMF) at the Bruce nuclear site in Kincardine, Ontario. Hydro Quebec and New Brunswick Power have their own interim storage facilities for L&ILW co-located at their reactor sites (CNSC 2015: 21).

3.4 The waste management strategy

The Adaptive Phased Management approach to Canada's nuclear waste is based on phased and adaptive decision making, incorporating flexibility to implementation to allow for a step-by-step approach, supported by research and public engagement at each interval along the way. The NWMO's 2005 report recommended that the Canadian Shield and areas underlain by Ordovician sedimentary rock (in southern Canada) could be considered for the ultimate centralised containment and isolation of used fuel in deep geological disposal. It also included an option for shallow storage at the central site prior to placement in the repository. The only geographic limitation on the final recommendation was that any developed facility should be sited within Ontario, Quebec, New Brunswick or Saskatchewan, i.e. in provinces that have benefited from the employment and electricity associated with the nuclear industry (Murphy 2009: 131). The report proposed that any deep geological repository be "sited only in a willing host community" (NWMO 2005: 6).

There are three phases of implementation under the Adaptive Phased Management approach: Phase I, Preparing for Central Used Fuel Management;

Phase 2, Central Storage and Technology Demonstration, and; Phase 3, Long-term Containment, Isolation and Monitoring (NWMO 2005: 24). It should be noted that the NWMO underscores there is no fixed timeframe to its APM with the timelines below "intended to be 'illustrative' only" as nominal time frames were "required in order to develop conservative cost estimates" (ibid.: 33). In other words, the actual period will be determined by society at the time.

Phase I

Canada is currently in Phase I which involves maintaining storage and monitoring of used fuel at reactor sites while engaging with citizens to develop a program for activities such as design of the process for choosing a site, development of technology and key decisions during implementation. Selection of a central site will be based on the rock formations suitable for shallow underground storage (optional) and an underground characterisation facility and deep geological repository. At the same time, continued engagement with regulatory authorities focuses on pre-licensing work and eventually initiating the licensing process, triggering the environmental assessment process. Safety analyses, underground characterisation and the development and certification of transportation containers and used fuel handling guidelines. Towards the end of the approximately 30-year timeline for Phase I, a licence will be obtained to construct the underground characterisation facility at the central site, including decision whether to proceed with construction of a shallow underground storage facility and if so to obtain a construction licence, followed by an operating licence for the storage facility (NWMO 2005: 24).

Phase II

The next 30 years then focuses on demonstrating central storage and technology and will involve transportation of used fuel from reactor sites to the central site if a decision and licence to operate shallow underground storage has been given. If not, used fuel will continue to be stored at reactor sites until a deep repository becomes operational. Research and testing at the underground characterisation facility will be conducted to demonstrate and confirm the suitability of the site and deep repository technology while engaging with citizens through site assessment, technology and timing of waste placement in the repository. Phase II will require complete final design and safety analyses to obtain required operating licence decisions along with the decision as to when to construct the deep repository. The plan notes that there may be a need to manufacture transportation containers and facilities, processing facilities for their loading and production facilities for storage containers along with processing facilities to transfer used fuel from transportation to storage containers (NWMO 2005: 24).

Phase III

Long-term Containment, Isolation and Monitoring then extends another 60 years and beyond. Used fuel at reactor sites would be transported to the central facility for repackaging, or if used fuel was already stored at a co-located shallow underground facility, it would be retrieved, repackaged into long-lived containers and placed into the deep geological repository for final containment and isolation. Continuous monitoring, including with the participation of citizens, for an extended time with the option of retrieval until a future generation would decide when to decommission the underground characterisation, close the repository and decommission surface handling facilities. During this period, nuclear waste owners would continue to be responsible for interim management at reactor sites and would assume responsibility of waste as it is transported to the central facility (NWMO 2005: 25). It should be noted that the timeframe for extended monitoring in NWMO's 2005 report estimated a possible 240-year period while a more recent report states that 70 years had been assumed for planning purposes (NWMO 2015a: 40).

Retrievability

The concept of retrievability has survived as a concept in Canada's nuclear waste management since the committee of waste owners presented their first study on storing and disposing of nuclear waste in 1972. At the time, given plutonium could be recovered from spent CANDU fuel and re-used in existing or future nuclear power plants, the committee did not exclude the possibility of reprocessing and storing fuel so, if one day necessary, it could be retrieved and (re)processed (Sims 1980: 146).

Canada's policy changed sometime after India's 1974 nuclear test. However, unlike the United States which announced in 1977 it would "defer indefinitely the commercial reprocessing and recycling of plutonium" (Carter 1977), Ottawa never made a formal policy announcement regarding reprocessing. Although commercial reprocessing has never been carried out in Canada[3] and there is little pressure from utilities to do so given uneconomical costs, reprocessing can overall reduce the volume of fuel waste by recycling and is therefore has been given attention as a topic for the occasional technical study throughout the years.

3 Chalk River Laboratories did reprocess spent fuel in the 1940s to 1960s to extract plutonium.

Reprocessing however does not obviate the need for geological storage as underscored by a background paper by the NWMO (NWMO 2003: 24).

With Phase III calling for a 70-year period of monitoring in which nuclear fuel is retrievable, technological and scientific advances in the decades or centuries to come may yield new ways to make reprocessing more economical. Canada's approach today to retrievability is one related to storage, whether interim at reactor sites or at a shallow storage facility co-located at the centralised site. Disposal however does imply permanence and irretrievability. This has been accepted since 1987 when the AECB first declared its preferred approach to long-term waste disposal was permanent disposal with no intention of retrieval (AECB 1987: 2).

4 The legal and institutional framework

4.1 The legal framework

Primary federal legislation regulating the nuclear industry and radioactive waste management include the Nuclear Safety and Control Act (NSCA) and the Nuclear Fuel Waste Act (NFWA). The nuclear industry is also subject to the Canadian Environmental Assessment Act, 2012 (CEAA 2012) and other Acts related to the environment such as the Canadian Environmental Protection Act (CEPA) and the Fisheries Act (FA).

The Nuclear Safety and Control Act (NSCA) passed by Parliament in 1997 entered into force on 31 May 2000 and established the Canadian Nuclear Safety Commission (CNSC) to implement the Act. The legislation provides for the limitation of the risks to human and environmental health and safety and national security associated with the development, production and use of nuclear energy, nuclear substances and prescribed equipment and information. The Act implements Canada's obligations pursuant to international treaties governing the development, production and use of nuclear energy, including the non-proliferation of nuclear weapons and nuclear explosive devices.

Under regulations issued by the CNSC, the possession and management of radioactive waste is subject to the same regulations and requirements for possession and management of any radioactive substance. The CNSC takes a consultative and graded approach, specifying technical requirements in licence conditions (versus issuing general technical regulations specific for waste management). A waste management facility is also subject to the same regulations as for other nuclear facilities, differing at the level of facility-specific conditions in licences. All licensees of nuclear facilities must provide information on the radiation protection program, environmental protection and monitoring program, waste acceptance criteria, operational program documents, public information program and conceptual decommissioning plans and financial guarantees for the costs of decommissioning (Ferch 2003: 39).

Two documents specific to waste management have been issued by the CNSC. Regulatory policy P-290, Managing Radioactive Waste (2004) and Regulatory Guide G-320, Assessing the Long-term Safety of Radioactive Waste Management (2006). The policy outlines the philosophy and principals underlying the CNSC's approach while the guide assists applicants in understanding the ways to meet NSCA obligations, CNSC Regulations, or other legally-enforceable standards.

The Nuclear Fuel Waste Act (NFWA) of 2002 formalised the polluter pays principle. The Act instructs nuclear utilities to form, operate and fund a waste management organisation which in turn proposes approaches to the Government of Canada for the long-term management of nuclear fuel waste, and to implement the approach selected by the Government. Essentially the Act entrenches a framework for decisions on the management of nuclear fuel waste, including the requirement for trust funds established by utilities and the AECL to finance implementation of the selected approach.

The Canadian Environmental Assessment Act designated that projects that are regulated by the CNSC or the National Energy Board automatically require an environmental assessment by those regulators (and therefore they must submit a project description and assessment to those regulators and the Canadian Environmental Assessment Agency). Other acts related to environment that nuclear projects have to take into account are: the Canadian Environmental Protection Act (sustainable development requirements for pollution prevention and protection of environment and human health): Fisheries Act (prevents fisheries resources from becoming threatened or polluted by proposed or existing projects); and the Migratory Birds Convention Act and the Species at Risk Act which prevent migratory birds and all wildlife species respectively from being threatened or endangered by a proposed or existing project.

4.2 *The institutional framework*

The main government actors involved in the management of Canada's nuclear waste are the Canadian Nuclear Safety Commission, the Nuclear Waste Management Organization and Natural Resources Canada (NRCan).

The Canadian Nuclear Safety Commission (CNSC) is responsible for the regulation of radioactive waste in Canada. It was established by the NSCA, thereby replacing the AECB which had been established in 1946 to regulate the development, production and use of nuclear energy and the production, possession and use of nuclear substances, prescribed equipment and information. As Canada's nuclear regulator, the CNSC licenses, regulates and monitors waste management facilities to ensure they operate according to approved plans. Under the Act, the CNSC is also mandated to implement Canada's international commitments on the peaceful uses of nuclear energy as well as to disseminate scientific, technical and regulatory information on its activities and on the effects of regulated activities on human and environmental health and safety.

Pursuant to the NFWA, the waste management organisation – the NWMO – was established in 2002 by the nuclear energy corporations of OPG, Hydro-Québec, and New Brunswick Power, along with AECL. The NWMO's mission is "to develop collaboratively with Canadians a management approach for the long-term care of Canada's used nuclear fuel that is socially acceptable, technically sound, environmentally responsible and economically feasible" (NWMO 2003: 6; NWMO 2005: 17). From 2002 to 2005, the NWMO studied various approaches to the long-term management of Canada's spent fuel, including the APM approach which was ultimately picked by the Government. In May 2010, the NWMO began the site selection process.

Natural Resources Canada (NRCan) is the lead government department responsible for developing and implementing uranium, nuclear energy and radioactive waste management policies. It also administers the NFWA and is responsible for funding and managing a number of programs in the areas of historic low-level radioactive waste, legacy waste and uranium mine and mill tailings.

There is also the Low-Level Radioactive Waste Management Office (LLRWMO) established in 1982 to carry out the responsibilities of the federal government for the management of historic low-level radioactive waste (LLRW) in Canada. The Office is operated by Canadian Nuclear Laboratories (CNL) through a cost-recovery agreement with Atomic Energy of Canada Limited (AECL), the federal Crown Corporation that provides funding, direction and priorities for the LLRWMO. The LLRWMO manages historic LLRW at numerous waste sites located throughout Canada and has successfully completed projects in Ontario, British Columbia, Alberta and the Northwest Territories. The Office responds to questions from the public, conducts consultation and stakeholder engagement activities and establishes partnerships to resolve LLRW issues within local communities. The LLRWMO has completed many successful remediation projects in its more than 30-year history (CNL 2017). As of 31 December 2013, the LLRWMO also publishes the definitive inventory of all radioactive waste in Canada (CNL 2013).

In addition, other Government of Canada departments and agencies involved in governing aspects related to the nuclear industry include: Health Canada (recommendations on radiological protection standards and monitors occupational radiological exposures); Transport Canada (in developing and administering policies, regulations and services for the Canadian transportation system, including the transportation of dangerous goods) and Environment Canada which is responsible for the administration of the Canadian Environmental Protection Act.

5 Siting procedures

Initiated in May 2010, the NWMO is currently in the site selection process. The organization is working with a number of communities to understand the implications of hosting a repository in their area by conducting multi-year preliminary assessments in a number of areas with communities that expressed initial interest in hosting the project. These assessments include field studies, involving geophysical surveys and analysis of geological features, followed by studies of environment, safety, engineering and transportation and assessments of whether the repository will be supported broadly and would enhance community well-being if sited in the study areas (CNSC 2015: 36)

The site selection process requires communities in an area initiating their interest to enter and move through the process that requires communities, First Nation and Métis communities and surrounding municipalities to work together in implementation. Communities are invited to learn more about the repository project initiative. This allows them to decide whether they would like to host the facility. Communities choose to be involved, have access to info and financial resources to help them make informed decisions and can withdraw from the process at any time. From a starting number of 22, there were nine areas, all located in Ontario still involved as a focus of study as of January 2017: Blind River; Elliot Lake and Area; Central Huron, Homepayne and Area; Huron-Kinloss; Ignace and Area; Manitouwadge and Area; South Bruce, and; White River and Area (NWMO 2017b).

To date there has been no recommendation by the NWMO nor a decision by any community to accept the repository.

5.1 Procedures and criteria for site selection

On technical aspects, the NWMO's criteria for site selection include (NWMO 2005: 233):

- location in suitable rock such as the crystalline rock of the Canadian Shield or the Ordovician sedimentary rock basins;

- absence of known potential economic resources at depth;
- sufficient surface area for receipt facilities and associated infrastructure;
- seismically stable region with low known or projected frequency and magnitude of earthquakes;
- low frequency of major groundwater-conducting fracture zones, features, or faults at repository depths;
- geotechnically suitable host rock formation below the surface for the optional shallow rock cavern vaults;
- geotechnically suitable host rock formation at least 200 metres below the surface with a preference for a suitable host rock formation between 500 and 1,000 metres below the surface for the underground characterisation facility and the deep geological repository
- geotechnically suitable (e.g. reducing) conditions in groundwater at repository depth
- evidence of rock mass homogeneity and stability at repository depth;
- low hydraulic gradient and low permeability; and
- diffusion-controlled transportation of dissolved minerals at repository depth.

It should be noted that the NWFA stated that the NWMO study must specify an economic region for implementation of its preferred approach. An 'economic region' is a defined grouping of complete census divisions created as a standard geographic unit for analysis of regional economic activity. In Quebec, economic regions are designated by law while in all other provinces they are created by agreement between Statistics Canada and the provinces concerned. Murphy and Kuhn (2009) point out that this requirement puts Canada as the only country predetermining a region using economic criteria (154).

However, the NWMO final report does not specify an economic region. The NWMO does take a regional approach but is not bounded by 'economic regions' as it maintains it would be inappropriate to select one specifically because 1.) storage on site would require implementation in a number of regions; 2.) centralised approaches, due to transportation, require implementation in more than one region, 3.) given actual site characteristics are important, it is difficult to propose an economic region prior to site investigation. 4.) pre-identification could screen out potential sites by prematurely identifying economic regions and eliminating others, and 5.) narrowing the number of economic regions at an early stage may unduly remove communities that might otherwise wish to be considered as potential host locations (NWMO 2005: 145). That said, a NWMO background paper specifies that out of the 76 economic regions in Canada, 21 are located on the Canadian Shield (Kuhn and Murphy 2003: 1).

Accordingly, the NWMO approach focuses on volunteer host communities, as developed in 1994 and in the Seaborn Panel's report. Instead of a top-down identification of areas, the process is a bottom up, transparent and voluntary process for a willing community to engage and eventually host the facility. The site must meet the scientific and technical criteria as noted above to ensure that multiple engineered and natural barriers will protect human beings, other life forms and the biosphere, sustain engagement of people and communities throughout the phased process of decision and implementation and will be responsive to advances in technology, natural and social science research. Aboriginal traditional knowledge and societal values and expectations are also taken into account (NWMO 2005: 44).

Critics of a policy of 'placelessness' suggest it has been a deliberate tactic of both the nuclear industry and government to avoid the controversy associated with identifying a particular site (Murphy 2009: 138). Murphy (2009) noted that "Canada's approach to nuclear fuel waste management has been placeless, focused on technical concepts rather than emplaced, site-specific decision-making" (130). Murphy and Kuhn argue the siting process will be tense and its success far from guaranteed (150). These arguments echo a number of conclusions from inquiries past and their accompanying recognition of the sensitivities and public challenges to nuclear waste management in Canada. The APM does not suggest, nor would regulators and the NWMO assume, the process will be free of discord, heated debate or financial and economic challenges along the way.

5.2 Compensation mechanisms and socio-economic impact

According to a NWMO background paper done by two consultancies, Golder Associates Ltd., and Gartner Lee Limited (2005), measures that can be put in place to avoid or minimise significant socio-economic effects on a community's way of life, including everything from establishing large buffer areas around main facility buildings to free or subsidised transportation to the facility for workers from surrounding communities; and from providing 'impact grants' to municipalities to offset costs associated with the expansion of infrastructure maintenance, capacity and government services that will accompany the project to local procurement of goods and services and compensations for potential loss in property value. The paper considers timelines further down the line, such as the repository's closure (planned or unanticipated) which may cause social disruption and psychological and financial strain for workers, their families and the local community. Suggested measures to mitigate closure include providing skills upgrading/re-training, improved early retirement (higher pensions than otherwise eligible) and external placement (ibid.: 19).

It is important to underscore that the NWMO is in the early stages of the first phase of the APM approach. Site selection requires the needs and expectations of communities and surrounding areas to be recognised and addressed in the design of implementation plans. Detailed study on community well-being is anticipated during Phase II with the goal to understand specific social and economic pressures that may occur and what planning will be required to address them. To this end, ongoing collaborative consultations with communities, and the eventual host community, will lead to the development of community-specific and regional-specific measures.

6 Information and participation

In 2002, the NWMO launched a three-year national consultation process as mandated by the NWFA in exploring and assessing the four disposal options that were studied in its report (NWMO 2005: 17). Methods of outreach included, among others, expert panels and workshops, e-dialogues, Aboriginal dialogues, and nuclear community dialogues (NWMO 2005: 61). According to the report, it was estimated that through 31 August 2005, more than 50,000 people had visited the NWMO website with conservative estimates that more than 18,000 citizens contributed to it, including more than 500 specialists (NWMO 2005: 62). Face-to-face conversations were held with more than 250 individuals and groups involved, including from communities currently storing used fuel, nuclear power plant workers, Aboriginal leaders, faith communities and so forth. Moreover, the NWMO states that 15 national, regional and local Aboriginal organisations conducted a range of dialogue initiatives (ibid.: 111).

Some critics of the NWMO's national consultation process argue that the process reinforced inequalities between the nuclear energy industry and a coalition of religious and environmental organisations and Aboriginal nations (Fuji Johnson 2009: 90). The criticisms at the inclusiveness of the process were not in terms of the actors and organisations approached, but the inclusion of all voices in its dialogues, reports and recommendations (Fuji Johnson 2009: 95). Darrin Durant and Anna Stanley argued that "we do not deny that extensive public consultation exercises have taken place, but we assert that the object of those consultations reflects continuing marginalisation of political dissent and stubborn refusal to accept that much public disquiet is about institutional interests rather than technology per se" (Durant and Stanley 2009: 36).

After the APM approach was adopted, the NWMO invited the interested public to contribute suggestions and ideas to the development of an inclusive site selection process. This yielded a draft document in August 2008, Moving Forward Together: Designing the Process for Selecting a Site which was presented for comments before being finalised as the May 2010 report, Moving Forward Together: Process for Selecting a Site for Canada's Deep Geological Repository for Used Fuel.

By the end of 2014, the NWMO and the nine communities mentioned in section 5 had entered into Step 3 to conduct a preliminary assessment of potential suitability. This assessment is to be conducted in two phases over several years with Phase 1 focusing on desktop studies for communities to learn and reflect; and Phase 2 focusing on fieldwork conducted with the involvement of communities in the area, including First Nation and Métis communities. Throughout, the potential for a safe site, fostering community well-being, and the community's continuing interest are all to be explored (NWMO 2017d).

Based on the community's geology and willingness, it will then move on to Stage 4 which would include detailed site evaluations. Steps 5 and 6 involve community confirmation of acceptance and willingness to host the repository. Step 7 follows with a full regulatory review and approval process and Step 8, construction and operation of an underground demonstration facility and then construction of the actual repository in Step 9 (NWMO 2017e). The APM is designed to be shaped by the pace of the communities involved throughout each step.

7 Costs and financing

All waste owners have financial guarantees for the long-term management of radioactive waste as required by the CNSC. A large portion of these guarantees exists in segregated funds, of which the NFWA is a defined portion. The funding of NWMO to finance long-term waste management is arranged in two stages: waste owners directly fund NWMO's annual costs until the receipt of a construction license for a deep geological repository. At that point, the NFWA trust fund will cover all further NWMO costs. As of 2011, the total lifecycle cost for 3.6 million fuel bundles was approximately CDN $17.9 billion (CNSC 2015: 14). By 2015, the cost estimate was approximately $22.8 billion (NWMO 2017c). A distinction should be made on these figures, recognising there is a difference between actual costs as spent over many decades (i.e. $22.8 bn) and present value of the required investment upon which current investments are based. When updated to 1 January 2017, present value, the APM costs to be funded under a scenario of 3.6 million lifecycle bundles is $7.5 billion. This covers liabilities from 2017 onwards (NWMO 2017f).

Starting from zero in 2002 when the trust fund was established, the four nuclear waste partners (Ontario Power Generation, NB Power Nuclear, Atomic Energy Canada Limited, and Hydro-Quebec) have been contributing annually to the fund which totalled $1.5 billion by 2009 (Kemp 2009); $2.8 billion as of December 2013 (OECD/NEA 2015) and $3.7 billion at the end of 2015 (NWMO 2016: 1). In 2016, the four waste owners were required to make a total deposit of $215 million. The amount of annual contribution for each waste owner is calculated based on the amount of fuel bundles it has produced to date (NWMO 2016: 3).

The NWMO maintains an ongoing obligation to assess the accuracy of the cost estimate, and the sufficiency of contributions to cover cash flow, for the entire life of the project (NWMO 2005: 39). Actual figures will depend on a range of factors such as the location of the host community, schedule, and timelines for implementation. Accordingly, the total cost will fluctuate but will follow the 'polluter pays principle' where nuclear waste producers are financially responsible for long-term management. While the funds in the trust fund today are far from the total estimate for the total project, the NWMO is confident that the current $3.7bn nest will cover 97 per cent of implementing APM from the start of the construction phase (NWMO 2016: 3).

8 Conclusions

The approach Canada has adopted to handling its nuclear waste may seem impaired by too many unanswered variables from location to cost to long-term acceptance and even viability. Yet, while it may seem ad hoc and more akin to 'finding the way as we go', the APM is an approach that has evolved from almost five decades of debate, study, and public discourse. It is designed to be shaped by the pace of Canadian communities involved while carrying forward scientific and technical study based on past and ongoing inquiries within Canada and abroad. It is crafted to be nationally-appropriate, reflecting Canada's nuclear history, provincial-federal make-up, and legislative and regulatory framework.

While a centralised, deep geological repository remains placeless for the time being in Canada, there is a pathway forward which allows for flexibility and agreement when needed. Nuclear projects are inherently controversial and usually among the top most expensive infrastructure projects, meaning the process will surely involve moments of tension and heightened emotion. As the years pass and new generations take over, the twinned approach to incorporating the technical and the social will require steady, dedicated efforts. The multi-generational character to permanent disposal of nuclear waste motivates flexibility while the enormity of the quest impels a comprehensive, continually-studied and reviewed approach. In the end, a geological repository will not only become the biggest, most expensive construction project done in Canada to date; it will also be one of the most publicly-vetted.

References

Aiken, A.M., Harrison, J.M. and Hare, F.K. (1977). *The Management of Canada's Nuclear Wastes*. (Report of a study prepared under contract for the Minister of Energy, Mines and Resources Canada, Report EP 77-6).

Atomic Energy of Canada Ltd. (AECL) (1983). *Guide to the Canadian Nuclear Fuel Waste Management Program*, Second Edition, Pinawa, Manitoba.

Atomic Energy of Canada Ltd. (AECL) (1994). *Environmental Impact Statement on the Concept for Disposal of Canada's Nuclear Fuel Waste*. Report AECL-10711, COG-93-1. 1994.

Atomic Energy Control Board (AECB) (1974). Guide for Licensing Radioactive Waste Management Facilities. In: *Information Bulletin,* 74(1). Ottawa.

Atomic Energy Control Board (AECB) (1987). *Regulatory Policy Statement: Regulatory Objectives, Requirements and Guidelines for the Disposal of Radioactive Wastes – Long-term Aspects*. Ottawa.

Canadian Environmental Assessment Act (CEAA) (2012). S.C. 2012, c. 19, s. 52.

Canadian Nuclear Laboratories (CNL) (2013). *Low-Level Radioactive Waste Management Office (LLRWMO) (2013)*. Inventory Summary Report.

Canadian Nuclear Laboratories (CNL) (2017). *Waste Management Program.* http://www.cnl.ca/en/home/environmental-stewardship/waste-management.aspx, last accessed 27 April 2017.

Canadian Nuclear Safety Commission (CNSC) (2004). *Regulatory policy P-290, Managing Radioactive Waste*. Ottawa.

Canadian Nuclear Safety Commission (CNSC) (2006). *Regulatory Guide G-320, Assessing the Long-Term Safety of Radioactive Waste Management.* Ottawa.

Canadian Nuclear Safety Commission (CNSC) (2015). *Responses to Questions from Peer Review of Canada's Fifth National Report for the Joint Convention on the Safety of Spent Fuel Management and on the Safety of Radioactive Waste Management.* Ottawa. May.

Canadian Nuclear Safety Commission (CNSC) (2017a). *Radioactive Waste.* http://nuclearsafety.gc.ca/eng/waste/index.cfm, last accessed 15 January 2017.

Canadian Nuclear Safety Commission (CNSC) (2017b). *CNSC Research on Geological Repositories*. http://www.suretenucleaire.gc.ca/eng/waste/cnsc-research/geologic-repositories/index.cfm, last accessed 3 March 2017.

Carter, J. (1977). *Nuclear Power Policy Statement on Decisions Reached Following a Review*. http://www.presidency.ucsb.edu/ws/?pid=7316, last accessed 20 January 2017.

Dewar, D.J. (1969). *Control of Radioactive Pollution from Canadian Nuclear Facilities and from Radioisotope Operations, AECB-1044*. The paper appeared in the 1969 Annual Review Number of the Northern Miner under the title "Disposa l Methods Used by Nuclear Facilities Must Win Atomic Energy Board Approval", November 1969.

Didyk, J.P.; Nurmsoo, U. and Power, J.C. (1975). *Regulating the Management of Radioactive Waste in Canada*, Atomic Energy Control Board Report AECB-1081, Ottawa. July.

Didyk, J.P. (1976). *Radioactive Waste Management in Canada*, AECB-1216, Ottawa. June.

Durant, D. and Stanley, A. (2009). An Official Narrative: Telling the History of Canada's Nuclear Waste Management Policy Making. In: Durant, D. and Fuji Johnson, G. (Eds.). *Nuclear Waste Management in Canada: Critical Issues, Critical Perspectives*, UBC Press, Ch.3.

Ferch, R. (2003). Regulating Nuclear Waste in Canada. In: *Public Confidence in the Management of Radioactive Waste: The Canadian Context*, Workshop Proceedings, Ottawa Canada, 14-18 October 2002. Organisation for Economic Co-operation and Development, Nuclear Energy Agency.

Fuji Johnson, G. (2009). The Darker Side of Deliberative Democracy: The Canadian Nuclear Waste Management Organization's National Consultation Process. In: Durant, D. and Fuji Johnson, G. (Eds.). *Nuclear Waste Management in Canada: Critical Issues, Critical Perspectives*, UBC Press, 90-129.

Garamszeghy, M. (2015). *Nuclear Fuel Waste Projections in Canada – 2015 Update*, Report No.: NWMO-TR-2015-19, Toronto.

Golder Associates Ltd., Gartner Lee Limited (2005). *A Review of Possible Measures to Avoid or Minimize Significant Socio-economic Effects on a Community's Way of Life*, Nuclear Waste Management Organization Background Papers, Assessments.

International Atomic Energy Agency (IAEA) (2017). *Canada. Power Reactor Information System*. https://www.iaea.org/PRIS/CountryStatistics/CountryDetails.aspx?current=CA, last accessed 17 April 2017.

Kemp, B. (2009). Storing nuclear waste a $24-billion problem. In: *CBC News*, 18 August 2009.

Kuhn, R.G. and Murphy, B.L. (2003). *An Examination of Economic Regions and the Nuclear Fuel Waste Management Act*, NWMO Background Papers, Executive Summary.

Murphy, B.L. (2009). Canadian Communities and the Management of Nuclear Fuel Waste. In: Durant, D. and Fuji Johnson, G. (Eds.). *Nuclear Waste Management in Canada: Critical Issues, Critical Perspectives,* UBC Press, 130-149.

Murphy, B. L and Kuhn, R. (2009). Situating Canada's Approaches to Siting a Nuclear Fuel Waste Management Facility. In: Durant, D. and Fuji Johnson, G. (Eds.). *Nuclear Waste Management in Canada: Critical Issues, Critical Perspectives*, UBC Press, 150-167.

Natural Resources Canada (NRCan) (2007). *Minister's Statement on the 2006 (NWMO),* Annual Report June 19, 2007. https://www.nwmo.ca/~/media/Site/Reports/2015/10/02/07/59/1373_ministersstatementonthe2006nwm.ashx?la=en, last accessed 5 January 2017.

Nuclear Waste Management Organization (NWMO) (2003). *Status of Nuclear Fuel Reprocessing, Partitioning and Transmutation*, NWMO Background Papers.

Nuclear Waste Management Organization (NWMO) (2005). *Choosing a Way Forward: The Future Management of Canada's Used Nuclear Fuel.* Final Study. Ottawa.

Nuclear Waste Management Organization (NWMO) (2015a). *Description of a Deep Geological Repository and Centre of Expertise for Canada's Used Nuclear Fuel.* Toronto.

Nuclear Waste Management Organization (NWMO) (2015b). *Progress Through Collaboration.* Annual Report.

Nuclear Waste Management Organization (NWMO) (2016). *Funding Canada's Plan for the Safe, Long-Term Management of Used Nuclear Fuel.* Backgrounder.

Nuclear Waste Management Organization (NWMO) (2017a). *How is it Stored Today?* https://www.nwmo.ca/en/Canadas-Plan/Canadas-Used-Nuclear-Fuel/How-Is-It-Stored-Today, last accessed 27 April 2017.

Nuclear Waste Management Organization (NWMO) (2017b). *Study Areas.* https://www.nwmo.ca/en/Site-selection/Study-Areas, last accessed 22 January 2017.

Nuclear Waste Management Organization (NWMO) (2017c). *Employment and Economics.* https://www.nwmo.ca/en/Canadas-Plan/About-Adaptive-Phased-Management-APM/Employment-and-Economics, last accessed 8 March 2017.

Nuclear Waste Management Organization (NWMO) (2017d). *Step 3: Preliminary Assessments of Suitability.* https://www.nwmo.ca/en/Site-selection/Steps-in-the-Process/Step-3-Preliminary-Assessments-of-Suitability, last accessed 8 March 2017.

Nuclear Waste Management Organization (NWMO) (2017e). *Steps 4 to 9: Site Confirmation, Construction and Operations.* https://www.nwmo.ca/en/Site-selection/Steps-in-the-Process/Steps-4-to-9-Site-Confirmation-Construction-and-Operations, last accessed 8 March 2017.

Nuclear Waste Management Organization (NWMO) (2017f). *Progress Through Collaboration.* Triennial Report 2014 to 2016. March.

Organisation for Economic Cooperation and Development (OECD)/ Nuclear Energy Agency (NEA) (2015). *Canada: National Nuclear Energy Context.* https://www.oecd-nea.org/rwm/profiles/Canada_profile_web.pdf, last accessed.

Porter, A. (1978). *A Race Against Time: Interim Report on Nuclear Power in Ontario*. Queen's Printer for Ontario, Toronto.

Ramana, M.V. (2013). Shifting Strategies and precarious progress: Nuclear waste management in Canada. In: *Energy Policy*, 61, 196-206.

Sims, G.H.E. (1980). *A History of the Atomic Control Board.* Minister of Supply and Services Canada, July.

Vestergaard, C. (2015). *Governing Uranium in Canada*, DIIS Report 2015: 12, Copenhagen.

A Democratic Deficit

The Challenges of Nuclear Waste Governance in Argentina and Brazil

Moïra Jimeno[1]

Abstract

In Argentina and Brazil, the management of nuclear wastes remains an unresolved question. Despite some relevant regulations issued after the restoration of democracy, like the Regulatory Requirements for the Mangement of Radioactive Wastes in Argentina (Law 25018, enacted in 1998), and Law 10308 on the Selection of Sites, Construction, Licensing, Operation, Monitoring, Costs, the Compensation, Civil Liability and Guarantees relating to Deposits of Radioactive Tailings, among other Provisions, enacted in Brazil in 2001, neither of these two countries have long-term disposal strategies for nuclear waste. Spent nuclear fuel is stored in interim facilities located at the nuclear power plants. In the case of Argentina, the National Atomic Energy Commission (CNEA) submitted the Strategic Plan of the National Program for Radioactive Waste Management (PEGRR) to the National Executive in 2014 for its evaluation and approval. The plan defines the characteristics for nuclear wastes sites. The PEGRR has been preliminary approved but has not obtained the final approval from the National Congress.

The difficulty in finding a solution resides in the limited involvement of relevant actors in the process of selecting a nuclear disposal strategy. The CNEA in Argentina and the National Nuclear Energy Commission (CNEN) in Brazil, together with their national governments, have generally exercised a centralized and low-transparency control over nuclear issues, including nuclear waste, leaving almost no room for social participation. This has resulted in great uncertainty that has defined the management of nuclear waste.

---

1 Moïra Jimeno, German consultant agency eclareon; E-Mail: mj@eclareon.com

A. Brunnengräber et al. (eds.), *Challenges of Nuclear Waste Governance*,
Energiepolitik und Klimaschutz. Energy Policy and Climate Protection,
https://doi.org/10.1007/978-3-658-21441-8_12

1 Introduction

Although the global use of nuclear power has been declining since 2006 (Brunnengräber and Schreurs 2015), two countries in South America have been experiencing a slow but progressive nuclear renaissance. The case of Argentina and, to a lesser extent, Brazil illustrate this pattern. In Argentina, a third nuclear power plant (NPP) entered into full operation during the second half of 2014. The Argentine-designed *small modular reactor*, Central Argentina de Elementos Modulares (CAREM-25), is as of November 2017, under construction and two additional NPPs are foreseen to start construction in 2018 and from 2020, respectively. Meanwhile, Brazil has two operating NPPs and one under construction. Argentina, and also Brazil, have also signed agreements with third countries, namely China and Russia, to construct new NPPs in the next years.

Since 2012, a number of high-level agreements have been signed by the Argentine and Chinese presidents. The agreements have addressed the construction of the fourth NPP, a Pressurized Heavy Water Reactor (PHWR) unit designed, built and operated by Nucleoeléctrica Argentina (NA-SA) and supported by China National Nuclear Corporation (CNNC) under long-term financing, and technology transfer for the manufacturing of nuclear components and fuel fabrication in Argentina. The latest agreements between Argentina and China were signed with the China National Energy Administration in May 2017, confirming the fourth and fifth reactors. Argentina also has some cooperation agreements with Russia, such as the agreement signed in April 2015 that establishes a framework for cooperation in construction of a 1200 MWe Water-Water Energetic Reactor (VVER) with Russian financing (World Nuclear Association (WNA) 2017a). Since Mauricio Macri assumed the presidency in December 2015, his administration has been more focused on cooperation with China than on advancing the agreements with Russia (Jimeno 2016).

An overview of all NPPs (under operation, construction, planning or proposed) in Argentina and Brazil is presented in Table 1.

Table 1: NPPs in operation, under construction, in planning, and proposed in Argentina and Brazil

Country	NPPs in operation	NPPs under construction	NPPs in planning	Proposed NPPs
Argentina	Embalse Atucha 1 Atucha 2	CAREM 25	Atucha 3 Atucha 4	CAREM 150 VVER-1200
Brazil	Angra 1 Angra 2	Angra 3		Northeast, Pernambuco Southeast, Minas Gerais

Source: Author's compilation from World Nuclear Association (2017 data for Argentina and Brazil).

As the level of ambition of Argentina's nuclear plan is higher than that of Brazil, this chapter focuses on the Argentenian case, but also includes a brief comparison with the Brazilian case in section 8.

2 Nuclear waste management policy

2.1 Historical background

Argentina started developing nuclear energy early in the 1950s around the same time as Brazil and Mexico. On 31 May 1950, President Perón issued the first decree (No. 10936) promoting nuclear energy and founded the National Atomic Energy Commission (CNEA). The decision to develop nuclear power was influenced by Austrian scientist, Ronald Richter who proposed to Perón the idea of constructing a nuclear energy pilot plant to produce controlled nuclear fusion (Montenegro 2005). The president, enamored by Richter's vision, authorized the construction of nuclear installations on Huemul Island in southern Argentina. In February 1951, Richter announced that he had achieved "thermonuclear reactions under controlled conditions at technical scale". A year later a scientific commission formed by the renowned physicists and scientists José Antonio Balseiro, Mario Báncora, Manuel Beninson, Pedro Bussolini, and Otto Gamba was formed to investigate the Huemul project. The commission found out that the Austrian scientist's project was a fraud (Montenegro 2005).

Despite this upset, Balseiro, Báncora, and other scientists began a new national nuclear program for Argentina. CNEA was entrusted with the direction and monitoring of all nuclear program activities (Jimeno 2015). The government allocated generous resources to CNEA to move forward with NPP construction. The first NPP, Atucha I, was constructed in 1973. Radioactive wastes (RW) from the plant were disposed of in a semi-contention system for solid wastes located in the Ezeiza Waste Management Area (AGE) at Ezeiza Nuclear Center (CAE).

The national policy of allocating significant economic resources to the nuclear program continued during the military dictatorship (1976-1983) but changed after the return of democracy in the mid- 1980s, when resources were drastically reduced and the nuclear program was suspended for a time. When President Nestor Kirchner (2003-2007) took office, the national government decided to reactivate the nuclear industry and allocate CNEA with more resources. The Strategic National Plan of the Nuclear Sector was launched in order to complete the construction of Atucha II, which had begun in 1981 but was suspended in 1994, and to relaunch other objectives of the suspended nuclear program (Jimeno 2015).

Under the Macri administration, CNEA signed an agreement with the German consortium TECNA-Siemens to advance the construction of CAREM-25. The prototype might be followed, in 2021, by a larger version (100 MWe or 200 MWe), located in the province of Formosa at the border with Paraguay. The new government has also confirmed the decision to extend the life of Embalse (Jimeno 2016).

Since its establishment, CNEA has been the unifying agent of the nuclear sector, participating in the subsequent creation of the national nuclear companies such as Invap, NA-SA, and Dioxitec. As the central nuclear energy body, CNEA is also the operating organization for the final management of spent fuel (SF) and radioactive waste, while the Nuclear Regulatory Authority (ARN) created in 1997 is responsible for regulating nuclear waste issues and control of nuclear safety activities. This regulatory institutional arrangement was created to comply with the safety provisions established in the Joint Convention on the Safety of Spent Fuel Management and on the Safety of Radioactive Waste Management ratified by law in 2000.

2.2 *The national inventory*

For the classification of radioactive wastes, Argentina has adopted the scheme proposed by the International Atomic Energy Agency (IAEA) that contemplates six classes of waste based primarily on considerations of long term safety and also by implications for disposal of the waste (Manzini, interview June 15, 2015). As Alberto Manzini[2] points out, "[t]he (referred) IAEA classification scheme is merely used to inform the inventories of nuclear wastes and organize the information. Yet, the acceptance of a waste for a particular disposal needs to be demonstrated in a safety assessment. The content limits for each radioisotope activity will be established based on the safety assessment of final disposal site after selecting the waste" (interview June 15, 2015). Note that uranium mining waste is considered a distinct class due to its different nature and the decision to treat it in the places where it originates. Table 2 shows the characteristics of the different nuclear waste categories.

---

2 Written interview with Alberto Manzini, member of the National Atomic Energy Commission (CNEA), responsible for the National Program for Radioactive Wastes Management (PNGRR). Manzini was interviewed on different issues concerning nuclear waste disposal.

Table 2: Classification of radioactive wastes in Argentina

Waste Category	Years	Disposal System	Institutional Control
Very Low Level	< 30	Outside surface system	50 years
Low and Intermediate-level	< 30	Monolithic disposal near to the surface with multiple barriers	300 years
High Level	> 30	Deep Geological Disposal (= 500 m)	Not applicable

Source: CNEA 2010.

The exact volume of radioactive waste and SF generated in Argentina is unclear. As required by the National Program for Radioactive Waste Management (PNGRR), CNEA provides an approximate value of the radioactive waste and SF in the country (see Table 3).

Table 3: Approximate volume of radioactive waste and SFs in 2013

Generator	Low and Intermediate-Level Waste	Spent Fuel stored (as of end 2013)	Annual Spent Fuel
Embalse + Atucha I	90 m^3/year	148,849 SF [Total uranium: 4,228,091 kg]	4,230 SF/year
Atucha II	No Data	No Data	No Data
Ezeiza Waste Management Area (research reactor RA-3 / radioisotope production)	25 m^3/year	176 SF [Total uranium: 204 kg]	13 SF/year

Source: Based on data provided by a member of the PNGRR (interview June 15, 2015).

Manzini observes that "[I]n practice, SF can not be stored together in contact with each other due to their criticality and thermal dissipation, but if all SF were stored in contact with each other an aproximate calculation of their volume in m^3 would be:

- Atucha I and Embalse: 36 m^3 of annual volume and 1,470 m^3 of total storage;
- Research reactor (RA-3): 0.1 m^3 of annual volume and 2 m^3 of the total stored volume. Yet, as the SFs cannot be stored together, the real volume occupied is greater than the above values" (interview 27 July 2015).

A national system to register, operate, and maintain radioactive waste (RW) data is expected to be implemented to ensure a complete tracking of current and future

RW from nuclear activities in the country. According to the Radioactive Waste Management Regime (Law 25018), CNEA, through the PNGRR, shall implement, maintain, and operate a data registration system to identify in a continuous and reliable manner the generators and transporters of nuclear wastes as well as other participants across the entire management cycle. The information and registration system shall contain an inventory of all existing RW in the country. When appropriate, copies of the documentation shall be sent to the competent authorities from the provinces and the city of Buenos Aires (Law No. 25018: Article 10). In practice, the AGE has a database that includes the generators and RW sent to and stored at this area, while each NPP has its own registration system with an inventory of their generated wastes (interview 15 June 2015).

2.3 Interim storage

Neither Argentina, nor Brazil, have a long-term disposal system for nuclear waste and instead store their spent nuclear fuel in interim facilities at the NPPs. At Embalse NNP (1985) there is a SF storage pool and a dry storage system. The stored inventory at the end of 2012 was 99,900 fuel elements in 185 silos. In 2013, the SF dry storage system (ASECQ) had a total of 248 silos. The SF stored in the ASECQ silos has been included, at the ARN's request, in the Ageing Management Program for Components and Systems of the Nuclear Power Plant Associated to Nuclear Safety. The surveillance plan for canisters, internal cladding and the concrete structure of all ASECQ system silos was incorporated in the framework of this program. Additionally, there is a periodic measurement of the content of aerosol and noble gases inside the silos (Autoridad Regulatoria Nuclear (ARN) 2013).

At both Atucha units (completed in 1974 and 2014) SF is stored in interim storage pools. Due to the limited storage capacity of the Atucha I interim pool, NA-SA developed a compact storage project in 2002, which created 1,360 new places for the SFs. Foreseeing that by 2015, the storage capacity would be fully reached, an interim dry storage system to store new SF generated from both Atucha NPPs was planned. Although the plan was to start operation of the interim dry storage facility by February 2015, construction is not expected to be finished until March 2018 (CNEA 2015). The facility will consist of a new building where vertical underground silos will be located; this new building is an extension of the SF storage building 1, located within the plant controlled area. Construction works, which are responsibility of the Construction Unit of Atucha II and Unit IV's Construction Group, began constructing a net for sustaining soil from excavation as well as pit foundation (ARN 2013). The interim facility will enable the dry storage of 2,754 SF elements (CNEA 2015).

Due to the construction delay, according to the National Nuclear Safety Report (ARN 2013), the Construction Unit of Atucha II is working on an alternative storage plan which will involved the transfer of SF from Atucha I to Atucha II. The fabrication of the necessary devices for the transfer operation was completed by the end of 2014. The transfer operation from the two pools of Atucha I to the four pools of Atucha II began in 2015 and more than 100 SF elements have been transferred thus far. The process should continue until all 620 SF elements have been transferred (Manzini, interview 28 March 2016).

Very low, low and intermediate-level waste including used fuel from research reactors, disused sealed sources, and production nuclear material are handled at Ezeiza Nuclear Center's (CAE) Waste Management Area (AGE), an eight hectare area located in the jurisdiction of Ezeiza, which is part of the Greater Buenos Aires area. The AGE has a central and specially designed interim storage facility where intermediate-level RW is stored in dumpsters and barrels awaiting conditioning and ultimately final disposal in a future intermediate-level repository.

Low-level waste (LLW) is treated, conditioned, and temporarily stored at AGE until a decision concerning transport to a future designated repository and/or final disposal is made. Until the early 2000s, LLW was disposed of in final disposal facilities (i.e. semi-containment and containment structures) located at the AGE. In 1999, CNEA conducted an analysis regarding the hydrological, meteorological, socioeconomic, and demographic changes taking place in Ezeiza that could affect the operation of AGE. Based on this analysis as well as the 1960s-era disposal design, CNEA decided to suspend the operation of solid, liquid, and structural LLW final disposal in order to conduct a safety reassessment of AGE. In 2006, CNEA decided to permanently discontinue the operation of the disposal systems independently of the results of the safety reassessment, as it concluded that these systems had already exhausted their operational phase. Concerns and complaints raised about pollution caused by LLW in Ezeiza's groundwater might have influenced this decision. CNEA is still waiting for a decision and authorization concerning the siting of a final disposal facility.

3 The waste management strategy

3.1 *Waste disposal and its financing*

The SF discharged from a nuclear reactor contains not only high-level radioactive wastes (HLW) but also fissile elements useful for making new nuclear fuel. For this reason, in Argentina, like in Brazil, the spent nuclear fuel is not considered a waste but a strategic energy resource with potential use. For now, spent nuclear fuel is stored in interim wet and dry facilities located at the NPPs. Whether Argentina will opt for an open cycle (skipping reprocessing of SF and focusing on DGD) or a closed cycle (nuclear fuel reprocessing and final disposal of remaining RW), remains an open question. The Strategic Plan of the National Program for Radioactive Waste Management (PEGRR) which has not obtained the definitive approval from the National Congress foresees "studies for the siting, construction, and operation of a future DGD. The decision on either reprocessing or direct DGD, will depend on the finalization of the studies for the location of the DGD, and studies must be completed no later than 2030" (CNEA 2011: 10).

For the final disposal of very low, low, and intermediate-level wastes, there is a project in planning that has two stages and three phases within the second stage. The first stage includes the selection of sites, the election of final disposal systems near the surface, conceptual design of the industrial complex, and public communication. The following near-surface final disposal systems have been chosen:

- For intermediate and particular LLW: Concrete disposal comprising of a set of multiple barriers in which wastes are packed in 200 liter barrels and/or special concrete dumpsters with an insulation period of 300 years. Similar examples are L'Aube in France and El Cabril in Spain (CNEA 2011, Section B: 9);
- For LLW: Concrete repository with a simple barrier (insulation period of 50 years) (CNEA 2010: 58);
- For very low-level waste: Surface systems with engineering improvements (CNEA 2010: 58).

Moreover, "the PEGRR plans to build a new disposal for very low and LLW. Works related to the search and selection of sites and areas to localize these repositories are currently ongoing" (CNEA, Section B: 9).

The second stage has three phases, including conceptual and basic engineering, as well as execution of the project. Under the conceptual engineering phase, the following steps are described: classification of RW, criteria of acceptance, characterization of the RW, establishing a matrix of the present inventory and forecasting future waste levels, addressing Atucha II wastes; development and testing of procedures for the treatment and conditioning of wastes, site characterization, engineering barriers, and issuing a preliminary safety report.

The basic engineering phase addresses the concrete plan for the project with a corresponding schedule of investments, their estimated costs and the issuing of the preliminary safety report. Finally, the execution of the project phase refers to the call for tenders, the construction and commissioning of the project, and the operation of the facilities.

Act 25018 foresees the creation of a Fund for the Management and Final Disposal of the Radioactive Wastes managed by the Argentine state and composed of the funds collected from the generators of RW. The fund has yet to be implemented. According to a representative of the PNGRR, to understand the financing of the management and disposal of RW and SF it is essential to note that Act 25018 was enacted in 1998, the same year that Decree No. 1390 which sets rules for the enactment of Act 24804 contemplating the possibility of the privatization of nuclear plants was approved. As there has been no privatisation, the national state is responsible for taking over the necessary investments related to nuclear waste issues (interview June 15, 2015).

The interviewee added that nuclear waste management activities are financed through CNEA's budget, which is approved annually by the National Congress. CNEA is under the Ministry of Federal Planning, Public Investment, and Services. The PNGRR representative's argument does not clarify why the fund has not yet been established.

Norma Morandini (2015) insisted that the National Executive explain why the fund has not yet been constituted, although already in 1998 the legal foundations for its establishment were set and radioactive waste generators and their waste levels are annually collected (following the enactment of Law 25018).[3] The only reference to the funding for nuclear waste management activities and SF found in official documents is the CNEA's report entitled "Gestión de los Residuos Radiactivos y de los Combustibles Gastados en la República Argentina: Informe al Honorable Congreso de la Nación Correspondiente al Ejercicio 2011 según lo prescripto por la Ley No 25.018". This report indicates that the funds used by CNEA are from various sources: the National Treasury ($ 6,008,015), Equity ($ 24,714), the National Treasury PRAMU ($ 3,850,117), World Bank Funds, PRAMU PPF ($ 915,827), and funds from services to third parties ($ 340,564) for a total of: $11,139,237. These figures do not include spending on staff (CNEA 2012: 26).

3 Norma Morandini is National Senator for the political party Civic Front of the Province of Córdoba.

4 The legal framework

The legal framework applicable to nuclear wastes in Argentina is based on the provisions of the National Constitution and the specific legislation issued to regulate nuclear activity such as the National Law on Nuclear Activity (Law 24804) and the Joint Convention on the Safety of Spent Fuel Management and Safety in the Management of Radioactive Waste (Law 25279 of July 2000), as well as Law 25018, which lays down the Radioactive Waste Management Regime.

The first legal regulation related to nuclear waste issues was Article 41 of the National Constitution, enacted in 1994 when the Constitution was reformed. Article 41 is relevant because it bans the entry of RW into national territory. It was drawn up by Juan Schröder, who until 1994 was the coordinator of the Greenpeace antinuclear campaign in the country (Greenpeace 2002). This legislation has been invoked by national environmental NGOs in their attempts to halt controversial agreements with international nuclear organizations, such as that with the Australian Nuclear Safety Organization (ANSTO) (see section 7).

The National Law on Nuclear Activity was issued in April 1997 to regulate nuclear activities, assigning the corresponding responsibilities to the National Atomic Energy Commission (CNEA) and the Nuclear Regulatory Authority (ARN). CNEA was tasked with the responsibility for the management of RW, the decommissioning of NPPs and other nuclear facilities, and the exercise of state ownership for the fissile elements contained in the SFs and developed or introduced in the country. The ARN was tasked with regulation and control of all activities related to radiological and nuclear safety. Law 24804 also regulates the siting for high, intermediate, and LLW repositories, assigning CNEA the duty to propose locations and the ARN and provinces the task of approving the proposed locations.

The Radioactive Waste Management Regime was set up in October 1998. It created the PNGRR, the Fund for the Management and Disposal of Radioactive Wastes, confirmed the responsibility of CNEA as introduced in Law 24804, and entrusted to CNEA the development of the PEGRR and the annual progress reports. The PNGRR has several duties and responsibilities related to the management of high, intermediate and low-level RWs. The most important of these are designing a RW management strategy, proposing a research and development (R&D) strategy for appropriate technologies and management methods, coordination and monitoring of these R&D projects, evaluation of new disposal sites or facilities, planning and operating the equipment and facilities, and managing nuclear wastes, including those generated by the closure of facilities, through uranium mining, and from abandoned mining sites or manufacturing facilities taken out of service. In addition, the PNGRR is expected to promote studies on safety and environmental preservation and assure that environmental impact is as low as possible through complementary sustainable development programs for the communities directly affected.

The decision of whether to reutilize the fissile material contained in SFs has to be made before 2030 (Manzini, interview 28 March 2016). The PEGRR states that by that time, "the installation of the underground laboratory must have been started to allow the design and construction of a DGD, which must be operative by the year 2060" (CNEA 2011: B1). Finally, as Argentina is a federal country, there are several subnational regulations from the provinces and city of Buenos Aires which are relevant to nuclear waste management.

5 The national institutional framework and the role of key actors

According to Law 25018, the National Atomic Energy Commission (CNEA) is the enforcement authority within the Radioactive Waste Management Regime; the state is obliged to provide it with necessary resources. It coordinates with the provinces and the city of Buenos Aires on all matters relating to the law's implementation. CNEA should submit the PEGRR and its updates to the National Executive, who after consultation with the ARN sends it to the National Congress for definitive approval by law. On 20 August 2014, the PEGRR was submitted to the former Energy Secretary (under the Federal Planning Ministry)[4] and it was preliminary approved by the ARN. Yet, as of this writing, the National Executive has not submitted it to the National Congress for definitive approval. Senator Morandini asked the National Executive "why the PEGRR has still not been sent to National Congress for its approval, as established in Article 9 of Law 25018, despite being already ellaborated by CNEA and approved by the ARN?" (Morandini 2015). Despite this prodding, the National Executive has never provided a concrete answer on the issue.

The CNEA establishes the acceptance criteria and conditions for the transport and final disposal of radioactive wastes, which must then be approved by the ARN. Additionally, the transport of radioactive waste, including SF elements, follows the schedule and procedures established by CNEA previously approved by the ARN.

However, the Macri Administration has started to implement some institutional changes. It has created an Undersecretary of Nuclear Energy and has

4 Note that when President Macri took office in December 2015, it was created the Ministry of Energy and Mining as a result of the higher rank of the Energy Secretary as well as the Secretary of Mining.

placed the CNEA under the competence of this undersecretary. The intention of the Undersecretary of Nuclear Energy is to create the Fund for the Management and Final Disposal of the Radioactive Wastes for which it has begun to consider the amendment of the Radioactive Waste Management Law (Law 25018) and the National Law on Nuclear Activity (Law 24804). The undersecretary's advisors are also considering the separation of the PNGRR from CNEA and the creation of a Radioactive Waste Agency. The latter must be approved by the National Congress. According to Manzini, the creation of a Radioactive Waste Agency will take some time due to the aprovals and expenses that involve this process (Manzini, interview 21 November 2017).

The ARN was established in 1997 as the successor authority of the National Nuclear Regulatory Body (ENRN). It is an autonomous entity that regulates matters regarding radiological and nuclear safety, the physical protection and control of the use of nuclear materials, licensing, and inspection of nuclear facilities as well as international safeguards and transport of nuclear materials. It is in charge of granting, suspending, and revoking licenses for construction, commissioning and operation. Additionally, it oversees decommissioning of NPPs as well as licensing uranium mining activities, security research reactors, facilities for waste management or radioactive waste, and nuclear applications used in medical and industrial activities. The ARN also conducts surveys and regulatory assessments on facilities; proposes to the National Executive the expiration, renewal or replacement of a use license for a nuclear installation; promotes civil or penal actions to the competent courts in case of a breach by a licensee; and applies sanctions when required. Finally, the ARN has to submit an annual report to the National Executive and National Congress on the year's activities and suggestions for actions to be taken for the public interest (Law No. 24804).

As a federal country, the provinces and the city of Buenos Aires also have a meaningful role. In activities related to the management of radioactive waste, CNEA must comply not only with the legal regulations established by the ARN but also with the regulations of the provinces and the city of Buenos Aires, especially regarding the siting of facilities. In order to install a new disposal site, locations must first be approved by law in the particular province or, in the case of Buenos Aires, the city (Law No. 25018: Article 12). The application of the "environmental impact as low as possible" principle is informed by the environmental impact assessment requirements not only of the provinces but also the city of Buenos Aires (Law No. 25018: Article 11).

6 Siting procedures

As noted above, the siting procedure for final disposals involves three actors: CNEA, the ARN, and subnational governments. The specific siting procedure for intermediate, and low-level RW disposals involves various phases from proposal through construction. First are studies for the selection of geologically favorable sites. This involves integrating Geographical Information Systems (GIS) and geological and geochemical data, interpreting satellite images, and making an inventory of the rock types suitable for RW disposals, i.e. granite, welded tuff (ignimbrite), and clays of geological basins and regions. This is followed by a determination of the selection criteria to be used in survey the most favorable areas according to the inventory of suitable geological regions, seismic activity, hydrogeology, volcanism, tectonics, and geomorphology as well as resources, and demographic conditions. Once a suitable site is selected, the project is to be presented to the authorities of the relevant province and a public hearing to inform the community is organized. Winning over a community may involve offering incentives, such as infrastructure developments. Once the approval of the ARN and the controlling agencies has been obtained, and construction has been approved by provincial parliament construction can begin. When the construction starts, public information about the details of the disposal plans and facility construction should be distributed in the concerned community (Law 25018: Article 12).

During the 1980s and the 1990s, CNEA undertook geological feasibility studies to evaluate different siting alternatives for DGD of HLW. At that time, CNEA's personnel participated in the IAEA program, "Training in and Demonstration of Waste Disposal Technologies in Underground Research Facilities - An IAEA Network of Centers of Excellence" to be trained on nuclear waste disposal issues. "For the first studies, a granite formation located in Sierra del Medio in Gastre, province of Chubut, was selected. Studies included drilling of up to 800 meters and a conceptual design for the engineering of the DGD. Yet, after more than 10 years of research, the project was finally dismissed in 1997 due to a large local demonstration opposing the HLW disposal" (CNEA 2010: 56) (see section 7). "After the failed attempt in Gastre (...), a few studies of favorable areas in sedimentary and volcaniclastic formations started to be conducted. Unfortunately, studies were not continued due to problems of public and political acceptance because most of the provinces where the studies were conducted declared their jurisdiction nuclear-free, stating expressly in their constitutions the prohibition of entry, transport, and storage of RW or the installation of a repository. It is expected that the approval of PEGRR will somehow allow the political support that is required to reactivate the studies" (Manzini, interview 27 July 2015).

7 Information and participation

In the late 1980s, the lack of information and transparency about efforts to construct radioactive waste repositories in Gastre and Salinas Grande spurred significant opposition from different social sectors and NGOs. In both cases, CNEA decided to cancel the feasibility studies in the early 1990s. These experiences contributed to the establishment of new regulations regarding nuclear waste issues and the formalization of public participation in Argentina's nuclear waste management in the late 1990s. In 1997, the ARN was created to control the safety of the nuclear sector and Law 24804 was issued requiring a formal authorization for the treatment and siting of nuclear wastes, as well as informing involved provinces, cities, and the general public. These processes were incorporated in 1998 in Law 25018. Subsequently, the siting procedure was formally regulated and the authorization of the ARN, as well as the affected subnational governments (provinces and the Buenos Aires city), were mandated. Before 1997, decisions to dispose of and store nuclear waste (such as the final disposal for very low or low-level RW and interim storage for intermediate-level RW, as well as SF from research reactors located in the Ezeiza Waste Management Area (AGE)) were made by CNEA without any authorization or public consultation.

Article 41 is particularly significant because of a controversial agreement with the Australian Nuclear Safety Organization (ANSTO). In 2000, Invap signed an agreement with ANSTO for the design and construction of a 20 MW OPAL research nuclear reactor to replace an old reactor located in Lucas Height, Australia. One of the agreed-upon clauses stated that Australia would send its spent nuclear fuel to Argentina for processing, after which it would be returned to Australia. The agreement with ANSTO was severely criticized by Greenpeace and the Environment Defense Foundation (FUNAM), who declared the clause anti-constitutional. Despite these complaints from environmental actors, the Argentinean and Australian Ministries of Foreign Affairs signed a Cooperation Agreement on the Peaceful Uses of Nuclear Energy on 8 August 2001, which strengthened the agreement between Invap and ANSTO (Greenpeace 2002). The national government imposed its will over the objectives of the environmental lobby, but the presence of antinuclear movements and NGOs was significant. After the return of democracy (1983), environmental NGOs and social movements have slowly but progressively helped to increase debate about nuclear activities in the country, especially regarding nuclear waste issues. This is particularly significant given that during the military dictatorship, "nuclear energy policy was imbued with great secrecy" (Jimeno 2015: 233).

NGOs and environmental activists such as Greenpeace, FUNAM, Chubut Anti-Nuclear Movement (MACH), "Unión de Asambleas Ciudadanas" (UAC), and Tierralerta have been influential in relation to nuclear questions. They were successful in campaigns to stop uranium mining activities as well as halt the construction of waste storage facilities and dumps. In 1989, the activism of FUNAM contributed to the closure of the uranium mine at Los Gigantes in the Province of Córdoba, due to the contamination of a lake basin located near the mine although it must also be noted that economic conditions also played a role; "it was cheaper to import uranium than produce it locally" (Jimeno 2015: 232). A few years later, the actions of local antinuclear movements halted the installation of planned uranium mining in Córdoba, located in the tourist area of Traslasierra. As it was a tourist area, the local community was able to obtain the political support necessary to officially declare Traslasierra a nuclear-free zone. Other Córdoba municipalities later endorsed this action.

During the 1990s, environmental and local movements helped to first stop the construction of a radioactive waste storage site in Salinas Grandes and then a HLW disposal site in Gastre, Chubut. Salinas Grandes is a large salt desert in central-northern Argentina within the provinces of Córdoba, Santiago del Estero, Catamarca, and La Rioja. A large storage site was planned for this region, however the actions of FUNAM and the local community halted the project. Similarly, later on denunciations of MACH and a large local demonstration led to the cancellation of the project to construct a repository for HLW in the Gastre basin (Palicio Lada 2016).

Local communities in the vicinity of the CAE have claimed since 2000 that the groundwater in Ezeiza has been contaminated because of the LLW managed and disposed of at AGE. These denunciations have been supported by several environmental NGOs like Greenpeace and FUNAM. Yet, the ARN has consistently denied these claims, arguing that monitoring has demonstrated there has been no such contamination of the water. In 2005, complaints reached a critical peak, when an independent laboratory was able to prove that the groundwater in Ezeiza was contaminated and elevated concerns about the safety of the AGE (Lipcovich 2005; Greenpeace and FUNAM 2005). CNEA and the government tried to calm citizens' worries on these issues but they failed in their attempts. Local community members claimed that the process "was marked by obscuring and silencing the real impact. A few years later the provincial government was forced to acknowledge the contamination values measured by independent laboratories, although official reports stated that there was no contamination from nuclear waste but just high radioactive background level [in Ezeiza's water]" (Damveld and Bannink n.d.: 4). CNEA subsequently suspended definitively LLW disposal in AGE and in 2010. A judge involved in the case ordered that LLW waste barrels be removed, re-packed and stored in a Repository for Long-term Storage, built for this purpose. The waste is to remain there until the construction of a final disposal facilities for LLW and intermediate-level waste.

8 A brief comparison with the Brazilian situation

Brazil has two operating NPPs (Angra 1 and Angra 2) and one under construction (Angra 3) (Table 1). Angra 3 was expected to be in operation by the end of 2015, but due to a corruption probe in mid-2015, the operating company Eletrobras suspended the installation contracts. In mid-2016, Eletrobras' subsidiary, Eletronuclear, became the focus of a corruption investigation. Funding ran out, halting the work and postponing the construction schedule beyond 2018. In January 2017, Eletronuclear formally even annulled the electro-mechanical contract, rejecting the companie's appeals. The unit is about 70 percent completed (World Nuclear Association (WNA) 2017b). Similar to the Argentinean case, Brazil has signed agreements with third countries, namely China, Russia and American company Westinghouse, to construct new NPPs in the next years. However, the Brazilian government has announced that construction of new plants after Angra 3 (the reactor under construction) would not commence until after 2020, due to difficulties in financing the nuclear projects.

Brazil began developing nuclear energy in 1951, when it was established the National Research Council. Five years later the country founded the National Nuclear Energy Commission (CNEN) with the purpose of regulating, licensing and supervising the production and use of nuclear energy as well as radioactive protection (CNEN, n.d.). CNEN as well as the state-owned companies, Brazilian Nuclear Industries (INB) and Nuclebrás Heavy Equipment (Nuclep), are under the Ministry of Science and Technology and Innovation (MCTI) and report directly to the ministry.

SF from Angra 1 and 2 is stored in interim facilities located at each NPP. RW generated by the nuclear fuel cycle facilities managed by INB, like that generated by uranium mining and the nuclear fuel factories, is stored in the same locations where it originates with the exception of waste generated from the exploitation and processing of monazite, which is stored at the USIN area in Sao Paulo. Unlike in Argentina, SF from Brazilian research reactors is stored at the Nuclear Energy Research Institute (IPEN) in Sao Paulo and the Nuclear Technology Development Center (CTDN) in Belo Horizonte, where it originates.

Similar to the Argentinean case, in Brazil, Law 10308 states that the Federal Government, through the CNEN, is responsible for intermediate and final disposal of all radioactive wastes produced in the country. As the focal point for management and disposal of radioactive wastes, CNEN is in charge of site selection, construction, installation, licensing, operation, supervision, costs, damages, liability and physical safety of radioactive repositories. As the interim storage facilities are not operated by CNEN, all costs related to their management are supported by authorized operators.

Brazilian law foresees a monthly compensation to the municipalities where initial, intermediate and/or final disposal sites are authorized. The compensation provided should not be less than 10 percent of the costs paid to CNEN by the operators of the installations that generate the waste. Yet, the law introduced a nuance when it says that it will be up to CNEN to monthly receive and transfer to the municipalities this amount owed by the holder of the authorization for the operation of the installation that generates the waste.

In Brazil, the legal framework applicable to radioactive wastes is defined by a series of laws (CDTN 2010). According to Law No. 7804, issued on 18 July 1989, the Brazilian Institute of Environment and Renewable Natural Resources (IBAMA [*Instituto Brasileiro do Meio Ambiente e dos Recursos Naturais Renováveis*]), is responsible for environmental licensing of all facilities with potential environmental impacts, including nuclear wastes facilities. Law No. 6189 was issued in 16 December 1974 and later modified by Law 7781, enacted on 27 June 1989. This laws as amended assigns to CNEN all competences in nuclear matters, including the transport, treatment, and final destination or disposal of radioactive wastes produced in the country.

Act No. 10308, issued on 20 November 2001, was a milestone in radioactive waste management and disposal; it defines the criteria, procedures, and rules for the construction, licensing, management, and operation of radioactive waste disposal and storage facilities. Act No. 10308 classifies radioactive waste disposal and storage facilities in different categories, i.e. initial, intermediate, and final. Initial disposal refers to interim storage normally located at the nuclear installation, while intermediate disposal refers to longer-term interim storage facilities located at specific and selected sites. For the design, construction, and operation of initial disposal facilities, the law assigns responsibility to the corresponding operator; for nuclear licensing, it assigns responsibility to CNEN and IBAMA. For site selection, design, construction, and operation of intermediate, and final disposal facilities, responsibility is assigned to CNEN, whereas both CNEN and IBAMA are in charge of nuclear licensing. CNEN is also expected to finance the costs related to site selection, design, construction, installation, licensing, management, operation, and physical safety of intermediate and final disposal facilities. In the case of initial disposal, these costs are borne by the corresponding operator. The law also highlights that the areas selected for final disposal facilities will be declared of public interest and utility and in case they are not already property of the Federal Government, the latter will expropriate them. Finally, Act No. 10308 forbids the entry of radioactive wastes into the country (Law No. 10308 2001).

The competences for nuclear licensing were defined earlier in Brazil than Argentina. In 1989, the responsibility for licensing activities and installations with significant environmental impact was assigned to the IBAMA, under the Ministry of the Environment, through Law 7804 (Art. 10). Since then, IBAMA, which is located under the Ministry of the Environment, has been in charge of environmental licencing for nuclear facilities, while CNEN remains the responsible agency for the radiation aspects of nuclear licensing. Eletrobrás Termonuclear (Eletronuclear) is responsible for the design, construction, installation, licensing, management, operation, financing, and physical safety of the interim storage facilities located at both Angra units, as Eletronuclear is the authorized operator of the Angra NPPs. Eletronuclear is a subsidiary of Eletrobrás.

As explained in the section below, Brazil started to regulate the management of nuclear wastes in 2001. However, in practice the country has failed to take a decision on a solution for for low or/and intermediate-level radioactive wastes.

9 Nuclear waste governance: A democratic deficit

In Argentina and Brazil regulation of nuclear waste management issues has started quite recently, respectively in 1998 with Law 25018 and in 2001 with Act 10308 (NOM-004-NUCL, NOM-020-NUCL, NOM-022/1/2/3-NUCL). Since then, while some further regulatory advancements in the field of radioactive waste management have been made, delays continue. In Brazil, a long-term solution for low and intermediate-level RW was foreseen by 2016. In Argentina, construction of LLW disposals was expected to start by 2015, followed by construction of intermediate-level waste repositories in 2018. Despite these former plans, long-term solutions for low and intermediate-level RW are not happening in none of these two countries. A decision on the final disposal for low and intermediate-level wastes is being delayed in both Argentina and Brazil. The identification and approval of a site for the disposal of very low, low, and intermediate level RW is not expected before 2020, and construction not before 2030.

Progress on site selection in Argentina may further be delayed, however, as previous studies conducted to select suitable sites for RW disposal were suspended and more than a half of all of the provinces in the country (Jujuy, Chaco, Formosa, Corrientes, San Luis, Mendoza, San Juan, Santa Fe, La Pampa, Río Negro, Chubut, Santa Cruz, and Tierra del Fuego) have legally declared their jurisdictions nuclear-free zones, banning the entry of radioactive waste and the installation of disposal facilities per law. Additionally, some municipalities like Traslasierras in the province of Córdoba have declared their territories nuclear-free zones. The federal component plays a large role in the country's governance, and it seems to be extremely difficult for CNEA to coordinate technical, geographical, social, and especially legal-political aspects. It remains unclear where CNEA will next propose as possible sites for final disposal facilities for the 2020 obligation and then whether or when approval will be finally obtained.

As is the case in other countries like France and Italy, nuclear issues in Argentina and Brazil (including the management of nuclear wastes) are treated centrally, with a few actors binding together and being responsible for all nuclear activities. CNEA and CNEN control not only waste management activities but also the national nuclear companies, which are often their subsidiaries. There is limited room for local and regional participation and decisions are characterized by low levels of transparency, although Argentinean legislation requires the formal authorization of subnational governments in siting nuclear waste disposal facilities. In Brazil, legislation does not require the permission of subnational governments, but rather foresees an eventual economic compensation for the municipalities where nuclear facilities are sited. Denunciations from local communities, environmental NGOs, and national senators, such as in the case of contaminated groundwater in Ezeiza and the still pending implementation of the Fund for the Management and Final Disposal of the Radioactive Wastes, as well as the still to be approved PEGRR, illustrate the low levels of transparency in nuclear activities.

The funding scheme for the management of radioactive wastes and SF is not well developed in Argentina or Brazil. Due to the fact that the national state finances all nuclear activities in both countries, it is not clear how, how much, and from where CNEA and CNEN will obtain funding resources. Financing of nuclear activities, including nuclear waste management, is managed in non-transparent ways in the rather secretive atmosphere that has historically characterized the nuclear sector.

National legislation for the classification of radioactive wastes as well as nuclear disposal and storage, licensing and approval rules, siting procedures, and definition of institutional arrangements is quite well developed. Yet, problems reside on the implementation side; there have been many delays in obtaining licensing, approval or authorization to construct a disposal at a particular site. As in other large and costly energy infrastructure plans, the implementation and application of rules and procedures often fail and experience great delays. Both countries are at a very early stage in the policy process concerning the management of SF and radioactive wastes. Currently, decisions on the "simplest" type of final disposals like very low, low, and intermediate-level waste repositories are experiencing delays. This does not bode well for the eventual decision to be made about a solution for HLW final disposal. Argentina is experiencing somewhat of a nuclear renaissance and new NPPs are either under construction or projected for the next years. If there is not a policy for the management of nuclear waste in place, how will the country be able to manage the increasing amount of SF and radioactive waste coming from the new NPPs and the additional planned nuclear reactors? These questions remain unanswered, unresolved, and even ignored.

References

Autoridad Regulatoria Nuclear (ARN) (2013). *National Nuclear Safety Report.* Argentinean National Report for the Convention on Nuclear Safety. http://www.arn.gov.ar/images/stories/informes_y_documentos/informe_nacional_de_seguridad/NATIONAL_NUCLEAR_SAFETY_REPORT_ARGENTINA_2013.pdf, last accessed 23 June 2015.

Brunnengräber, A. and Schreurs, M. (2015). Nuclear Energy and Nuclear Waste Governance: Perspectives after Fukushima Nuclear Disaster. In: Brunnengräber, A.; Di Nucci, M. R.; Isidoro Losada, A. M.; Mez, L. and Schreurs, M.A. (Eds.). *Nuclear Waste Governance: An International Comparison*, Wiesbaden: Springer VS, 47-78.

Centro de Desenvolvimiento da Tecnología Nuclear (CDTN) (2010). *Gestión de Desechos Radiactivos en Brasil.* Presentation by Clédola Cássia Oliveira de Tello for the Regional Training Group in the framework of IAEA program "Desarrollo de un Almacenamiento en Superficie de Residuos Radiactivos RBMA", 22-26 March 2010, Córdoba, Spain. https://www.iaea.org/OurWork/ST/NE/NEFW/WTS-Networks/DISPONET/disponetfiles/Lanzamiento_Proyecto_Almacenamiento_en_Superficie-Cordoba_Espana_Marzo2010-Spanish/Brasil.pdf, last accessed 23 June 2015.

Comisión Nacional de Energía Atómica (CNEA) (2010). *Gestión de Residuos Radiactivos en la República Argentina.* Presentation by Alberto Manzini for the Regional Training Group "Desarrollo de un Almacenamiento en Superficie de Residuos Radiactivos RBMA" in the framework of OIEA Disponet program, 22-26 March 2010, Córdoba, Spain. http://forosocialantinuclear.blogspot.de/2017/03/marco-legal-para-residuos-radioactivos.html, last accessed 25 November 2017.

CNEA (2011). *Joint Convention on the Safety of Spent Fuel Management and on the Safety of Radioactive Waste Management.* Fourth Argentinean National Report. http://www.cnea.gov.ar/sites/default/files/4_Informe_Nacional_a_Convencion_Conjunta-2011.pdf, last accessed 23 June 2015.

CNEA (2012). *Gestión de los Residuos Radiactivos y de los Combustibles Gastados en la República Argentina: Informe al Honorable Congreso de la Nación Correspondiente al Ejercicio 2011 según lo prescripto por la Ley No 25.018.*

CNEA (2015). *Joint Convention on the Safety of Spent Fuel Management and on the Safety of Radioactive Waste Management.* Fifth Argentinean National Report. http://www.cnea.gov.ar/sites/default/files/5_RESPUESTAS_A_PREGUNTAS_RECIBIDAS-2014.pdf.pdf, last accessed 23 July 2015.

CNEN (n.d.). *Quem Somos*. http://www.cnen.gov.br/quem-somos, last accessed 28 January 2017.

Damveld, H. and Bannink, D. (n.d.). *Management of spent fuel and radioactive waste. State of affairs: A worldwide overview*. http://www.nirs.org/mononline/nm746_48.pdf, last accessed 23 July 2015.

Greenpeace (2002). *El Informe de Greenpeace sobre Invap*. http://www.greenpeace.org/argentina/Global/argentina/report/2006/3/informe-greenpeace-sobre-invap.pdf, last accessed 15 June 2015.

Greenpeace and FUNAM (30 March 2005). *Contaminación con uranio en Ezeiza: Denuncian manipulación de datos*. http://argentina.indymedia.org/news/2005/03/279005.php, last accessed 24 June 2015.

Jimeno, M. (2015). *Explaining Divergent Energy Paths: Electricity Policy in Argentina and Uruguay*. Dissertation zur Erlangung des akademischen Grades eines Doktors der Politikwissenschaften (Dr. rer. pol.) am Fachbereich Politik- und Sozialwissenschaften des Otto-Suhr-Institutes der Freien Universität Berlin.

Jimeno, M. (2016). Argentina: From an Energy Stalemate Towards Shale Gas Expansion and Creating a Renewables Market. In: *Sustainable Energy in the G20: Prospects for a Global Energy Transition*, 19-25. http://publications.iass-potsdam.de/pubman/item/escidoc:1906910:2/component/escidoc:1906911/IASS_Study_1906900_3.pdf, last accessed 21 January 2017.

Law No 7781 (1989). *CNEN Competences*. http://www.planalto.gov.br/ccivil_03/leis/L7781.htm, last accessed 23 June 2015.

Law No 7804 (1989). *National Environmental Policy* (IBAMA). http://www.planalto.gov.br/ccivil_03/leis/L7804.htm, last accessed 24 June 2015.

Law No 24804 (1997). *National Law for Nuclear Activity*. file:///Users/moirajim/Desktop/%20NUCLEAR%20WASTE%20PAPER/Ley%20N°%2024.804%20del%2023_04_97.htm, last accessed 23 June 2015.

Law No 25018 (1998). *Radioactive Waste Management Regime*. http://www.cnea.gov.ar/sites/default/files/REGIMEN-DE-GESTION-DE-RESIDUOS-RADIACTIVOS.pdf, last accessed 23 June 2015.

Law No 10308 (2001). *Site selection, construction, licensing, operation, supervision, costs, damages, civil responsability and security for radioactive waste disposals*. http://www.planalto.gov.br/ccivil_03/leis/L7804.htm, last accessed 23 June 2015.

Lipcovich, P. (19 March 19 2005). Cerca del Centro Atómico de Ezeiza el agua está contaminada con uranio. In: *Pagina 12*. http://www.pagina12.com.ar/diario/sociedad/3-48664-2005-03-19.html, last accessed 23 June 2015.

Manzini, A. (2015). Written interview on June 15, 2015; member of the Comisión Nacional de Energía Atómica [Nuclear Energy Commission] (CNEA), responsible for the responsible of the National Program for Radioactive Wastes Management (PNGRR).

Manzini, A. (2017). Written interview on November 21, 2017; member of the Comisión Nacional de Energía Atómica [Nuclear Energy Commission] (CNEA), responsible for the responsible of the National Program for Radioactive Wastes Management (PNGRR).

Montenegro, R.A. (2005). The Nuclear Program of Argentina and the Creation of Nuclear-Free Zones for Reducing the Risks of Nuclear Facilities. In: *Updating International Nuclear Law*. http://www.updatingnuclearlaw.at/fileplus/abstract_montenegro.pdf, last accessed 24 July 2015.

Morandini, N. (2015). *Actividad nuclear en la Argentina: demasiadas preguntas*. http://www.normamorandini.com.ar/?p=10494, last accessed 24 July 2015.

Palicio Lada, P. (June 21, 2016). *A 20 anios de la gran marcha a Gastre*. http://www.elchubut.com.ar/nota/171454/, last accessed 21 January 2017.

World Nuclear Association (WNA) (2017a). *Nuclear Power in Argentina*. http://www.world-nuclear.org/information-library/country-profiles/countries-a-f/argentina.aspx, last accessed 19 November 2017.

World Nuclear Association (WNA) (2017b). *Nuclear Power in Brazil*. http://www.world-nuclear.org/information-library/country-profiles/countries-a-f/brazil.aspx, last accessed 19 November 2017.

Disposal and Contamination

Nuclear Waste Governance in South Africa

David Fig[1]

Abstract

Since the Manhattan project, South Africa's links with the global nuclear industry have been profound, starting out as a provider of uranium. From 1965 it commissioned a research reactor, and since 1984 has been producing nuclear electricity. There was a secretive nuclear weapons programme between 1978 and 1990. Together with considerable waste resulting from the mining of uranium, it is surprising that the first policy document on radioactive waste appeared only in 2005. Government was forced to do this because of civil society litigation, yet the policy was issued with almost no public input. The industry has seldom felt accountable to the public, and the regulator has struggled to meet its obligations, given shortages of expert staff and budgets. There is concern about current government plans to add six to eight more reactors to the existing two, given the need to ensure that the institutions governing and regulating the industry are competent and viable. This chapter aims to uncover the history of nuclear waste in South Africa, providing some detail on the inventories, the legal and institutional background, and current controversies.

1 Research Associate, Chair in Bio-Economy, University of Cape Town, and Fellow, Transnational Institute, Amsterdam, Box 533, Wits 2050, South Africa; E-Mail: davidfig@iafrica.com

A. Brunnengräber et al. (eds.), *Challenges of Nuclear Waste Governance*,
Energiepolitik und Klimaschutz. Energy Policy and Climate Protection,
https://doi.org/10.1007/978-3-658-21441-8_13

1 Introduction

Although South Africa has only possessed two nuclear power reactors since the 1980s, it has a long record of generating radioactive waste. Primarily, this has been as a result of the uraniferous content of the ore in the gold-bearing reefs around Johannesburg. From 1886, the deep sitings of the gold reef meant the generation of huge amounts of waste, mostly seen as tailings dams or waste dumps adjacent to the mines, often close to residential areas. These dumps have significant radioactive content.

The venture during the apartheid era into making nuclear weapons of mass destruction has also had an impact on the generation of radioactive waste. Current policy to procure a further six to eight new reactors means that the problems associated with the management and safe disposal of nuclear waste are likely to multiply. This is extremely perturbing given the fragility of the current nuclear regulatory apparatus, already a victim of austerity measures and the inadequate development of human expertise.

2 Nuclear waste disposal in South Africa

2.1 Historical background

South Africa was integrated into the global nuclear industry in the latter part of the 1940s. Since the 1920s geologists had pointed out that the gold-bearing Witwatersrand reef was also rich in uranium (Cooper 1923: 90ff). At the time the uranium had no known application, and hence little commercial value. It was treated as part of the waste stream of the gold mining industry, one of the largest in the world.

During the Second World War, the United States and the United Kingdom had collaborated in the Manhattan Project, responsible for developing the bombs dropped on Hiroshima and Nagasaki (Alperovitz 1995). Instead of internationalizing the nuclear weapons, as scientists like Einstein and Oppenheimer advocated, those nations which possessed nuclear weapons felt that they could gain a strategic military edge in the impending Cold War. Both the US and the UK sought to cultivate nuclear arsenals in their postwar competition with the Soviet Union. The Soviets developed their own weapons, and the first of their tests took place in 1949 (Bird and Sherwin 2005).

The Western Allies had no really secure source of supply of uranium, which had been relatively scarce during the war. The material for the Manhattan Project originated from the Belgian Congo and Canada. Other known sources included

Czechoslovakia, but access to supplies here receded behind the Iron Curtain by the end of the 1940s.

In a global search, the US and UK approached a number of friendly countries in an effort to seek a reliable and plentiful supply of uranium (Fig 2009a: 182). The South African Prime Minister Jan Smuts responded positively, and the gold mining industry agreed to re-mine their tailings dams (waste dumps) for uranium. South Africa signed a secret agreement whereby the US would receive two-thirds of its uranium exports and Britain the remaining one-third. This was destined for their bomb programmes, as the commercial nuclear energy industry was not yet under way (Fig 2005: 183). Seventeen uranium separation plants servicing 26 gold mines on the Witwatersrand made clandestine arrangements to provide the uranium to the US and UK weapons programmes. The secret treaty absorbed South Africa's output of uranium for a period of a decade (Fig 1999: 78).

Smuts also set about creating an Atomic Energy Board through the passage of legislation. Smuts lost his seat and his party the control of government in the May 1948 elections, in an electoral process which excluded the black majority from the vote. Power then accrued to the National Party, which advocated apartheid, an intense form of racial subjugation of black inhabitants by a small white minority. This policy combined strict personal, residential, educational and job segregation, the confinement of black people to 13 percent of the land, and their disenfranchisement. It also intensified a harsh system of labour migration, particularly on the mines.

From 1949, the newly created Atomic Energy Board presided over nuclear research and development. Scientists were sent abroad, particularly to the US, West Germany and the UK for nuclear training (Cervenka and Rogers 1978). It took some time before South African scientists began to develop an interest in reactor research and other parts of the fuel chain.

By the early 1960s, the research establishment had been moved out of cramped offices in downtown Pretoria, the capital, to land on the western edge of the city, at a farm called Pelindaba. Under the US Atoms for Peace programme, South Africa received a small research reactor which was installed at Pelindaba in 1965 (Newby-Fraser 1979).

As resistance to apartheid intensified, the ruling regime felt more compelled to militarise itself. A sophisticated conventional arms industry was developed in the face of global sanctions. The burgeoning nuclear establishment also offered the state the chance of having its own nuclear weaponry.

From the 1970s, the Atomic Energy Corporation (AEC, successor to the AEB) along with its twin, the Uranium Enrichment Corporation (Ucor), a newly created state-owned enterprise, secretly proliferated in conjunction with Armscor, the state's agency for weapons procurement and manufacture. At first, gun-type nuclear weapons were developed on site at Pelindaba. Manufacture was later moved to Advena, a complex a few kilometres closer to Pretoria, and more control over the process passed to Armscor (Albright 1994: 13).

Meanwhile on a site adjacent to Pelindaba, South Africa built a conversion plant and enrichment facilities. Ostensibly these were constructed to serve a pair of reactors commissioned by the electricity utility, Escom. However, in reality, the enrichment was aimed mainly at producing weapons grade uranium for the weapons programme. An estimate given was that each of the bombs required an estimated 55 kg of 90 percent-enriched uranium.[2]

While proliferating, South Africa also encouraged the construction of two civilian nuclear power reactors at a site outside Cape Town called Koeberg. Built by a consortium organised by Framatome, the precursor to Areva, two pressurized water reactors went into service in 1984. The reactors together can produce a maximum of 1844 MW, although the maximum output is not always reached. This amounts to just over five percent of South Africa's electricity production, historically dominated by over 90 percent coal burning. Part of the logic for construction of Koeberg was its substantial distance from the country's coalfields. The alternative was to experience inefficient losses of coal-generated power during long-distance transmission. To date, the nuclear power reactors at Koeberg are the only ones operating on the African continent (Fig 2005: 54-59).

By the mid-1980s, apartheid South Africa had, despite political and economic sanctions, acquired most of the links in the nuclear fuel chain (except for reprocessing, although there were plans to develop a site for experimentation at Gouriqua). It had also embarked on a plan to proliferate nuclear weapons.

Apartheid South Africa faced additional security problems at this time. The independence of Angola, Mozambique and Zimbabwe, all under Marxist-leaning governments, brought the front line ever closer. South Africa was involved in numerous cross-border invasions and incursions, as well as in trying to repress growing socio-economic discontent at home.

The purpose of South Africa's weapons was, President F.W. de Klerk later suggested, deterrence. There was no logic in utilizing it on the local population, hostile neighbours, or battlegrounds. Instead, if apartheid was ever to be threatened vitally, the existence of the weapons would have been revealed to the world in an attempt to defend the life of the apartheid regime (Purkitt and Burgess 2005).

Instead it was apartheid's own contradictions that defeated it. A combination of sanctions, disinvestment and a loss of confidence slowly strangled the apartheid economy. Border wars against liberation movements proved costly and disruptive, the low point being the defeat of conventional South African forces by Cubans and Angolans at Cuito Cuanavale. Soon after, South Africa had to

2 Personal interview with David Albright, the founder and current president of the non-governmental Institute for Science and International Security, on 30 December 1993.

concede the independence of Namibia, whose territory it controlled between 1915 and 1990.

At the same time, civil unrest grew to a peak, and the apartheid regime began to realize that its days were numbered. In 1990 the African National Congress and other proscribed parties were unbanned, and Nelson Mandela released. Discussions with the ANC leadership led to the negotiation of a transitional executive, democratic elections, and the rewriting of the constitution to reflect democratic and non-racial values.

De Klerk foresaw the ANC gaining a significant share of political power. To prevent the incoming party obtaining the knowhow of nuclear weapons, he decided to dismantle the programme, shred all evidence of it, and to see that South Africa rejoin the IAEA and the NPT. The identities of the 1000 people involved in the nuclear weapons programme have never formally been revealed. Nor were the manufacturers of weapons of mass destruction ever held accountable for their actions when the "Truth and Reconciliation Commission", under Archbishop Tutu, held its deliberations.

At first the ANC, which had in its years of exile been critical of South Africa's nuclear programme, gave indications that it would not want to repeat apartheid proliferation, or the secrecy in which nuclear decisions were made, and favoured a review of the nuclear establishment (EMG 1994: 3-5).

However, once in power, the ANC continued to budget for the continuation of nuclear research, although on a much-reduced basis. There was a further renaming in 1999 of the state's nuclear research establishment from the Atomic Energy Corporation to the South African Nuclear Energy Corporation (NECSA). The conversion, enrichment and fuel fabrication plants at Pelindaba were dismantled, with the zirconium cladding manufacturing facility of the fuel fabrication process being sold to China. Work on molecular laser enrichment was also discontinued (World Nuclear Association 2017).

The state began to bankroll plans to build a demonstration plant for a small-scale high-temperature helium reactor. Called the Pebble Bed Modular Reactor (PBMR), the aim was to allow for technological innovation while at the same time keeping happy a number of the scientists and engineers who had been involved in the bomb programme (PBMR Company 2002).

The problem was that the PBMR programme was far too ambitious for the South African nuclear industry to sustain. Dates of completion of the demonstration plant were placed further and further into the future, costs ballooned, and there was a decided absence of other investors or clients. It was surmised that Eskom, whose attitude to the PBMR programme was only lukewarm, would have been obliged to take delivery of future reactors to justify the expenditure on it (Thomas 2004; Fig 2010a).

By 2010, the PBMR company said that it needed a minimum of an extra R31 billion to complete the demonstration plant. The state, however, having already devoted R9 billion with nothing to show for it, became impatient, and put an end to the programme.

A growing electricity crisis (Fig 2010b) has propelled the government into paying more attention to the commissioning of a new fleet of nuclear reactors. It was argued that 9.6 GW of additional nuclear power would provide base load electricity and help diversify from coal, which provided over 90 percent of the country's electricity needs. Land had been earmarked at coastal sites in the Eastern and Western Cape, as well as adjacent to Koeberg. More recently, other sites closer to the coal-mining complex at Witbank, and in the province of KwaZulu-Natal have also been proposed (Mathews 2017).

The country's formal nuclear policy foresees the putting in place of the whole gamut of the nuclear fuel chain, reconstructing conversion, enrichment, and fuel fabrication plants, but also having pretentions to the undertaking of reprocessing as well. The policy also envisages the creation of a specialized nuclear police force to safeguard installations and the movement of nuclear materials.

Should the construction of a new fleet go ahead, it will occur despite considerable opposition. This arises not only from the anti-nuclear and environmental movements, but also from faith-based organisations, trade unions, human rights advocates, community-based groups, and, increasingly, the private business sector and the business media. Their arguments are that the nuclear fleet will be too costly, create a debt problem, crowd out investment in cleaner and cheaper renewables, and force dependency on rigid inflexible energy provision in the face of a contracting economy and decreasing demand.

The new fleet would multiply the waste stream coming out of reactors, adding an extra five times the existing amount. This poses a very large challenge to the underfunded and already overstretched regulatory process.

2.2 The national inventory

2.2.1 Spent fuel at Koeberg nuclear power station

Koeberg is the site of two 922 MW pressurized water power reactors built by Framatome (now Areva). Low- and intermediate-level wastes are removed for disposal at Vaalputs, a site 700 km distant from the reactors. Spent fuel is stored on site in cooling ponds, or in four interim dry storage casks (Castor X/28F) above ground.

Table 1: Inventory of spent fuel stored at Koeberg nuclear power station, 2014

LOCATION	STORAGE	NUMBER OF SPENT FUEL ASSEMBLIES
Koeberg reactor 1	Wet storage	1009
Koeberg reactor 2	Wet storage	996
Castor X/28F casks	Dry storage	112
	Total	2117

Source: NNR 2014b: 125, table 3 (A2-1).

2.2.2 Spent fuel at Pelindaba precinct (NECSA)

Pelindaba houses the national nuclear research facility, which includes a research reactor installed in 1965 (Safari-1).

Table 2: Inventory of spent fuel stored at Pelindaba complex, 2014

LOCATION	STORAGE	NUMBER OF SPENT FUEL ELEMENTS	NUMBER OF USED CONTROL RODS
Safari-1 reactor	Wet storage	175	31
Thabana pipe storage facility	Dry storage	867	134
	Total	1042	165

Source: NNR 2014b: 125, table 4 (A2-2)

2.2.3 Storage of non-reactor wastes at Pelindaba

Pelindaba is home to a number of research and production facilities. It was also the original site at which nuclear weapons were manufactured. When the weapons programme was discontinued in 1990, an estimated 385 kg of dismantled weapons-grade enriched uranium was returned to Pelindaba and is held under IAEA safeguards.[3] Reporting on the inventory for each of the Pelindaba waste sites has not been comprehensive, but is supplied where known.

3 Personal interview with David Albright on 30 December 1993.

Table 3: Inventory of non-reactor wastes stored at Pelindaba

PURPOSE	NAME	LOCATION	INVENTORY
Decontamination of uranium contaminated metals	Decontamination services Area 26 Decontamination	V-A8 Area 26	
De-heeling of UF_6 cylinders	UF_6 De-heeling facility	Area 27	
Liquid effluent treatment	LEMS Uranium bearing liquid effluent treatment facility	P-2400 A8	Volume of 13,300m^3
Predisposal operations	Volume reduction facility	Area 14 Pelstore	
Historic liquid waste facilities	Pelindaba East CaF2 Pans BEVA Pans ABC & 1-14	Pelindaba East Thabana Pelindaba West	
Storage of post-reactor test fuel pin related waste	Cell 3 NTP Hotcell complex	P-1701	
Low- and intermediate-level waste drummed & containerised sold waste storage facilities	Pelstore Thabana stores UF_6 cylinders Decontamination services Bus Shed waste drum store Drum store	Area 14 Thabana Areas 21, 16 V-A8 Pelindaba East A-West	56,099 x 210 litre drums Storage capacity of 14,900 drums Medical waste Storage capacity of 5000 drums
Storage of non-clearable decontaminated metals	Quarantine store	Quarantine store	
Disused sealed radioactive sources	Pelindaba East	Area 24	7,790 sealed sources
Historic disposal trenches	Thabana Radio-active Waste Storage Facility	Thabana (trenches) & P-5100	
Mixed use	Dorbyl Camp	Area 18	
Metal smelter	Test smelter	Area 25	
Storage of confiscated radiological material	Illicit Materials Store	BEVA E-1	

Source: NNR 2014b: 127, Table 5 and narrative pp. 127-33.

2.2.4 Storage of waste at Vaalputs National Radioactive Waste Disposal Facility

Vaalputs (situated in a remote part of the Northern Cape province) was opened in 1986, two years after the completion of the Koeberg nuclear power station. It received most of the low- and intermediate-level radioactive waste produced at Koeberg and other nuclear installations. The distance from Koeberg to Vaalputs is 700 km. The amount of low- and intermediate-level waste stored in drums was 13,882 m^3 by volume, recorded in 2014 (NNR 2014b: 137, Table 11).

2.2.5 Storage of waste at mines, scrapyards, smelters and research facilities

The National Nuclear Regulator has compiled figures for these sources.

Table 4: Inventory of radioactive waste at mining and mineral processing facilities, scrap dealers and smelters, and at small users (e.g. laboratories and refurbishers). 2014 (in tonnes).

WASTE TYPE	METHOD OF STORAGE	MINING	SCRAP & SMELTERS	SMALL USERS & LABS
Scrap metal	Salvage yard	6.76E+06	1.04E+05	00E+03
Tailings	Tailings storage facility	2.16E+08	2.22E+02	0.00E+00
Waste rock	Waste rock dumps	5.12E+07	2.04E+01	1.35E+05
Other waste	Salvage yard	1.78E+05	3.55E+04	6.99E+02

Source: NNR 2014b: 140, Tables 14-16.

2.3 The storage sites

2.3.1 Vaalputs

With the commissioning of two nuclear power reactors in 1984 at the opposite end of the country, it became clear that the amount of low- and intermediate-level waste generated at Koeberg would have to be accommodated at a dedicated site, preferably where population density was low.

It was imperative for the nuclear industry to show that it was capable of managing the additional waste. The state commissioned studies on the location of a new disposal site. After rejecting proposals for the Kalahari and Richtersveld areas, the Atomic Energy Corporation selected a site at a place called Vaalputs, in the arid region of Namaqualand. Vaalputs topped the list because of its low average annual rainfall (74 mm), deep ground water (>50 m) and geological stability. Levels of public consultation were minimal and will be dealt with in section 5. Deliveries of Koeberg's nuclear waste began in 1986, yet it took over 17 years before the regulator communicated with local people in September 2003.

The Vaalputs site was engineered to accept low- and intermediate-level wastes and it was envisaged that at a later stage high-level wastes could be stored in specially crafted above-ground silos. However a plan for disposal of high level wastes has never emerged and continues to elude the industry.

Low-level waste, according to the US Nuclear Regulatory Commission, consists of contaminated protective shoe covers and clothing, wiping rags, mops, filters, reactor water treatment residues, equipment and tools, luminous dials, medical tubes, swabs, injection needles, syringes, and laboratory animal carcasses and tissues (US NRC 2015). At Koeberg, these wastes are compressed into steel drums with the same dimensions as petrol drums.

The intermediate-level waste is mainly made up from parts of the reactor building that have been contaminated, filters, as well as resins, chemical sludge and metal nuclear fuel cladding. The Canadian Nuclear Safety Commission defines intermediate-level waste as "waste that has been exposed to alpha radiation, or that contains long-lived radionuclides in concentrations that require isolation and containment for periods beyond several hundred years" (CNSC 2015). These materials are mixed with concrete and stored in sealed concrete drums, which are much larger and heavier than the drums used for low-level waste.

At Vaalputs, drums from Koeberg are stored in trenches about 10 m deep. When full, the trenches are covered with compacted soil from the vicinity. Rehabilitation plans include replanting of the covered trenches with local flora.

Accusations of mismanagement of Vaalputs surfaced in June 1997, when it was revealed that drums of intermediate level waste had been leaking radio-activity for a considerable time. This was ascribed to some wear and tear while the drums stood in open trenches for two to three years awaiting soil cover (Fig 2005: 63).

The operators of the facility, the AEC, were faced with having to suspend their activities at Vaalputs until they could conform to licencing conditions. It turned out that on an inspection visit in September 1996, the regulator noted 55 violations of the licence (CNS 1997: 23). Brian Hambleton-Jones, then head of the AEC's nuclear waste division, blamed the corrosion on the cold weather ("extreme cold and heavy rains could have been responsible for hairline cracks in the concrete blocks containing radioactive waste"), while Tony Stott at Koeberg attributed the problem to the extended times between deliveries of waste (SAPA 1997).

According to the regulator, however, the AEC had failed to implement quality controls and training programmes, develop emergency planning, maintain proper records, control radioactive effluent, maintain records of staff radiological protection, check instrumentation, and audit safety procedures. Even worse was the realization that the process of storing the drums was inadequate, and that there were no arrangements to store or retrieve records on the disposed waste. The AEC was accused of a general lack of management and supervision at all levels and having no effective mechanism of complying with licence conditions. The document detailing these allegations took a decade to reach the public domain (Fig 2009b: 83).

In 2009 an official working for the regulator publicly admitted to an International Atomic Energy Agency conference that the Vaalputs technology was outdated and ageing. Former workers at Vaalputs have openly voiced their criticism of the waste dump's operations. Hendrik Fortuyn said: "When I went to work at Vaalputs, they (management) told me that the underground water was contained in such a way that it could not reach the surface, and could not become contaminated, no matter what. Yet we noticed that the water level did not drop but continued to climb. We measured this regularly. What it tells me is that the Vaalputs management was lying to me. I am prepared to testify to this in court." Another worker, Willem Ghaal, testified to leakages: "There was a leak in one of the drums containing nuclear waste. People came from Koeberg to see the drums. At that time, they closed the drums and reburied them in a square metre of earth set aside for this purpose."[4]

2.3.2 Pelindaba including Radiation Hill (Thabana)

From the time of the installation of an Allis-Chalmers experimental reactor (SAFARI-1) at Pelindaba in the mid-1960s, the nuclear waste that was generated was stored on the property in a section known as Radiation Hill. Drums of nuclear waste were haphazardly placed in trenches and covered over with soil, without much planning or regulation. Radiation Hill was also the site which accepted the waste of South Africa's bomb programme (1978-90).

In 1990, as a result of a nuclear incident, condensers from SAFARI-1 were contaminated and placed in Trench 7 at Radiation Hill. In 1995, workers were ordered to retrieve them and each was issued with a pickaxe, a shovel, a pair of boots, an overall and a paper mask to cover their mouths and noses. According to

4 Interviews conducted by Helena Kingwill for the film Buried in Earthskin, 2010.

one of them, Bennie Masemola, "Very soon we unearthed mountains and mountains of drums, many of them were rusty and full of cracks. I remember the dust and the smell, and when I went home, the smell would still hang around me. It was very bad. As we dug deeper, we were covered with blue, red and green dust" (Kanhema 1996). A senior manager at Pelindaba confessed: "Look, we screwed up, made several mistakes, infringed on many of our licence conditions and no waste disposal records were kept for Radiation Hill." In a formal statement the regulator said that these occurrences are indicative of a degradation of safety culture at the AEC" (CNS 1996).

In May 2005, Shaun Smillie, a journalist on South Africa's Independent group newspapers, was included in a media tour of Pelindaba. He was told by senior manager Arie van der Byl that Radiation Hill had not been operated up to international standards when it came to nuclear waste management. "In 1997 we decided that we should rename the facility. We would like to forget that legacy. It is no longer seen as acceptable to bury waste," Van der Bijl said. (Smillie 2005). From 1997, the site was renamed Thabana Hill. In 1998 the AEC commissioned its interim retrievable dry store at Thabana Hill for weapons-grade spent fuel from SAFARI-1 after receiving the necessary safety and safeguards approvals from the International Atomic Energy Agency in Vienna. As a result, 120 spent fuel elements were transported from the spent fuel racks in the pool of SAFARI-1 research reactor to the retrievable dry store.

2.3.3 Koeberg's spent fuel rods

Since no site in South Africa is yet licenced to receive high-level waste for final disposal, the high-level power reactor wastes, in the form of depleted fuel rods, have to date been stored in ponds on site at the existing Koeberg nuclear power station. This was meant to be a temporary solution. The aim was initially to store the fuel rods for up to a decade while they cooled off, to permit later transport to a site dedicated to longer-term disposal. However no such site has ever been commissioned. The fuel rods, amounting to 32 tonnes of waste per year, have increasingly populated the ponds at Koeberg. Eskom claims to have densified the racks to allow storage for the 40-year life of the reactors (Eskom 2017).

In August 2015, it was announced that the operators of the Koeberg power station, Eskom, would order seven new reinforced metal casks, supplied by US manufacturer Holtec Interntional, at a cost of R200 million (12.8 million Euro). The casks would store waste from Koeberg's spent fuel.

By 2019, a further 30-40 dry storage casks would be ordered, constituting an interim on-site storage facility. By 2025, the casks would all be transported to an off-site "central interim storage facility" (Reuters 2015).

2.3.4 Pelindaba smelter

In 2003, at least a decade after the decommissioning of NECSA's enrichment plants, the corporation applied for environmental authorisation of a new test smelter (and two subsequent 4 tonne induction smelters) within Area 26 of its premises at Pelindaba East, aimed at melting down 14,000 tonnes of uranium-contaminated scrap metal used as equipment in the former enrichment programme and stored at Pelindaba. NECSA had judged that smelting was the best way to get rid of this equipment.

The smelting was aimed at separating the uranium from the molten metal, and concentrating it in the slag. This would be collected, drummed and managed as radioactive waste. NECSA's estimate is that 1400 tonnes of material will be smelted each year, allowing for an operations life of 10 years, after which it would be decommissioned (NNR 2014a: 70-71).

However, strenuous objections were received by a number of NGOs, including the Pelindaba Working Group, an association of concerned local residents. They feared that emissions from the smelter would lead to radioactive contamination of the area. When the minister of environmental affairs finally authorized the smelters, on 31 August 2007, he rejected local complaints about the emissions, the failure to have conducted an independent assessment of the smelter technology, and the exclusion of local communities from the inadequate consultation process. Objectors also feared that once the smelters had been allowed to operate, the corporation would start to accept imported radioactive waste. However, the Record of Decision accepted NECSA's safety and other assurances, but insisted that authorisation would fall away if not acted upon during the following five years.

Nothing was done to commence work on the smelters within the five years, and when the authorization lapsed on 31 August 2012, Earthlife Africa, a concerned environmental justice organisation, took legal steps to ask for a high court injunction to prevent NECSA from going ahead with their construction (Earthlife Africa Johannesburg 2014). Without taking into account that the matter was *sub judice,* the National Nuclear Regulator board approved of the smelters in April 2013, and issued a licence to that effect on 2 July 2013.

After much negotiation between the parties, and a series of postponements, the matter came up before Pretoria High Court. On 7 March 2017, Judge Rabie dismissed Earthlife Africa's application, allowing the smelters to go ahead (Naidoo 2017). The court did not impose any obligation on NECSA to undertake an Environmental Impact Assessment (EIA) process with respect to the climatic impact of the smelters. This judgment therefore forecloses further opportunities for public participation in expressing concerns about the smelters.

2.3.5 Radioactive waste in the environment

For around 130 years, the area around Johannesburg has been at the centre of the mining of the richest underground gold and uranium bearing reefs in Southern Africa. The mining has occurred over an arc of approximately 75 kilometres, along a geological ridge of white waters, or the Witwatersrand. This also constitutes the continental divide with different rivers flowing into the Indian and Atlantic Oceans (Turton et al. 2006: 316).

Today the gold and uranium are mostly mined out, and most mines have closed, along with the companies that operated them. An extensive legacy of environmental pollution goes unchecked, since little obligation was ever placed on the mine owners to clean up the damage. Part of this damage to the environment has been the mismanagement of tailings dams, the huge piles of solid and liquid wastes discarded by the extraction process. There are 270 tailings dams across 400 square kilometres in the Witwatersrand, for the most part unlined and unvegetated. In 1997, it was calculated that of 468 tons of mining waste, 221 or 47 percent came from the gold and uranium complex, and thus constituted the largest single source of pollution in the country (DWAF 2001).

Additionally, bomb-grade 90 percent-enriched uranium from the dismantled apartheid nuclear weapons is being held on site at Pelindaba under IAEA safeguards. Some of this material has provided fuel for the Safari-1 research reactor. However, the US exerted pressure on South Africa to run Safari-1 on blended down uranium at a much lower level of enrichment.

The bulk of radioactive waste is generated by the mining and mineral extraction industries and is situated in tailings dams in the mining areas. The resulting dumps are not always revegetated, nor is the public removed from their vicinity. This results in air and water borne contamination, and additional health problems which are said to affect at least 1.7 million inhabitants living adjacent to the dumps in the environs of Johannesburg.

Increasingly, attention is also having to be paid to the problem of acid mine drainage (AMD), consisting of flows of polluted water from the mines into the surface watercourses and underground aquifers surrounding the former mines. To prevent this, water was pumped out of the mines, but since most have closed, the pumping has also ceased, and the water table is rapidly rising towards the surface. In the West Rand basin, the AMD has already entered the environment and has been polluting watercourses for some years. This has led to some farmers being prevented from allowing their cattle to drink water with elevated levels of radionuclides. It has also led to the blinding of local hippopotami in nearby nature reserves affected by the flows of AMD into their waterholes.

Worst of all, a population estimated at 1.7 million lives in areas subjected to elevated radioactivity, without being informed of the dangers. Most are poor inhabitants of informal settlements located close to tailings dams, who continue to grow crops and allow their children to play on contaminated land (the Tudor Shaft example will be cited below, on page 317). Some pregnant women have developed a custom of ingesting earth comprised of the waste in order to satisfy a craving for minerals, without any awareness of the dangers of doing so. The waste is bagged and sold all over the townships of the Witwatersrand. There is no public education programme aimed at reversing this practice.[5]

It is likely that the Witwatersrand is the most radioactive urban area in the whole of Africa. It would take a major public investment to rehabilitate the region so that it causes no further harm to humans, animals and the environment. Meanwhile the government struggles to deal with the consequences of AMD, which it has characterized as a "potential environmental catastrophe" (DWAF 2009: 22).

2.4 *The waste management strategy*

Until obliged to do so by the courts in 2005, there had been no specific government policy on the question of radioactive waste management. The policy document subsequently produced (with almost no public participation) proved to be agnostic on the question of long-term disposal of high-level radioactive wastes. It produced a checklist of options (deep-level disposal, vitrification, etc.) from which future policy makers might select. Therefore the policy is a shallow one on paper, and does not specify a long-term strategy (DME 2005).

In the meantime, wastes that are being generated by power stations are simply being held on site, mostly in cooling ponds. Justification is being given to an increase in above-ground storage of some of this waste. Low- and intermediate-level reactor waste is being contained in concrete and metal drums and transported from the power station to a disposal site 700 km away. Some of the wastes generated at and around Pelindaba find themselves being stored above ground in varied configurations at the site.

The management of the national waste disposal facility has not always gone smoothly, and the site has had to close temporarily before conforming to licencing conditions. More recently it has been placed in the charge of a new radioactive waste disposal entity, the National Radioactive Waste Disposal Institute, whose problems are highlighted in a subsequent section. Management of wastes at the former Radiation Hill (Thabana) at Pelindaba have also been criticised by the media, the public and the regulator.

5 Mudanalo Tshishibedi, informant, 25 September 2015.

Imports of enriched uranium and the transport of wastes place a very high burden on South Africa's already dangerous road system. Not enough information, skills and budgets are available to adjacent municipalities to perform emergency services when required. The state should review its radioactive waste management policy, particularly in the light of its promotion of a fivefold increase in nuclear electricity in the near future.

3 The legal and institutional framework

3.1 The legal framework

The Atomic Energy Act of 1949 established an Atomic Energy Board, which became the main state body to control the production of and trade in uranium. In 1959 the Act was extended to allow the Board to undertake research, development and use of nuclear technology. The AEB began with offices in downtown Pretoria, but by the mid-1960s had moved into a dedicated precinct on a farm outside the capital. The farm was known as Pelindaba, Zulu for "the talking has ceased", which was an injunction for researchers to respect the secrecy associated with the nuclear industry.

In 1970, Prime Minister B.J. Vorster announced in parliament that the Board was being replaced by an Atomic Energy Corporation (AEC) and a separate entity, the Uranium Enrichment Corporation (Ucor), which would establish facilities for the conversion and enrichment of uranium on land at Valindaba ("the matter is closed" in Zulu), adjacent and linked to Pelindaba. In 1985, the two corporations merged under the AEC. The siting and the management of the Nuclear Waste Disposal Facility for low and intermediate level waste was put under the charge of the AEC.

The Nuclear Energy Act No. 131 of 1993 created the Council for Nuclear Safety (CNS), a new statutory body which took over the regulatory function from the former AEC Licencing Branch. The Council drew its staff almost exclusively from former AEC employees. Theoretically the CNS had more independence, but its small budget and staff establishment limited its effectiveness. There was no provision for public participation in its work. The same Act governed nuclear operations and nuclear regulation.

To correct this, the state created two new laws in 1999, separating out the research and development establishment from the regulatory functions. The Nuclear Energy Act 46 of 1999 transformed the old Atomic Energy Corporation into the South African Nuclear Energy Corporation (NECSA). Under section 45 of the Act, the Minister of Minerals and Energy was given responsibility for the management of radioactive waste and the storage of irradiated fuel:

> (1). The authority over the management and discarding of radioactive waste and the storage of irradiated nuclear fuel vests in the Minister. (2) The Minister in consultation with the Minister of Environmental Affairs and Tourism and the Minister of Water Affairs and Forestry may make regulations prescribing the manner of management, storage and discarding of radioactive waste and irradiated nuclear fuel. (3) The Minister must perform that function with due regard to the provisions of the National Nuclear Regulator Act 47 of 1999 (South Africa, 1999).

It was envisaged in section 45(2) that something would be done in an undefined future to establish regulations for the disposal of radioactive waste. Glazewski (2000: 585) suggests that "this is irresponsible and that until a clear and satisfactory policy is established for dealing with nuclear waste and regulations are in place in this regard, nuclear processing should not continue."

The companion act, the National Nuclear Regulator Act 47 of 1999 separated out the functions of the regulator from the research and development function. The Act created the National Nuclear Regulator (NNR), a successor to the CNS, and gave it full licencing authority over nuclear facilities. In addition, the NNR was made responsible for managing the safe transport of radioactive materials in line with IAEA regulations.

In 2008, new legislation was passed to create the National Radioactive Waste Disposal Institute, aimed at taking over the management of nuclear waste from the AEC. It took another six years before the Cabinet approved the first Board of the NRWDI, whose term of office began on 1 January 2014.

3.2 The institutional framework

From the beginning of uranium mining in the 1940s, no government has seriously considered a policy on radioactive waste, despite this being envisaged in the legislation of 1999.

It should be mandatory for every country with nuclear facilities to install a nuclear regulatory authority. In South Africa's case, no such body existed prior to 1988. This is at least 40 years after the creation of a uranium mining industry and the AEB, whose brief included the conduct of nuclear research and development.

Until 1988, the operation of nuclear installations was inspected by the licencing branch of the AEC. While licencing is an important part of the regulatory process, a full-scale regulator has to monitor a much wider range of contingencies.

Because this internal licencing process ran against global practice, the government of the day decided to carve the licencing branch out of the AEC and declare it to be the new regulator. Between 1988 and 1999 it was known as the

Council for Nuclear Safety (CNS). During this phase the same piece of legislation covered the regulation as well as the promotion of the industry.

Even the management of the CNS found this situation untenable and began to lobby for a discrete law to provide for regulation. The outcome was successful and in 1999 a new law created the National Nuclear Regulator, successor to the CNS.

Part of the governance problem has not been resolved. Although the regulatory functions have been placed in a separate Act, the energy minister now has political responsibility for both the promotion and the regulation of the industry. This may present a conflict of interest for the minister when she is in a position to decide between promotional and regulatory priorities and has created a problem for democratic governance in the sector. The NNR should have been granted more independence, falling under a different department of government, or even becoming a "chapter nine" institution (an autonomous body) under the provisions of the South African Constitution. This would have removed the potential for any conflict of interest.

The NNR operates under both budgetary and human resource constraints (NNR 2008: 4). Together with its limited political leverage, this has made it extremely difficult for it to take on the well-entrenched and powerful mining industry with its huge legacy of indiscriminately depositing radioactive waste in the form of mine tailings and slurries in the Witwatersrand and other uranium mining areas. Under enormous public pressure, it has only since 2007 teamed up with the Department of Water Affairs in addressing the massive contamination of the Wonderfonteinspruit, a tributary of the Vaal River on the West Rand, by establishing a steering committee to implement a remediation action plan.[6]

The scientific reporting of the NNR has turned out to be inadequate and unreliable. This was demonstrated when its reporting on the levels of contamination of the Tudor Shaft area in the Wonderfonteinspruit catchment, outside Johannesburg, where communities had built informal housing on radioactive land close to mine tailings. The NNR reported in a publication (NNR 2010) that it could find no evidence of a threat to the public. Diligent NGOs encouraged the intervention of an independent scientist who verified that the contamination was a serious threat to health (Busby 2011). Worrying mistakes were identified in the NNR report, for example, on the calculation of dose impacts. Licking its wounds, the NNR claimed that it was "not a research body" and it sheepishly undertook to revise its findings. Clearly its level of scientific and radiological expertise is poor (Fig 2010a: 23-25).

6 Interview with Mariette Liefferink, chief executive officer of the Federation for a Sustainable Environment, 16 June 2011.

As such it is generally a weak protector of the public interest, and often fails to understand its mandate. For example, instead of taking the lead on questions of radioactive contamination, it waits for other government agencies to become active. It fails to recognise its broad responsibility and argues that it is only accountable for overseeing licenced nuclear facilities. Therefore it cannot function effectively to reduce radioactive contamination outside these facilities, which, in the case of South Africa, is quite extensive. Thus the NNR has played a very ineffectual role with respect to the seepage of waste water from abandoned mines (known as acid mine drainage) into the environment. This waste water is highly contaminated due to the presence of radioactive and toxic radionuclides.

The NNR's record with respect to transparency and openness has been somewhat contradictory. Despite its mandate to include representatives from labour and affected communities on its Board, it acted very slowly in doing so. Until late 2003, it had appointed a former Eskom (utility operating Koeberg) employee as its community representative, despite many approaches from civil society organisations denouncing this person as an illegitimate spokesperson for community interests. Only after this did the NNR advertise the position more broadly so as to attract a representative of affected communities and labour. However the voices of these representatives have not carried huge weight in the Board, and in some cases, have experienced being silenced by confidentiality provisions. In recent times, community representative Mariette Lieferrink has resigned from the board after being blocked from playing a representative role and pressing for the body to become more effective. Instead the NNR chose a community representative from a government-aligned body, who has had no interest in making the body more accountable.

One compromise of the Board's position was the appointment as chief executive officer of the NNR in April 2005 of Maurice Magugumela, a former safety and licencing manager of the PBMR company. The appointment was made by the minister over the heads of Board members, a practice in clear violation of the NNR Act. Later this was given full Cabinet approval. This showed the impunity with which the Minister acted to create a "revolving door" between the industry and the regulator, at a time when the regulator needed to be seen as scrupulously impartial and to reject special interests. During Magugumela's three years in office, there was a notable downturn in the release of information to the public. The NNR refused numerous applications by civil society organisations for documents to which they were entitled under the Promotion of Access to Information Act (PAIA) (Fig 2009b).

As government policy foresees an immense expansion of the nuclear industry, so the question of the competence of the regulator can no longer be seen as a side issue. The NNR needs a major revision of its core competencies in order to be able to safeguard the population from radiation risks and regain public confidence. It needs to ensure that its mandate to regulate is broadly implemented, and conduct appropriate scientific investigations. It needs to become more independent of the strongly promotional Department of Energy and more accountable directly to the public and to parliament. It needs to function free of licencing fees, on which it relies for the bulk of its income. This gives it a built in stake in the expansion of the industry. There needs to be a complete review of the effectiveness, transparency, objectivity and neutrality of the NNR and a complete overhaul of its skills and its understanding of its legal mandate.

Former NNR board member Liefferink, who developed a strong critique of the institution, stated the following: "You judge the future from the current. I judge the NNR on what they are doing now. And I judge it from physical or real evidence. And I despair. Because if we cannot even manage or monitor our low dose radioactive waste from the mining industry, let alone the radioactivity from nuclear facilities, then I really despair."[7]

However, the NNR played a role in exposing the wholesale breach of licencing conditions by the AEC at its disposal site for low and intermediate level radioactive waste in Namaqualand.

In 2002, environmental watchdog organisation Earthlife Africa challenged the government's acceptance of a flawed Environmental Impact Assessment (EIA) of the now defunct Pebble Bed Modular Reactor (PBMR) (Andrews and Pole 2002; Wynberg and Fig 2014: 332ff). Earthlife listed its objections to the facts that the state had not demonstrated the need for the reactor, grossly underestimated the safety of its design, had not taken full costs into account, did not have a policy on radioactive waste nor an integrated energy plan. The EIA process had also violated Earthlife's right to be heard, as its submission had never been considered before the Record of Decision was issued. Judge Dennis Davis of the Cape High Court reversed the state's positive Record of Decision, on the grounds that the decision maker should be "fully informed of the submissions made on behalf of interested parties and ... should properly consider them" (Earthlife Africa Cape Town 2005: para. 76).

Eskom set about instituting another EIA. To pre-empt some of the potential Earthlife objections, it announced that the next design would include a containment building (to redress safety concerns) and that the government would be requested to put in place a policy on the disposal of radioactive waste.

At the behest of these pressures, the Department of Minerals and Energy set about putting together a policy document on radioactive waste. Not having its own expertise on the matter, it based the policy on the advice of a French consultant close to his country's nuclear industry.

---

7 Personal interview, 16 June 2011.

The national policy and strategy document is a minimal intervention in setting the way forward for the management of radioactive waste. It sets out a list of principles within which the policy is established, notably drawn from the IAEA and obligations under national legislation, such as the existing nuclear energy and nuclear regulatory acts, and the national environmental management act (*op. cit.*: 8-11). It foresees NECSA handing over responsibility for radioactive management to a new body that would continue to rely on NECSA's infrastructure (*op. cit.*: 15-6) and creates an interdepartmental executive committee to discuss waste management (*op. cit.*: 15). It reviews existing practice, but does not comment on its efficacy. The Vaalputs site will continue to be used for storage of low- and intermediate-level wastes. It notes that spent fuel is currently stored on site in ponds at Koeberg and Safari 1, also in canisters at the latter site. It states that this could continue for the lifetime of Koeberg, and that this should continue in the interim, but that "storage of spent fuel at a reactor site is not sustainable over long periods of time. Government shall … investigate the options for safe management of these wastes in South Africa" (*op. cit.*: 26).

However the policy made no choice of disposal technology, whether above ground, reprocessing, deep geological disposal or transmutation of the wastes. The document leaves the choice of options to future decision makers. Therefore it is questionable whether the document provides any fresh policy recommendations at all with respect to the management of radioactive waste.

In one of the forewords to the document, the deputy minister of minerals and energy said that "the radioactive waste management policy was founded on the belief that nuclear resources are a national asset and the heritage of the entire people and should be managed and developed for the benefit of present and future generations in the country as a whole" (*op. cit.*: 6).

To remove the onus on NECSA for managing the country's radioactive waste, the government legislated the creation of a National Radioactive Waste Disposal Institute with Act 53 of 2008. In 2009, the Act was promulgated by parliament.

It took a further five years before the state-owned institute was launched. In January 2014 the minister of energy appointed its first board members. The NRWDI was given the function of managing all the country's radioactive wastes. These include wastes from the use of ionizing radioactivity emanating from medical, mining and energy-related facilities. Despite claims to the contrary, there was scarcely any public consultation on the establishment of the institute. The NRWDI took over the management of Vaalputs, the current repository for low- and intermediate-level radioactive wastes, and will have to construct, manage and monitor any future radioactive waste disposal facilities (Tancott 2014).

No country has yet developed and licenced a high-level radioactive waste deep disposal facility. One of the roles of the NRWDI will be to research and develop such a facility.

Thirteen months after the launch of the NRWDI, the minister of energy announced that she was setting up a task team to investigate "serious mismanagement in relation to corporate governance and management issues" at NRWDI and NECSA. The task team consists of representatives of the departments of Energy, Mineral Resources, International Relations and Cooperation, as well as the National Energy Regulator of South Africa (NERSA).

Concerns of the task team were said to include that NECSA was in the red to the extent of R147.8 million (9.5 million Euro). Although it was supposed to report back by 31 March 2015, no report had entered the public domain as of April 2017.

Conflicts within the Board continue to dog NECSA (Mail & Guardian 2015; Business Day 2015). "If left unattended," the minister stated, "they (the concerns) may adversely impact on the effectiveness of the board in overseeing and guiding NECSA and NRWDI operations" (Engineering News 2015).

In view of the forthcoming procurement of six to eight new nuclear reactors, the minister cannot afford to have such a serious governance crisis in its key nuclear institutions.

4 Siting procedures

4.1 Procedures and criteria for site selection and compensation mechanisms

In South Africa it is mandatory to obtain authorisation for the siting of nuclear installations. The applicant for a licence must provide an up-to-date safety assessment, including a risk assessment, and this must take into account feedback from local and international experiences. The regulator receives a site safety report which characterizes the site and shows how plant design will meet safety standards. The site safety report must take into account the ecology of the surroundings, population distribution and growth, land and sea use, the location of transport and infrastructure facilities, meteorology, oceanography, hydrology, geology, seismology and cooling water supply. It would also outline impacts of natural and human-generated hazards, and the steps taken to ensure site control and emergency services. If the regulator concludes that the site is unviable or unsuitable, no authorisation will be granted.

As pointed out earlier, siting of the Vaalputs disposal site in the 1980s, only took into account the presence of "white" settlement, ensuring a 50 km buffer zone. However, settlements of indigenous Nama people were not given this consideration, and some found themselves living within 24 km of the site.

There is as of yet no plan for the implementation of deep-level disposal of nuclear waste. Storage of various types of waste has expanded at the Pelindaba site. In addition, spent fuel at the existing Koeberg power station has had to be "re-racked" in cooling ponds with the installation of high-density storage racks. Pressure to extend the life of the reactors, and the storage configuration itself, have led to consideration of storage of some spent fuel on site on an interim basis in above ground dry storage casks of the Castor X/28 type (NNR 2014b: 66).

No mechanism for granting compensation to affected parties has been envisaged, for example, in a case involving forced population removal. The state usually provides for occupational compensation, but some hundreds of Pelindaba workers who claim to have been irradiated while working in the complex in the 1980s, are today still struggling to obtain compensation (PMG 2007; Gilbert 2008). Their case is currently with the Office of the Public Protector, whose report is said to be imminent.

During the resolution of this case, it transpired that NECSA had not been keeping reliable, retrievable medical records for its personnel. This clearly undermined the possibility of the workers being able to satisfy strict conditions proving occupational exposure. A committee appointed by NECSA to resolve this matter ended up spending its budget on its own meetings and failing to report back.

Planning should take into account the full life of the nuclear facility in order to assure that emergency planning arrangements for the site are viable throughout this period. Technology has enabled the extension of the lifetime of reactors, and the Koeberg nuclear power station is installing technology that will extend its life from 40 to 60 years.

Most household insurance clauses exclude liability for very serious nuclear accidents. In the secret agreement with Russia, it was revealed that South Africa would give Rosatom considerable control over the local nuclear energy industry without holding Rosatom liable in any way for any accidental damage.

4.2 Socio-economic impact

The social and economic impacts of nuclear waste cannot be entirely distinguished from those arising from the nuclear industry in general. Nevertheless, there are a few specific elements that pertain to the impact of waste.

The radioactive residues arising from the gold and uranium mining industries, left in the environment, have a profound impact on the health of 1.7 million people living adjacent or in contact with these wastes. The toxic and radioactive nature of the wastes creates epidemic levels of cancer, elevates levels of genetic disturbance, and also contributes to weakening resistance to other diseases such as tuberculosis and the HIV. These impacts range across the Witwatersrand basin, into the North-West and Free State provinces.

Nuclear installations that produce waste are not big employers, and contribute less than other sources of energy generation to combatting the scourge of unemployment. Investment in nuclear also crowds out potential investment in cleaner energy technologies. A study conducted by the Energy Research Centre at the University of Cape Town predicts a loss of 75,000 jobs should the nuclear build programme go ahead and should costs be high (ERC 2015: 14-15).

The waste has to be kept on site or transported long distances on South Africa's notoriously accident-prone highway system. The waste passes through a number of impoverished local authorities whose scarce resources impede them from developing accident prevention and management strategies. This imposes a significant risk burden on those who benefit little from the industry, and increases their poverty.

Nuclear procurement of another 9.6 GW - because of the enormous cost - creates opportunities for corruption in a society whose anti-corruption measures are either absent, poorly implemented or deliberately overlooked. There is widespread mistrust in the incumbent president, who has sidestepped corruption charges, and continues to favour members of his family and his wider entourage. Electricity, in order to reach the masses, has to be affordable, and nuclear is proving to be one of the most expensive sources. Despite a considerable expansion of electrification, levels of poverty mean that the poor and disadvantaged struggle to afford it. In turn, this places a burden on women and the environment, learners' chances to do homework at night, and the option of running small businesses. Affordability is also a consideration for large electricity users. This is why nuclear technology has been rejected by the Electricity Intensive User Group, which includes the three dozen corporate users of bulk electricity, including mines, cement producers, smelters and other firms.

In the fiscal year 2017/2018, budgetary allocations for the National Radioactive Waste Disposal Institute (NRDWI), the state-owned company legally charged with managing the country's nuclear waste, amount to R20 million (1.4 million Euro). It is unclear whether the substantial wastes stored at Pelindaba fall under the budget for the NRDWI or that of NECSA, which totals R64.8 million (4.5 million Euro). If the latter, only a portion of the total would cover the waste management at Pelindaba (Kubayi 2017).

Radioactive wastes have to be managed for long periods of history, so the burden of risk has to be shared across numerous generations, most of which will have no say in the process. Ongoing costs of management have to be assumed by the state. Hence falling on all citizens, not just the users of nuclear electricity.

Even were there to be a moratorium on future development of the nuclear industry, there is already a significant legacy which will have to be addressed for a number of centuries to come.

As a country which still has to overcome massive levels of inequality, poverty and unemployment, South Africans should acknowledge more clearly that this burden is unaffordable financially and environmentally.

5 Information and participation

Prior to the advent of democracy (1994) and the formalization of an inclusive Constitution (1996), the previous apartheid regime paid little attention for most of its history to environmental questions, and was dedicated only to the participation of white citizens. The whites were accorded the vote but the regime did not make allowances for much day-to-day participation. Key political decisions were conducted behind closed doors, and often in consultation with elitist Afrikaner nationalist secret societies such as the Broederbond.

Given the already secretive nature of the nuclear industry, and its clear linkages with a weapons programme, very little consultation even occurred with elected representatives of whites. For example, the municipality of the city of Cape Town was never consulted with regard to the siting of the Koeberg Nuclear Power Station, located 23 km to its north-west. Very little political space existed for objections to the state's nuclear energy plans. When the reactors were switched on, the city provided its ratepayers with supplies of iodine tablets, which were thought to be helpful in building the thyroid's resistance to radio-activity in case of any leaks. The city's Medical Officer of Health, Dr Reg Coogan, made a public show of his decision to retire to an area outside the municipal boundary in the opposite direction to Koeberg.

The most glaring violation of people's rights came with the siting of the Vaalputs nuclear waste disposal facility, 700 km north of Cape Town, in the district of Namaqualand. In choosing the location, researchers had also drawn no-go areas in a radius of 50 km around what were then white-run municipalities in the district. The indigenous Nama people, who occupied remaining land, were neither consulted nor included as part of any buffer zone. The settlements of Paulshoek, Nourivier and Leliefontein were each less than 50 km from Vaalputs.

"Must we prepare our own gallows?" community leader Oom Oulak asked when subsequently informed of the decision. Another revered community elder, Oom Japie Bekeur, went on to question the justice of locating Vaalputs so close to Nama settlements: "Why should we accept that radioactive waste be buried close to our homes when we are not even supplied with electricity?" In a local community newspaper, *Namaquanuus*, journalist Elizabeth Beukes confirmed that there was no consultation whatsoever with the people of the area. "For the AEC," she claimed, "our people do not seem to exist." Apartheid planning refused to take the voices of indigenous communities into account (Fig 1991: 122-123).

Communities had to rely on the voices of activists to keep them informed. In Kommaggas, the Rev Peter Grove and his wife Terry made sure that their congregants were aware of developments. The Namaqualand branch of the National Union of Mineworkers, advice centres, the *Namaquanuus*, and environmental organisations based in Springbok and Cape Town all contributed to greater public understanding of what was at stake (*ibid.*).

The secrecy around Vaalputs has also alarmed the local population. "We regard it as dangerous because they are keeping it secret and because they don't want to consult with us," according to local resident Petrus Rosseuw. He goes on to say, "What I still don't understand is that after all these years most people in those small communities still don't know what is going on there. The people they employ there – and I am not scared to say this – must shut their mouths in order to keep their jobs. And even when meetings are open to the public, the information policy of the AEC has upset the locals."[8]

Oom Wolfie Waldeck of Rooifontein village agrees: "I have attended their meetings, where they only tell the people about the positive aspects of nuclear waste, but the dangers are never explained to you. If the government thinks it isn't dangerous, then why can't Koeberg bury the waste in its own back yard?"[9]

In Cape Town, attention was focused on the siting of the reactors. There was enough public interest to establish a credible anti-nuclear campaign, Koeberg Alert, which originally formed a number of locally based committees, including one which was based in a dormitory industrial working class area close to the Koeberg site, called Atlantis. The campaign raised funds, held public debates, engaged with the media, the universities, and the growing wave of anti-apartheid civic and trade union movements. A number of its activists were detained in state clampdowns, particularly around the 1986 State of Emergency. The climate of repression was such that activists had to exercise great caution. For example, the offices of Koeberg Alert in Community House, Salt River, which also housed faith-based organisations, labour groups, community arts projects and the United Democratic Front, was bombed by security police agents in August 1987.

By the end of the 1980s, it was clear that the apartheid regime was morally and financially exhausted. As the popular anti-apartheid movement gained

8 Interviewed by Helena Kingwill for the film Buried in Earthskin, 2010.

9 Interviewed by Jenny Hunter for the film Uranium Road, 2008.

traction, the new wave of environmentalism also took the question of human rights extremely seriously. Its focus was no longer simply on the protection of species or landscapes, but had a stronger accent on reversing environmental injustices which had disproportionately affected the poor. As power passed to the liberatory African National Congress, so the activists were drawn into projects to democratize environmental governance. A strong national organisation, the Environmental Justice Networking Forum, with 500 affiliated groups at its height, participated in multi-stakeholder processes to craft new democratic environmental laws, containing ample provisions for public participation. EIA regulations, informal during apartheid, were codified, honouring the right of public participation.

This window lasted to the end of the 1990s, when the ANC assumed its governance role more exclusively, downplaying an independent role for activists, while incorporating some of them into structures of government. Bureaucratic alliances were made, not with progressive elements, but mostly with the residue of former public servants, who had survived under a compromise agreement called the "sunset clause". Activists found that they were excluded from access to the state, and began to play a role as a progressive opposition, without having political party representation. Whether campaigning on issues of human rights, land, health, education, jobs, social provision, environment or information, the activists were faced with a conservatizing state and ruling party which had embraced a neo-liberal ideology.

In relation to the nuclear industry, the ANC government was at first suspicious, applying more stringent spending constraints to the research function, and commissioning studies to address the future of Koeberg and Pelindaba. However, by the time of the 1996 Energy Summit, which involved the participation of various social sectors and affected communities, the state no longer pursued an agenda of closing the industry down. Gradually Pelindaba was allowed to propose and receive funding for research into a high temperature Pebble Bed Modular Reactor.

Public participation took the form of opening up of anodyne public fora at both Koeberg and Pelindaba. Often the industry used its insiders to block effective proposals from the activists. Similarly the 1999 law establishing the National Nuclear Regulator allowed for both a community and a labour movement representative on its board. In practice, these representatives have to abide by confidentiality clauses that prevent them from speaking out in public. A key activist who became a community representative on the NNR board found that she could not uphold her community commitments by continuing to remain silent. This provoked her representation, and she was replaced by a government loyalist from a moribund civic federation.

Activist movements have been more effective intervening by means of litigation or making legal challenges to EIA decisions. This strategy was helpful in finally getting government to develop a policy on nuclear waste, and contributed to government reconsidering its commitment to the PBMR. It has not prevented NECSA from building smelters, as we have seen. The judgment is still awaited on the case challenging government's secret agreement with Russia to favour Rosatom's bid to develop more nuclear power reactors for South Africa.

The latter case, although launched by environmental NGOs, has attracted universal interest across all social sectors, who recognise the centrality of the nuclear procurement in the plans of President Zuma. Because this would be the most expensive infrastructural project in the country's history, and because the President is implicated in favouring corruption of the procurement process, the nuclear industry is no longer considered marginal but a central issue in contemporary South African politics.

The issue has united not only the environmental movement, and its faith-based allies, but anti-corruption and human rights groups, progressive trade unions, most of the business sector and financial press, in a loose cross-class alignment which is gathering support. This may in the future help to defeat the president, his loyalist beneficiaries and the nuclear lobby, while ushering in a new phase of energy democracy for the country.

The regulator makes little provision for stakeholder involvement in the licencing process for nuclear installations. However, the law states that Environmental Impact Assessments (EIAs) have to be conducted for nuclear installations. EIAs allow for the registration of interested and affected parties (I&APs) who must be invited to public meetings and allowed a reasonable time to submit written comments on proposed developments. The applicant usually employs a consultant firm to manage public participation. Nuclear installations are deemed to be the purview of national-level authorities, so the municipal authority is only listed as an I&AP and the EIA is not held under its auspices.

Additionally, there was no broad stakeholder public consultation when the Department of Minerals and Energy set about putting together a policy document on radioactive waste. The period allotted for public comment on the national Radioactive Waste Management Policy and Strategy was so short that the minister of minerals and energy was forced by public opinion to extend it for an extra ninety days (DME 2003). Very few of the concerns expressed by the public found their way into the final policy document (DME 2005).

In practice, I&APs have complained that documentation is not provided in accessible language, that the zero (non-development) option is seldom addressed, that deadlines are often set at inconvenient times (e.g., during summer holidays). On occasion, I&APs have been obliged to turn to litigation to ensure that there is fair application of administrative justice.

6 Conclusions

As has been demonstrated, South Africa has a weak legacy of radioactive waste management. Yet despite that, President Zuma strongly favours a fivefold expansion of its nuclear power generation capacity. This would necessarily mean a multiplication of the amount of nuclear reactor waste that has to be managed.

Estimates for the overnight cost of the additional procurement of six to eight reactors, generating 9600 MW of electricity, have ranged between R450 and R1200 billion (29 - 77.3 billion Euro). This would make it by far the largest infrastructural investment ever undertaken by the South African state.

The state envisaged a procurement process to get vendors to bid for the new build programme. Potential vendors included Areva (France), China Guangdong, KEPCO (South Korea), Rosatom (Russia), and Toshiba/Westinghouse (US/Japan). Of these, both Areva and Westinghouse are experiencing serious financial setbacks (Prof. S. Thomas, quoted in BBC 2014; Cardwell and Soble 2017), while Rosatom will lose its state subsidies and bailouts by 2020, according to its chief executive officer (Alimov et al. 2017: 4; World Nuclear News 2016). As a result of a pre-bidding request for information (RFI) in 2017, 27 nuclear-related companies expressed interest, but these are probably not all reactor vendors. Due to offers by Rosatom to fund the construction of the reactors, and strong links between Presidents Zuma and Putin many felt that Russia might have the edge in the bidding process. South Africa has sought to sign nuclear co-operation agreements with all the vendor countries. Most of these did not contain contractual commitments. However, the agreement with Russia did contain an agreement that Rosatom would build the new nuclear fleet, get control of South Africa's nuclear establishment and would be exempt from any liability for nuclear accidents or damage. The agreement was not released to the South African public, and only discovered later on the website of the Russian Ministry of Foreign Affairs (Earthlife vs Minister of Energy 2017).

Two NGOs, Earthlife Africa and Southern African Faith Communities' Environmental Initiative, sought relief from the courts to contest the legality of the nuclear build, the procurement process, and especially the content of the secretive agreement with Russia. On 26 April 2017, the Western Cape High Court pronounced judgement. The procurement process and its allocation to Eskom was declared illegal, as were the nuclear agreements with Russia, the USA and South Korea. The national electricity regulator was berated for rubber stamping the government's procurement decision without ensuring a process of public inquiry.

This judgement has come as a huge blow to the state's nuclear plans, forcing the procurement process back to the drawing board. It is unlikely that anything can materialize within the final two years of Zuma's presidential mandate. His aspiration to see the nuclear procurement as part of his legacy to the nation has thus been thwarted by the courts.

There has been substantial public disquiet expressed over the procurement programme, with key misgivings being the enormous cost (renewables are now cheaper than coal and nuclear), the potential for corruption, the overinvestment in a rigid centralized form of power, the dangers posed by the technology, the lack of a solution to the safe disposal of the waste, security over the nuclear material, and the potential for future proliferation of nuclear weapons. Power cuts have played a role in softening the public up for acceptance of the deal, however, despite this, serious opposition has been expressed by the private sector, the business press, the intensive users of electricity (constituting 44 percent of consumption), faith-based groups, and environmental and climate campaigns. Some parts of government, including the National Planning Commission, and research groups like the Energy Research Centre (University of Cape Town) and the CSIR have raised concerns. Only the Presidency, the utility Eskom, tender beneficiaries and the small but well resourced nuclear lobby seem to be in favour.

The nuclear procurement process has also been cited as a reason for the dismissal of Minister of Finance Pravin Gordhan on 31 March 2017, as well as one of his predecessors. Minister Gordhan had been a well-known opponent of the nuclear-build programme on financial grounds, and this had antagonized President Zuma, who has been a strong proponent. The cabinet reshuffle has been extremely controversial, and provoked the investment ratings companies to declare that South Africa had been downgraded to junk status.

Because of these consequences, Zuma's ambitions have placed the question of South Africa's nuclearized future at the centre of politics. For the present, it is difficult to see the procurement succeeding in the short term. Zuma's political successors are unlikely to champion this procurement so strongly.

However, while vested interests still exist, it is too early to write the programme off entirely. The abiding question is whether South Africa possesses sufficient technical skills, finances and commitment to manage the resulting waste effectively and carefully regulate an expanded nuclear industry.

References

Albright, D. (1994). *South Africa's Secret Nuclear Weapons*, Washington: Institute for Science and International Security.

Alimov, R.; Bayram, D.; Haverkamp, J.; Berger, A. and Tiainan, O. (2017). *Rosatom Risks: Exploring the Troubled History of Russia's State Nuclear Corporation*, updated April 2017. http://www.greenpeace.org/hungary/PageFiles/636986/Rosatom-Risks-Report-2017_Update.pdf, last accessed on 30 April 2017.

Alperovitz, G. (1995). *The Decision to Use the Atomic Bomb*, New York: Alfred A. Knopf.

Andrews, A. and Pole, A. (2002). *Submissions on the Pebble Bed Modular Reactor and Associated Fuel Manufacture Draft Environmental Impact Reports: Analysis of Legal Compliance,* Cape Town: Legal Resources Centre.

Bird, K. and Sherwin, M.J. (2005). *American Prometheus. The Triumph and Tragedy of J. Robert Oppenheimer*, New York: Alfred A. Knopf.

British Broadcasting Corporation [BBC] (2014). *Nuclear Deal 'at Risk of Collapse' Due to Areva Crisis*, 20 November. http://www.bbc.com/news/uk-england-somerset-30125622, last accessed on 30 April 2017.

Busby, C. (2011). *Radioactivity in the Upper Wonderfonteinspruit Catchment Area.* Review of National Nuclear Regulator Surveillance Report TR-NNR-10-001. Unpublished. http://earthlife.org.za/www/wp-content/uploads/2011/01/Prof-Busbys-review-of-NNRs-surveillance-report.pdf , last accessed on 1 October 2015.

Business Day (Johannesburg) (2015). *NECSA Latest State Entity To Be Hit by Boardroom and Financial Disarray.* 1 October. http://www.bdlive.co.za/business/energy/2015/10/01/necsa-latest-state-entity-to-be-hit-by-boardroom-and-financial-disarray, last accessed on 1 October 2015.

Canadian Nuclear Safety Commission (2015). *Radioactive waste* (page updated 16 February). http://www.nuclearsafety.gc.ca/eng/waste/low-and-intermediate-waste/index.cfm#Intermediate-level, last accessed 1 October 2015.

Cardwell, D. and Soble, J. (2017). Westinghouse Files for Bankruptcy, in Blow to Nuclear Power. In: *New York Times*, 29 March. http://www.nytimes.com/2017/03/29/business/westinghouse-toshiba-nuclear-bankruptcy.html, last accessed on 30 April 2017.

Červenka, Z. and Rogers, B. (1978). *The Nuclear Axis. Secret Collaboration between West Germany and South Africa*, London: Julian Friedman.

Cooper, R. A. (1923). Mineral Constituents of Rand Concentrates. In: *Journal of the Chemical, Metallurgical and Mining Society of South Africa,* 24, 90-96. Reprinted in the *Journal of the Institute of Mining and Metallurgy*, 94(3) (1994), 47-51.

Council for Nuclear Safety (CNS) (1996). *Nuclear Occurrence at AEC's Pelindaba Complex, statement by R F Botha, Minister of Minerals and Energy.* http://www.polity.org.za/polity/govdocs/pr/1996/pr0225a.html, last accessed on 30 July 2015.

Council for Nuclear Safety (CNS) (1997). *Annual Report 1996/7*, Pretoria: CNS.

Department of Minerals and Energy (DME) (2003). *Press Release*, 16 October.

Department of Minerals and Energy (2005). *Radioactive Waste Management and Strategy for the Republic of South Africa*, Pretoria: Government Printer.

Department of Water Affairs and Forestry (DWAF) (2001). *Waste Generation in South Africa.* Water Quality Management Series, Pretoria: DWAF.

Department of Water Affairs and Forestry (2009). *Water for Growth and Development Framework* (Version 7), Pretoria: DWAF.

Earthlife Africa Cape Town v. Director General of Environmental Affairs & Tourism and Eskom Holdings Ltd. (2005). *Case number 7633/2003*. Judgment in the High Court of South Africa, Cape Town, 26 January.

Earthlife Africa Johannesburg v. Minister of Environmental Affairs and Others (2014). *Case number 51505/14*. Application in the High Court of South Africa, Gauteng Division, Pretoria.

Energy Research Centre (ERC) (2015). *South Africa's Proposed Nuclear Build Plan: An Analysis of the Socio-Economic Risks*, Cape Town: ERC, University of Cape Town.

Engineering News (Johannesburg) (2015). *Task Team to Investigate NECSA and NRDWI Allegations*. 23 February. http://www.engineeringnews.co.za/article/task-team-to-investigate-necsa-nrwdi-allegations-2015-02-23, last accessed on 1 October 2015.

Environmental Monitoring Group (EMG) with Western Cape ANC Science and Technology Group (WCANCSTG) (1994). *The Nuclear Debate. Proceedings of the Conference on Nuclear Policy for a Democratic South Africa*, 11-13 February 1994. Cape Town: EMG/WCANCSTG.

Eskom (2017). *Waste and Re-Racking*. http://www.eskom.co.za/whatweredoing/ElectricityGeneration/KoebergNuclearPowerStation/Pages/Waste_Reracking.aspx, last accessed on 30 April 2017.

Fig, D. (1991). Flowers in the Desert - Community Struggles in Namaqualand. In: Cock, J. and Koch, E. (Eds.). *Going Green. People, Politics and the Environment in South Africa*, Cape Town: Oxford University Press, 112-128.

Fig, D. (1999). Sanctions and the Nuclear Industry. In: Klotz, A. and Crawford, N.C. (Eds.). *How Sanctions Work. Lessons from South Africa*, Basingstoke: Macmillan, 75-102.

Fig, D. (2005). *Uranium Road. Questioning South Africa's Nuclear Direction*, Johannesburg: Jacana.

Fig, D. (2009a). A Price too High - Nuclear Power in South Africa. In: McDonald, D. A. (Ed.). *Electric Capitalism. Recolonising Africa on the Power Grid*, London: Earthscan, 180-201.

Fig, D. (2009b). In the Dark - Seeking Information about South Africa's Nuclear Energy Programme. In: Allen, K. (Ed.). *Paper Wars. Access to Information in South Africa*, Johannesburg: Wits University Press, 246-252.

Fig, D. (2010a). *Nuclear Energy Rethink? The Rise and Demise of South Africa's Pebble Bed Modular Reactor*, Pretoria: Institute of Security Studies.

Fig, D. (2010b). Darkness and Light: Assessing the South African Energy Crisis. In: Freund, B. and Witt, H. (Eds.). *Development Dilemmas in Post-Apartheid South Africa*, Pietermaritzburg: University of KwaZulu-Natal Press, 116-136.

Gilbert, D. (2008). *Fact Sheet and Overview on the Nuclear Workers' Health Study*. Unpublished Press Kit. Copy in author's possession.

Glazewski, J. (2000). *Environmental Law in South Africa*, Durban: LexisNexus Butterworths.

Kanhema, N. (1996). Articles on Radiation Hill. In: *Sunday Independent* (Johannesburg), 3 and 10 March.

Kubayi, M., Minister of Energy (2017). *Budget Vote Speech*, South African National Assembly, 19 May.

Mail & Guardian (Johannesburg) (2015). *More Nuclear Fallout at NECSA*. 31 August. http;//mg.co.za/article/2015-07-30-more-nuclear-fallout-at-necsa, last accessed on 1 October 2015.

Masondo, S. (2017). Nuclear Ball Rolls in June. In: *City Press* (Johannesburg), 9 April. http://www.news24.com/SouthAfrica/News/nuclear-ball-rolls-in-june-20170408-2, last accessed on 30 April 2017.

Mathews, C. (2017). Eskom Moots Nuclear Plans for Witbank. In: *Business Day,* 30 March. http://www.businesslive.co.za/bd/companies/energy/2017-03-30-eskom-considers--nuclear-power-plants-for---witbank, last accessed on 25 April 2017.

Naidoo, R. (2017). *Nuclear Waste Smelting to Commence after Earthlife Loses in Court.* 8 March. http://infrastructurene.ws/2017/03/08/nuclear-waste-smelting-to-commence-after-earthlife-loses-in court/, last accessed on 2 April 2017.

National Nuclear Regulator (NNR) (2008). *Annual Report 2007/8*, Centurion: NNR.

National Nuclear Regulator (2010). *Radioactivity in the Upper Wonderfonteinspruit Catchment Area.* Surveillance Report TR-NNR-10-001, Centurion: NNR.

National Nuclear Regulator (2014a). *Annual Report 2013/4*, Centurion: NNR.

National Nuclear Reguator (2014b). *Third South African National Report on the Compliance to Obligations under the Joint Convention on Safety of Spent Fuel Management and on the Safety of Radioactive Waste Management*, Centurion: NNR.

Newby-Fraser, A. R. (1979). *Chain Reaction. Twenty Years of Nuclear Research and Development in South Africa,* Pretoria: Atomic Energy Board.

Parliamentary Monitoring Group (PMG) (2007). *Minutes of the Environment and Tourism Parliamentary Portfolio Committee* (L. Zita, chair), 20 June.

Pebble Bed Modular Reactor (PBMR) Company (2002). *Report on the Proposed Demonstration Model and Potential Commercialisation of the PBMR*, Centurion: PBMR Company.

Purkitt, H. and Burgess, S. (2005). *South Africa's Weapons of Mass Destruction*, Bloomington: Indiana University Press.

Reuters (2015). R200m for Nuclear Waste Storage. In: *The Times* (Johannesburg), 13 August.

Smillie, S. (2005). Infamous 'Radiation Hill' Given a Miss. In: *Sunday Independent*, Johannesburg. 6 May. http://www.iol.co.za/news/south-africa/infamous-radiation-hill-given-a-miss-1.240443#.Vg0riukxH-Y, last accessed on 1 October 2015.

South Africa (1999). *Nuclear Energy Act, No. 46 of 1999*, Pretoria: Government Printer.

South African Press Association (SAPA) (1997). *Weather May Have Caused Leak at N-Cape Nuclear Dump Site*. 27 June.

Tancott, G. (2014). National Radioactive Waste Disposal Institute Launched. In: *World Nuclear News*, 4 April 2014.

Thomas, S. (2004). *The Economic Impact of the Proposed Demonstration Plant for the PBMR Design,* London: Public Services International Research Unit, University of Greenwich.

Turton, A. R.; Schultz, C.; Buckle, H.; Kgomongoe, M.; Malungani, T. and Drakner, M. (2006). Gold, Scorched Earth and Water in Johannesburg. In: *Water Resources Development,* 22(2), 313-335.

United States, Nuclear Regulatory Commission (US NRC) (2015). *Low-level waste* (page updated 8 April). http://www.nrc.gov/waste/low-level-waste.html, last accessed 1 October 2015.

World Nuclear Association (2017). *Nuclear Power in South Africa.* (page updated April 2017). http://www.world-nuclear.org/information-library/country-profiles/countries-o-to-s/south-africa.aspx (scroll to Fuel Cycle, Historical R & D), last accessed on 30 April 2017.

World Nuclear News (2016). *Rosatom Chief Outlines Commercial Vision*, 3 March. http://www.world-nuclear-news.org/C-Rosatom-chief-outlines-commercial-vision-08031701.html, last accessed on 30 April 2017.

Wynberg, R. P. and Fig, D. (2014). Realising Environmental Rights. Civic Action, Leverage and Litigation. In: Langford, M.; Cousins, B.; Dugard, J. and Madlingozi, T. (Eds.). *Socio-Economic Rights in South Africa. Symbols or Substance?* Cambridge: Cambridge University Press, 310-340.

Contributors

Achim Brunnengräber is an adjunct professor (Privat-Dozent) at the Department of Political and Social Sciences, Freie Universität Berlin. He is the coordinator of the "Multi Level Governance-Perspective on Management of Nuclear Waste Disposal" research project at the Environmental Policy Research Centre (FFU). Key teaching and research areas include: global and multi-level governance; international political economy; and international environmental, energy and climate policy.
Email: achim.brunnengraeber@fu-berlin.de

Maria Rosaria Di Nucci is a senior researcher at the Environmental Policy Research Centre (FFU) of the Freie Universität Berlin and advises - as an independent consultant - the European Commission and European research organisations. Di Nucci holds a master's degree in Economics from the University of Rome and a PhD from the University of Sussex. Formerly, she held an executive position at the Ministry of Environment of the Land Berlin and was a lecturer in Industrial Economics at the Berlin Technical University. Her research interests include: energy policy, in particular smart grid and smart metering, and comparative nuclear policy. She is a member of the "Multi-Level-Governance-Perspective on Management of Nuclear Waste Disposal" research project at the FFU.
Email: dinucci@zedat.fu-berlin.de

David Fig is a research associate in the Chair of Bio-Economy, University of Cape Town, as well as in the Society, Work and Development Institute, University of the Witwatersrand, Johannesburg. He is a Fellow of the Transnational Institute in Amsterdam. He has been writing about the South African nuclear industry since the 1970s. His research includes energy, environment, extractives,food sovereignty and corporate behaviour in Southern Africa and Latin America. He has strong links with the South African environmental justice movement. He chairs the Biowatch South Africa Trust. Much of his writing is available at eeu.academia.edu.
Email: davidfig@iafrica.com

Ana María Isidoro Losada is a research fellow at the Environmental Policy Research Centre at the Freie Universität Berlin since 2013. Currently she works in the research project "Methods and measures to deal with socio-technical challenges in storage and disposal of radioactive waste". Her primary fields of interest include theory of democracy, research on participation, nuclear waste

A. Brunnengräber et al. (eds.), *Challenges of Nuclear Waste Governance*,
Energiepolitik und Klimaschutz. Energy Policy and Climate Protection,
https://doi.org/10.1007/978-3-658-21441-8

governance, multi-level governance and spatial politics. She holds a Diploma in Landscape Architecture and Planning and studied Political Science, Sociology, and History at the Leibniz Universität Hannover and at the Université Paris VII-Denis Diderot/France.
Email: isidoro.losada@fu-berlin.de

Felix Jaitner holds a MA in Political Science from the University of Vienna and a BA in European Studies from the University of Bremen. Currently, he writes a PhD on resource-extractive industries and re-industrialisation strategies in Russia. He works as an external lecturer at the Department of Political Science, Vienna University. His key teaching and research areas include: international political economy; international environmental, energy and climate policy and development theories.
Email: F.Jaitner@gmx.de

Moïra Jimeno is a Project Manager in the Policy Department of the German consultant agency eclareon. She takes part in different projects aiming at analysing finantial, legal and policy framework conditions of RES in EU and Latin American countries. Jimeno holds a PhD from the Environmental Policy Research Centre (FFU) of the Freie Universität Berlin. Her dissertation topic addressed questions on why and how South American countries adopted different energy policies to deal with similar problems in the electricity sector. She also looked into the electricity structures and policies of France and Germany. Beside the PhD, Jimeno holds a Master's Degree in Territorial and Environmental Planning from the Universidad de Barcelona, Spain.
Email: mj@eclareon.com

Zsuzsanna Koritár currently works as Program director at MEHI Hungarian Energy Efficiency Institute and was a senior energy policy expert at the Hungarian Energiaklub Climate Policy Institute and Applied Communications. She holds a master's degree in Geology from the Eötvös Loránd University Budapest, specializing in Environmental Geology and Hydrogeology. She holds a postgraduate diploma in Environmental Sciences and Policy from the Central European University. Her main research area is energy policy, in particular nuclear policy, with a major focus on radioactive waste and spent fuel management.
Email: koritar@mehi.hu

Sung-Jin Leem is a Professor in the School of Social Sciences and at Jeonju University in South Korea. He is also the Director for the Institute of

Environmental and Energy Policy (IEEP) at the university. He graduated from Free University in Berlin, Germany (Dr. rer. Pol.). Since 1997 he has served a lot in Seoul as an expert witness in areas of environmental, energy and technology policy. He was a Chairman of the Antinuclear Committee of the Korean Federation for Environmental Movement.
E-Mail: leemsj@gmail.com

Lutz Mez is a senior adjunct professor (Privat-Dozent) of Political Science at the Department of Political and Social Sciences, Freie Universität Berlin. In April 1986, he co-founded the Environmental Policy Research Centre (FFU) and served as its executive director until April 2010. Since November 2009, he has been the coordinator of the Berlin Centre for Caspian Region Studies (BC CARE). In 1993/94, Mez was a visiting professor at the Department of Environment, Technology, and Social Studies at Roskilde University, Denmark. He holds a doctoral degree and a diploma in political science from Freie Universität Berlin. His major research area is comparative environmental and energy policy, with particular reference to nuclear and electricity policy.
Email: lutz.mez@fu-berlin.de

Peter Mihók is a researcher at the Research and Innovation Centre at the Faculty of Economics of the Matej Bel University in Banská Bystrica. His key recent research and implementation activities were a part of the two EURATOM's FP7 projects acronymed IPPA (2011 - 2013) and PLATENSO (9/2013 - 8/2016). Mihók is a member of the Nuclear Transparency Watch network under an affiliation of the Slovak NGO CEPTA, in which he has been active since his PhD studies in a field of Public Economics and Services (10/2006 - 8/2009). In 3/2003 – 6/2006, Mihók worked for international networks of NGOs CEE Bankwatch Network and the Friends of the Earth, focusing on social consequences of earlier NPP closure, public control of the European Bank for Reconstruction and Development administered Bohunice International Decommissionign Support Fund, and research of financial deficit at the Slovak national nuclear fund.
Email: peter.mihok@umb.sk

Lila Okamura is an associate professor at the Department of German Studies and a Member of the Institute of Human and Environmental Symbiosis Research at Dokkyo University in Japan. She holds a master's degree in Political Science from Gakushuin University in Tokyo and a doctoral degree from the Environmental Policy Research Centre (FFU) of the Freie Universität Berlin. Her teaching and research areas are comparative environmental policy, energy and climate policy, waste management, and nuclear policy.
Email: lilaokamura@dokkyo.ac.jp

Olexi Pasyuk coordinates activities of the CEE Bankwtach Network work in Eastern Europe and Central Asia and serves as a campaigns coordinator in Ukrainian Center of Environmntal Initiatives 'Ecoaction'. Pasyuk holds a master's degree in Environmental Management from the University of Amsterdam, a specialist degree in Analytical Chemestry from the Kyiv Taras Shevchenko State University and was a lecturer for Environmental Impact Assesement at National University of 'Kyiv-Mohyla Academy'. Since 1996, he follows nuclear energy issues for environmental non-govermental organisations. His areas of interst include energy policy, public participation and international financial flows.
Email: opasyuk@bankwatch.org

Povilas Poskas is a Head of the Nuclear Engineering Laboratory at the Lithuanian Energy Institute. Since 1994 he is involved into research related to the radioactive waste and spent nuclear fuel management, and decommissioning of nuclear facilities. His special interests are on safety and environmental impact assessment problems during treatment, handling, storage and disposal of the radioactive waste and spent nuclear fuel, and during decommissioning of the nuclear facilities. He holds a doctoral degree from Lithuanian Energy Institute. Email: povilas.poskas@lei.lt.

Miranda A. Schreurs (PhD, University of Michigan) is Professor of Environment and Climate Policy at the Bavarian School of Public Policy, Technical University of Munich. Her research focuses on climate change, low carbon energy transitions, and environmental movements. She is currently Vice Chair of the European Environment and Sustainable Development Advisory Councils and co-chairs the National Committee to Monitor the Nuclear Waste Management Site Location Search in Germany. She was also a member of the Ethics Committee for a Secure Energy Future set up by Chancellor Angela Merkel in 2011 in reaction to the Fukushima nuclear accident.
Email: miranda.schreurs@hfp.tum.de

Leo Šešerko is an Associated Professor at the Environmental Protection College /Visoka šola za varstvo okolja/ at Velenje, Slovenia that he co-founded in 2007. He was previously an associate professor of philosophy at the Faculty of Social Sciences of University of Ljubljana 1987 – 1997. In 1989 he was elected member of the Parliament of Slovenia (1989-96) and vice-premier of the slovene government (1989-92). In the years 2003-2006 he was member of a research group Sustainability Strategy of the EU. He studied sociology and philosophy at

Ljubljana University and received a PhD in Philosophy and is member of International Hegel Society and Slovenian Sociological Society. He was founding member of a scientif periodical Časopis za kritiko znanosti (Periodical for critique of sciences), Ljubljana in 1973 and its editor.
Email: leo.seserko@gmail.com

Eva Sternfeld is a consultant and researcher specialised in China's resource economy and environmental governance. Previously she has been Visiting Professor at the Institute for East Asian Studies at Freie Universität Berlin and Head of the Center for Cultural Studies on Science and Technology in China at Technische Universität Berlin. From 2000 to 2008 she worked as a foreign expert at the Center for Environmental Education and Communication (CEEC) of the Chinese Ministry for Environmental Protection in Beijing. She is editor of the *Routledge Handbook of Environmental Policy in China* (2017).
Email: eva.sternfeld@gmx.net

Cindy Vestergaard is a senior associate at the Stimson Center in Washington D.C. Previously she was a senior researcher at the Danish Institute for International Studies (DIIS) in Copenhagen, Denmark. Before DIIS, she worked on non-proliferation, arms control and disarmament at Canada's Department of Foreign Affairs and International Trade. She currrently leads Stimson's Nuclear Safeguards Program, with a focus on the global governance of natural uranium. Her portfolio also includes chemical weapons disarmament, biosecurity and import/export controls.
Email: cvestergaard@stimson.org

Made in the USA
Monee, IL
13 September 2026

62973773R00193